WITHDRAWN FROM LIBRARY

The Library
Exeter College
Oxford.

Time Sequence Analysis
in Geophysics

Time Sequence Analysis in Geophysics

Third Edition

E.R. Kanasewich

The University of Alberta Press
1981

First published by
The University of Alberta Press
Edmonton, Alberta, Canada
1973

Third revised, reset, and enlarged edition
1981

Copyright © 1981 The University of Alberta Press

ISBN   0-88864-074-9

The Library
Exeter College
Oxford.

Canadian Cataloguing in Publication Data

Kanasewich, E.R., 1931 -
   Time sequence analysis in geophysics

   Bibliography: p.
   Includes index
   ISBN 0-88864-074-9

   1. Time-series analysis. 2. Spectral theory
(Mathematics). 3. Geophysics - Mathematics.
I. Title

QC806.K36 1981   551'.01'519232   C81-091301-1

All rights reserved. No part of this publication may be produced, stored in a retrieval system, or transmitted, in any form or by any means, electronic, mechanical, photocopying, recording, or otherwise, without the prior permission of the copyright owner.

Printed by Hignell Printing Limited
Winnipeg, Manitoba, Canada

Contents

	Preface to the third edition	xi
1.	**Introductory observations**	1
	1.1 Historical notes	1
	1.2 Applications of time sequence analysis	2
	1.3 References	18
2.	**Convolution of a time series**	22
	2.1 Introduction	22
	2.2 Convolution as a digital integration	24
	2.3 Convolution as a geometric operation of folding	25
	2.4 Convolution as a geometric operation of sliding	26
	2.5 Convolution as an algebraic operation - z transformation	26
	2.6 Convolution of analog signals	29
	2.7 Convolution algorithm	29
	2.8 References	30
3.	**Fast Fourier transforms**	31
	3.1 Fourier series	31
	3.2 The discrete Fourier transform	34
	3.3 The principle of the fast Fourier transform	43
	3.4 Fast Fourier transform using matrix notation	45
	3.5 Fourier transformation of very long data sets	51
	3.6 Simultaneous computation of two Fourier transforms	53
	3.7 Two-dimensional Fourier transforms	54
	3.8 Fast Fourier computer programs	58
	3.9 References	61
4.	**Laplace transforms and complex s plane representation**	63
	4.1 Introduction	63
	4.2 The Laplace transform	65
	4.3 Examples of Laplace transformation	66
	4.4 Translation of a function	71
	4.5 References	73

Contents

5.		**Impulse response, convolution and the transfer function**	74
	5.1	The impulse response	74
	5.2	Convolution in the frequency domain	79
	5.3	The transfer function	80
	5.4	References	82
6.		**Correlation and covariance**	83
	6.1	Definitions	83
	6.2	Properties of the autocovariance function	85
	6.3	Vibratory sources	91
	6.4	Multivariate regression	94
	6.5	References	97
7.		**Wiener-Khintchine theorem and the power spectrum**	98
	7.1	The periodogram	98
	7.2	Wiener-Khintchine theorem	99
	7.3	Cross-power spectrum	102
	7.4	Compensation for instrument response	106
	7.5	References	109
8.		**Aliasing**	110
	8.1	The Nyquist frequency	110
	8.2	Aliasing	111
	8.3	References	117
9.		**Power spectral estimates**	118
	9.1	Introduction	118
	9.2	Effects of data truncation	118
	9.3	Tapering the data	120
	9.4	Daniell power spectral estimate	121
	9.5	Bartlett's power spectral estimate	124
	9.6	Prewhitening	126
	9.7	Matrix formulation of the power spectrum	127
	9.8	Spectral estimates using the autocovariance	130
	9.9	References	130
10.		**Joint spectral analysis on two data sets**	131
	10.1	Cross covarience with the fast Fourier transform	131
	10.2	Cross-spectral analysis and coherence	132
	10.3	Coherency	134

Contents

	10.4 Bispectral analysis	137
	10.5 Spectral estimation of a moving average process	139
	10.6 References	141
11.	**Maximum entropy spectral analysis**	142
	11.1 Data adaptive spectral analysis methods	142
	11.2 The concept of entropy	146
	11.3 Maximum entropy spectral analysis	147
	11.4 Order of the AR process	162
	11.5 The autocovariance function	165
	11.6 Extension of maximum entropy method	167
	11.7 References	173
12.	**The maximum likelihood method of spectral estimation**	177
	12.1 Introduction	177
	12.2 Comparison of MEM and MLM methods	180
	12.3 References	183
13.	**Minimum delay and properties of filters**	184
	13.1 Introduction	184
	13.2 Z transform	185
	13.3 Dipole presentation of filtering	188
	13.4 Normalized wavelet	190
	13.5 Minimum delay wavelet	192
	13.6 Maximum delay wavelet	194
	13.7 Mixed delay filter	195
	13.8 The partial energy	196
	13.9 References	197
14.	**Deconvolution**	198
	14.1 Inverse filtering	198
	14.2 Truncated deconvolution	199
	14.3 Least squares deconvolution dipole	200
	14.4 General equations for deconvolution	204
	14.5 Z transform of inverse filter	208
	14.6 Levinson recursion for inverse filter	210
	14.7 An inverse shaping filter	216
	14.8 Rice's inverse convolution filter	223
	14.9 The Wiener-Hopf optimum filter	225
	14.10 Alternate methods of deconvolution	228
	14.11 References	233

15. Band pass filters — 237
15.1 Ideal filters and their truncated approximation — 237
15.2 Recursion filters — 240
15.3 Zero phase shift filters - cascade form — 246
15.4 Rejection and narrow pass filters — 247
15.5 Butterworth filters — 252
15.6 Number of poles required for a given attenuation — 253
15.7 Singularities of a low pass Butterworth filter — 254
15.8 Frequency transformation for high pass filter — 258
15.9 Frequency transformation for a band pass filter — 260
15.10 The bilinear z transform — 266
15.11 Other digital filters — 277
15.12 References — 279

16. Wave propagation in layered media in terms of filter theory — 281
16.1 Introduction — 281
16.2 Reflection and transmission at an interface — 281
16.3 Reflection in a multi-layered media — 284
16.4 Synthetic seismogram including dissipation — 291
16.5 Reverberation in a water layer — 291
16.6 Removal of ghost reflections — 294
16.7 References — 297

17. Velocity filters — 299
17.1 Introduction — 299
17.2 Linear velocity filters with box-car shaped pass bands — 300
17.3 Recursive relation for a velocity filter — 304
17.4 Non-linear velocity filters — 309
17.5 References — 318

18. Velocity spectra — 319
18.1 Common depth point and correlations — 319
18.2 Vespa and Covespa — 323
18.3 References — 326

19. Polarization analysis — 328
19.1 Introduction — 328
19.2 Theory of polarization — 328
19.3 Time domain polarization filters — 334
19.4 The remode filter — 341
19.5 Surface wave discrimination filter — 344

Contents

	19.6 Polarization states and unitary matrices	347
	19.7 References	353
20.	**Homomorphic deconvolution**	355
	20.1 Cepstrum and quefrency analysis for echoes	355
	20.2 Homomorphic filtering	357
	20.3 References	360
21.	**The Hilbert transform**	361
	21.1 Minimum phase and delay	361
	21.2 Hilbert transform equations	362
	21.3 Envelope of a function	367
	21.4 The phase and modulus of causal filters	368
	21.5 References	371
22.	**Acquisition of digital data**	372
	22.1 Components of a digital system	372
	22.2 Conversion between number systems	377
	22.3 Bipolar and other binary codes	379
	22.4 Reconstruction of analog signals	381
	22.5 References	383
23.	**Walsh transforms and data compression**	384
	23.1 Introduction	384
	23.2 Walsh function	385
	23.3 Hadamard matrix	387
	23.4 Walsh and Hadamard transforms	389
	23.5 Paley transform	390
	23.6 Data compression	392
	23.7 Algorithm for Walsh transform	395
	23.8 References	398
24.	**Generalized linear inverse method**	399
	24.1 Introduction	399
	24.2 Discrete linear inverse problem	399
	24.3 Inverse for an overconstrained system	403
	24.4 Inverse of the underdetermined case	408
	24.5 Conclusions	410
	24.6 References	411

Appendix 1 — 413
  1.1 Fourier series and integrals — 413
  1.2 References — 429

Appendix 2 — 430
  2.1 The Dirac delta function — 430
  2.2 References — 433

Appendix 3 — 434
  3.1 The Wiener-Hopf equation — 434

Appendix 4 — 447
  4.1 Stationary time series and white noise — 447
  4.2 References — 455

Appendix 5 — 456
  5.1 The Hanning and Hamming windows — 456
  5.2 The Bartlett window — 458
  5.3 The Parzen window — 460
  5.4 References — 461

Appendix 6 — 462
  6.1 Stability of filters — 462

Appendix 7 — 465
  7.1 Cooley-Tukey fast Fourier transform — 465

Author Index — 471

Subject Index — 477

Preface to the third edition

Time sequence analysis is carried out extensively by earth scientists in industry and in research laboratories. Courses on the subject are being introduced in every department that pursues geophysics. This text gathers together material from many sources and organizes it in a unified presentation that is suitable for the training of a professional geophysicist at the level of a senior or graduate level course. There are now many books published on various aspects of the subject. Some are entirely devoted to one special topic like deconvolution. Some are written at a very elementary level for the technician or as an introduction to specialists in another field like geology and a few are suitable only for the advanced graduate student. Because of the success of the first two editions it was felt the third edition in this rapidly evolving field should be at the same mathematical level. It incorporates many new developments that were only being researched when the earlier edition was being written but still retains the important introductory sections which are vital for an appreciation of the field.

The title given to this discipline is *time sequence analysis* since many of the applications involve wave phenomena from well-defined temporal sources and it is desired to study relations between a sequence of data points or a sequence of signals in order to determine the physical properties of the earth. Some of the processes involved in contributing to the observed data will be deterministic while others will be random. The origin time of the wave pulse is often known or is a quantity to be determined. The use of the phrase *time series* is usually reserved for a study of random events or for data whose properties are independent of translation of the origin of time.

Care has been taken to relate the subject material to the fields of physics, electrical engineering and geophysics. Original authors have been given credit for their published research discoveries and the references are included. The contents of this book have been used in two semester courses during the past ten years; the material being split between a senior undergraduate level and a graduate level. Additional material, from the fields of seismology and geomagnetism as applied in industry, are assigned to the student as subjects for research projects, seminars and computer modelling experiments. The lecturer will find this material readily available in journals of the Society of Exploration Geophysicists or the European Association of Exploration Geophysicists. This book is intended to present the fundamental background necessary for digital processing of geophysical or other types of experimental data. Much active research is still being done in this field but it is felt that the subjects presented here will assist students in reading the current literature.

Readers familiar with the previous editions will find a large amount of similarity in the topics covered but most of the chapters have been revised extensively. This is particularly true of the chapters on the fast Fourier transform, correlation and covariance, power spectral estimation, maximum entropy analysis, deconvolution, band-pass and polarization filters. A few computer programs are included either as a tutorial aid or to augment subroutines not available in the open literature. The emphasis is a physical understanding of the subject rather than a technical facility in manipulating subroutines. The chapters on spectral analysis were revised extensively to reflect current methods of computation. Most of the material on spectral windows has been relegated to the appendix so that it can be consulted when reading the earlier literature. There are new chapters on the acquisition of digital data, Walsh transforms and data compression, Hilbert transforms, Cepstral analysis, homomorphic deconvolution and linear inverse methods.

For an undergraduate course large sections of Chapters 1 to 3, 6 to 9, 13 to 15 and 21 may be covered in a semester. For a graduate course the various chapters on Fourier, Laplace, Hilbert and Walsh transforms can be reviewed as a section followed by a thorough treatment of spectral analysis (Chapters 6 to 12). This may be followed by a study of filter and deconvolution (Chapters 13 to 15 and 22).

The entire text including the mathematical equations has been set with the aid of a computer using a University of Alberta Computing Services language called TEXTFORM. The computer algorithm carrying out this operation was written by the author and is called BOOKED. The author wishes to acknowledge the assistance of Mr. Dave Holberton in the programming. Members of the Univerty of Alberta Press designed the layout. The help of Ed Grams in exploring the full capabilities of the phototypesetter was invaluable. I would especially like to thank Mr. James Dymianiw for his assistance in formatting the mathematical equations and for preparing the entire text into computer files. Mr. C. H. McCloughan assisted with the computer programs and the figures.

I am greatly indebted to my colleagues for suggestions and encouragement and to Professor F. Abramovici with reference to inverse methods. The many letters and comments from readers were much appreciated and are too numerous to be acknowledged in this preface. I would like to acknowledge the assistance of the students taking my courses on the subject

Preface to the third edition

and my graduate students, particularly Messrs. C. D. Hemmings, J. F. Montalbetti, K. F. Sprenke, G. Mann and Drs. R. G. Agarwal, T. Alpaslan, R. M. Clowes, P. R. H. Gutowski, A. C. Bates, J. Havskov and D.C. Ganley.

E. R. Kanasewich

Department of Physics
The University of Alberta
Edmonton, Canada

1 Introductory observations

1.1 Historical notes

Time sequence analysis involves the extraction of a flow of messages from a background of random noise and unwanted signals and a study of their statistical distribution and relationship in time. The analysis will often involve the use of Fourier integrals, autocorrelation, cross correlation, power spectral studies and convolution or filtering. Because of the speed and convenience of digital computers the signals are most often analyzed in discrete form and are then called either a time series or a time sequence. The time series is represented by a row vector whose coefficients are usually real but may be complex functions of time or space.

$$\mathbf{x} = (x_1, x_2, \ldots, x_n) \qquad 1.1\text{-}1$$

The theoretical basis for these studies was formulated independently by Norbert Wiener (1930, 1949) and A. N. Kolmogorov (1939). This theory was developed for the harmonic analysis of functions of time which did not die out as time approaches infinity. Since the signals were neither periodic nor transient they could not be dealt with directly by the Fourier series or the Fourier integral. Of particular interest is a relation which is analogous to the Euler-Lagrange equation in the calculus of variations and is called the Wiener-Hopf equation (Wiener, 1949). It formulates the solution to an integral equation which gives the optimum linear filter for extracting desirable signals if we are given the shape of the desired signal and a noisy signal which forms the input to the filter.

More recent studies of special significance to geophysics include the work of Blackman and Tukey (1958) on the measurement of power spectra and Burg's (1967) maximum entropy spectral analysis. In the development of digital filters the pioneering research of Hurewicz (1947) on the z transform has not been acknowledged adequately. Also of note is the concept of minimum phase operators by Bode (1945) and minimum delay operators by Robinson (1962); the development of inverse convolution filters by Rice (1962); recursion filters by Golden and Kaiser (1964); and the introduction of the fast Fourier transform by Cooley and Tukey (1966). Of note is the formulation of the theory of digital multichannel filtering by Robinson (1963), Wiggins and Robinson (1968) and Treitel (1970). The development of wave equation computations using finite difference techniques by Boore (1962, 1972) and Claerbout (1970) has been of importance in the inversion of earth structure. More complex models have been studied by Trorey (1970) and Hilterman (1970) with a numerical approach to Huygen's principle using the Helmholtz equation.

Previously Schuster (1898) had invented a *periodogram* analysis for finding hidden periodicities in meteorological and economic data. This involved the determination of a power spectrum by a Fourier analysis. The periodogram has a very irregular appearance because no spectral window has been used for smoothing. The variance of the spectral density is independent of sample size so increasing the number of observations does not reduce it. Taylor (1920) introduced the autocorrelation function

$$a(L) = \frac{[x_t \cdot x_{t+L}]}{[x^2]} \qquad 1.1\text{-}2$$

(in which the brackets indicate a summation) to study changes of pressure in the atmosphere. He showed that the function was always even and applied it to diffusion by turbulent fluid motion. Wiener defined the correlation function without the normalizing factor in the denominator.

$$a(L) = \lim_{T \to \infty} (2T)^{-1} \int_{-T}^{T} x(t) \cdot x(t+L) dt \qquad 1.1\text{-}3$$

Wiener added mathematical rigour to previous work on harmonic analysis and contributed many important theorems which were useful in designing optimum filters for signal detection. Of particular importance was his discovery that the autocorrelation function and the power spectrum are Fourier transforms of each other.

1.2 Applications of time sequence analysis

The earliest example of an outstanding use of power spectral analysis is in the field of physical oceanography. The results are summarized by Munk, Miller, Snodgrass and Barber (1963) but the original papers are of interest also (Barber and Ursell, 1948; Barber, 1954). In one study a tripartite array of pressure transducers was located on the ocean bottom off the coast of California and recorded digitally. Pressure fluctuations decrease exponentially with the ratio of depth of water to the wavelength. Therefore, the water can be used as a convenient filter for the attenuation of short wavelengths which are generated locally by the surf. A complex cross spectrum was obtained from pairs of stations and used to determine the direction and phase velocity of surface waves generated by distant storms. In some studies spectral analyses have been used to detect waves generated 14,000 kilometers away and correlated with storms in

Applications of time sequence analysis

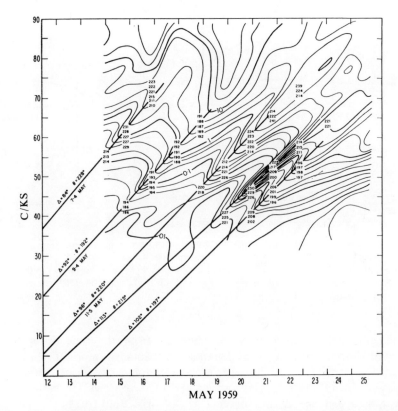

Figure 1.1 Contours of equal power density, E(f,t), on a frequency-time plot. The contours are at equal intervals of log E(f,t). Heavy contours correspond to 0.01, 0.1, 1.0 and 10 cm²/(c/ks). On the time axis the ticks designate midnight U.T. The ridge lines correspond to the dispersive arrivals from a single source, and their slope is inversely proportional to the distance from the source. For each ridge line we have marked the estimated direction, Θ, of the source (clockwise from true north) and the great circle distance Δ in degrees. The small numbers give some computed directions of the optimum point source, Θ_0 (f,t). (From Munk et al., figure 13, 1963, courtesy of the Royal Society, London.)

the Indian Ocean. The waves may have an amplitude of only one millimeter and a wavelength of one kilometer. The detection was possible through the study of the non-stationarity of the time series. In the power spectra of many four-hour-long records they found a peak on successive records corresponding to the wave. The frequency became higher with time indicating a dispersive train of waves. Theoretically the group velocity should be

$$U = g/4\pi f \qquad \text{1.2-1}$$

where f is the frequency. The longer waves travel faster than those with a short

Introductory observations

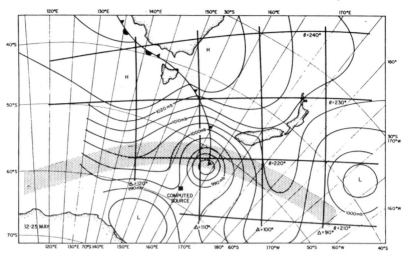

Figure 1.2 Portion of the Southern Ocean Analysis for 0600 U.T., 12 May. The great circle routes bearing Θ=210, 220, 230, 240° from San Clemente Island, and the distances Δ = 90, 100, 110, 120° are indicated. The computed source was for 12.4 May at the position shown in the central figure. Shaded band indicates path of storm. (From Munk et al., figure 36, 1963, courtesy of the Royal Society, London.)

wavelength. The distance may be obtained from the time difference in recording shorter and longer wavelengths. An example of a power density plot is shown in figure 1.1. The contours are at equal intensity of log E where E is the power density. The ridges correspond to dispersive arrivals from a source at a great circle distance, Δ, and direction Θ clockwise from true north. The date correlates with a storm travelling along the shaded path (shown in figure 1.2) south of New Zealand. This example illustrates the use of spectrum analyses to a non-stationary time series since the waves were discovered only because of the characteristic non-stationarity.

The discovery of free oscillations in the earth provide another notable example of the utility of spectral analysis. A theoretical formulation of the problem for a uniform sphere had been made by Lame' (1853), Lord Kelvin (1864) and Lamb (1882). Love (1911) solved the problem for a uniform sphere of the same size and mass as the earth, taking both gravity and compressibility into account, to obtain a period of 60 minutes for the oscillation of the lowest mode, $_0S_2$. The free oscillations are described, principally, by spherical harmonics

$$S_l^m = e^{im\phi} \cdot P_l^m(\cos\theta) \qquad 1.2\text{-}2$$

where P is the associated Legendre polynomial, θ is the colatitude, ϕ is the

Applications of time sequence analysis

longitude, l is the angular degree and m = -l, ..., +l is the angular order. The number of nodal surfaces is given by the radial order, n. Oscillations having a radial displacement are called spheroidal and are denoted by the symbol $_nS_l^m$. Oscillations with only a horizontal component, for which the dilatation is always zero, are called toroidal and identified by $_nT_l^m$. The order number, m, is

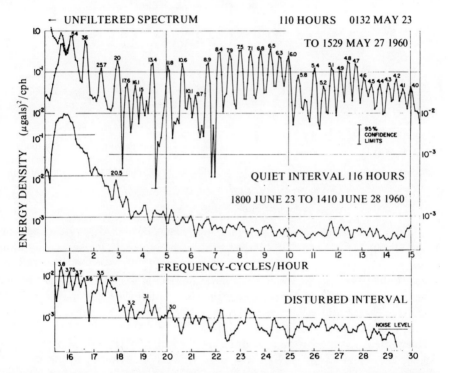

Figure 1.3 Power spectrum of gravity record after Chilean earthquake (22 May 1960) in comparison with that for a corresponding quiet period. (From Ness, Harrison and Slichter, 1961, courtesy of the J. Geophys. Res.)

usually omitted because the frequency, ω_{ln}, is independent of the $2l+1$ values of m for a non-rotating sphere. The fundamental modes are obtained by setting n = 0, the lowest, $_0S_2$, of which is described as the *football* mode because the earth is distorted alternatively from an oblate to a prolate spheroid. The amplitude of the modes depends upon the nature of the earthquake source causing the excitation.

Benioff (1958) observed a wave with a period of about 57 minutes on the Isabella, California, strain seismometer record following the magnitude 8.25 Kamchatka earthquake of 1952. A convincing identification of the modes

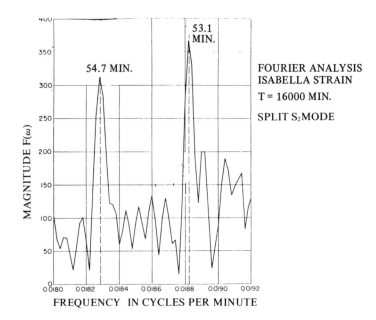

Figure 1.4 High resolution Fourier analysis of Isabella strain seismograph to show split spectral peak. A total of 16,000 minutes was used of the Chilean earthquake of May 1960. (From Benioff, Press and Smith, 1961, courtesy of J. Geophys. Res.)

of oscillation was only possible when more numerical computations had been extended to treat a layered spherical earth. These conditions were met when the next large earthquake occurred on 22 May 1960. Power spectral studies were carried out on a tidal gravimeter recording (figure 1.3) of this 8.25 magnitude Chilean shock. Studies were also done by Benioff et al. (1960); Alsop et al. (1960) and Ness et al. (1960) on observations of strain and pendulum seismographs. Over 70 modes of oscillation were identified with the aid of theoretical computations of Alterman, Jarosch and Pekeris (1959).

Several of the spectral lines were reported by Benioff, Press and Smith (1960) (figure 1.4) to be split and Pekeris (1960) conjectured that this was due to the rotation of the earth since Lamb (1932) had shown theoretically that in a rotating circular basin of water the waves advancing in the direction of rotation had a longer period than waves moving in the opposite direction. The very difficult theoretical problem of free oscillations in a rotating earth was solved by Backus and Gilbert (1961). For the spheroidal mode the frequency interval between the lines is given by

$$\Delta\omega = \beta_{nl}\Omega m \qquad 1.2\text{-}3$$

where Ω is the earth's angular velocity of rotation and β is a complicated

Applications of time sequence analysis

integral function of the displacement field from the center of the earth to the surface. Backus and Gilbert predicted periods of 52.33, 53.10, 53.89, 54.71 and 55.55 minutes for m = +2,1,0,-1,-2 respectively. Since the observed values (figure 1.4) are a doublet at 54.7 and 53.1 minutes they assumed that orders ±1 were excited preferentially. This would be expected if the earthquake source was approximated by a vertical force. It is amusing to note that one obtains an independent confirmation of the rotation of the earth on its axis from a rather sophisticated observation of spheroidal oscillations set up by a large earthquake.

The magnitude 8.4 Alaskan earthquake of 24 April 1964 and the magnitude 7.2 Columbian earthquake of 31 July 1970 are the only other events after which significant free oscillations were observed. An analysis of these has produced a large amount of detailed information about wave velocities, density and attenuation (Q) in the earth. Dziewonski and Gilbert (1972) have established conclusively that the inner core, with a radius of 1,229 km, is solid, having a high Poisson's ratio of 0.44 and an average shear velocity, unknown to that time, of 3.6 km/sec. Their observational results, obtained from a spectral analysis of the records of 84 World Wide Standard Network Seismograph Stations, are shown in figure 1.5.

An average earth model has been difficult to obtain from body waves because the waves arise in an anomalous earthquake belt and are recorded on a continental crust which is structurally different from the more widespread ocean crust. Using 67 hours of free oscillation observations following the Columbian earthquake, Dziewonski and Gilbert have obtained an earth model that is a better representation since the waves have traversed inside the earth many times and sampled it more uniformly. They have concluded that the models being used currently have errors of 4 seconds in S or shear wave travel times and 2 seconds in P or compressional wave travel times.

A determination of the physical properties as a function of radius of the earth as obtained at the earth's surface has been called the *inverse problem of geophysics*. It may be considered a problem in time series analysis in which one desires to find an optimal solution if one is given sets of inaccurate gross earth data. Three techniques have been advanced to solve this inverse problem:

   (1) the Monte Carlo technique
   (2) the hedgehog technique
   (3) the general linear inverse technique.

The Monte Carlo technique was introduced by Press and Biehler (1964) in America and by several groups in the USSR (Buslenko and Schreider, 1961; Asbel, Keilis-Borok and Yanovskaya, 1966; Levshin, Sabitova and Valus, 1966). This involves a random search to find permissable models using a large computer. Press (1968) has made the most comprehensive use of it, testing

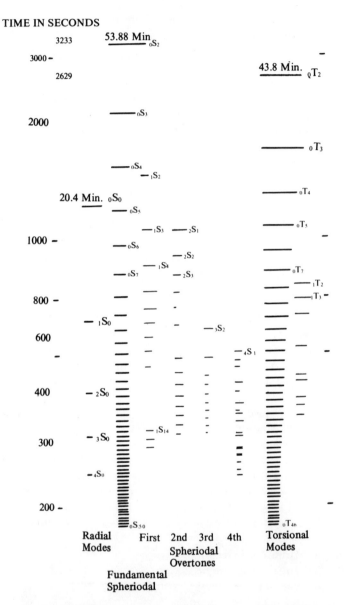

Figure 1.5 Observed spheroidal and torsional modes for the Alaskan earthquake, 28 March 1964. Periods are plotted against amplitude as tabulated by Dziewonski and Gilbert (1972). Data was obtained from 84 World Wide Standard Network Seismograph Stations. Identification of the eigenperiods was by comparison with earth models and by displacement polarizaton.

Applications of time sequence analysis

some 5 million models on data consisting of 97 free oscillation periods, travel time of compressional and shear waves, mass and moment of inertia of the earth. Only six fairly similar models were found to satisfy all the observational tests.

The hedgehog technique was introduced by Valus (1968) and has been described briefly by Levshin, Sabitova and Valus (1966) and also Keilis-Borok and Yanovskaya (1967). A number, N, of parameters (velocities, densities, layer thicknesses, etc.) are chosen together with an upper and a lower band and a step size. This defines an N-dimensional network in which the nodes are possible earth models. The initial determination of an acceptable model is made by a Monte Carlo technique but subsequent models are obtained by a systematic search to find acceptable models. These produce a *hedgehog* pattern in parameter space. A program which incorporates body and surface wave data has been produced under the guidance of Leon Knopoff and V. I. Keilis-Borok and a current version has been implemented by Edo Nyland at UCLA and the Universities of Cambridge and Alberta. The method has the greatest potential for solving ill-determined problems involving a heterogeneous earth. It may be expensive computationally unless controlled through an interactive graphical device.

The third and most successful method for solving inverse problems is the general linear inverse technique which was introduced by Backus and Gilbert (1967, 1968, 1970) in a series of brilliant but difficult papers. More lucid variants of the method have been given by Smith and Franklin (1969), Parker (1970), Jordan and Franklin (1971) and Wiggins (1972). The theory has analogies with the general equations for deconvolution (Rice, 1962) in which the square of the difference between the desired and actual output from a filter is minimized by setting partial derivatives with respect to filter coefficients equal to zero. Consider a set of n model calculations, A_j, and a set of n observations, O_j. The model is computed from a set of m model parameters, M_i. It is desired to minimize the difference between the observations and the model calculations. An essential assumption is that we know an initial starting model well enough so that it is possible to linearize the problem by expanding the computed values in a Taylor series, discarding all second and higher order terms.

$$O_j - A_j = \sum_{i=1}^{m} \frac{\partial A_j}{\partial M_i} \Delta M_i \qquad j=1,2,\ldots,n \qquad \qquad 1.2\text{-}4$$

There will be a systen of n linear equations to solve in order to find the unknown corrections, ΔM_i, to the parameter. This involves the generation of a m by n matrix which relates small changes in the model to small changes in the data, $\partial A_j / \partial M_i$. The normal mode data described previously has been used by both Gilbert, Brune and Dziewonski (1972) and Anderson and Jordan (1972) to obtain new gross earth models with revised P and S wave velocities and a solid inner core.

Time sequence analysis has made a profound change in the analysis and interpretations of geophysical data used in mineral exploration. An early attempt was by Wadsworth, Robinson, Bryan and Hurley (1953) in the Geophysical Analysis Group at MIT. Perhaps the most successful application

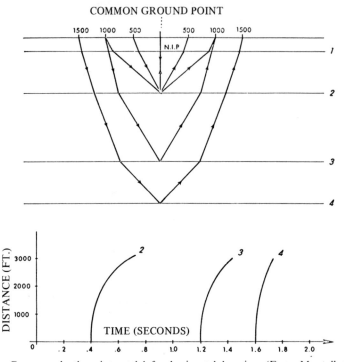

Figure 1.6   Common depth point model for horizontal layering. (From Montalbetti, 1971, courtesy of the Canadian Society of Exploration Geophysicists.)

has been in reflection seismology. Elastic waves may be generated from chemical sources and recorded at nearly vertical incidence to yield reflection seismograms. A large amount of structural mapping has been carried out by geophysical companies in this manner because of the high resolution possible with this technique. Elaborate computer filtering and correlation of digitally recorded data have enhanced the signal to noise ratio so that even stratigraphic traps can be explored for petroleum.

Another example is the automatic computer determination of seismic wave velocities. The feasibility of obtaining velocities was explored by Schneider and Backus (1968) with the use of reflection data acquired by the

Applications of time sequence analysis

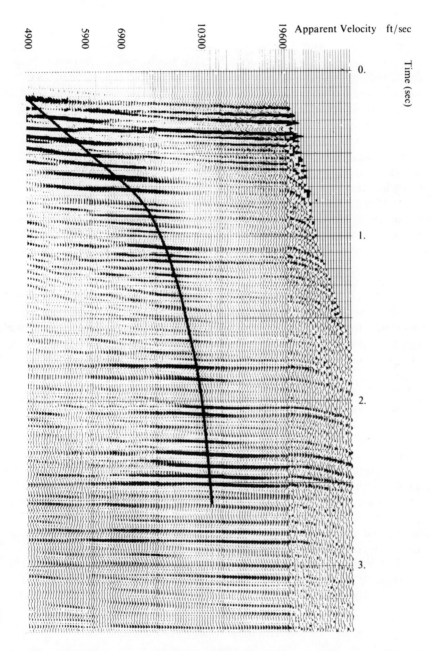

Figure 1.7 Example of a constant velocity stack. The 24-fold input record is shown, followed by the stacked traces for each velocity used. This display demonstrates use of the coherency measure. (From Montalbetti, 1971, courtesy of the Canadian Society of Exploration Geophysicists.)

common depth point (CDP) method pioneered by Mayne (1962). In a CDP reflection section (figure 1.6) the shots and receivers are spaced symmetrically about a central point so that ray paths have a common subsurface reflection point. Coherant seismic energy reflected from a layer at depth is identified by a cross correlation of data at all receivers and the velocity is determined from the time lags having the highest coherency measure. Figures 1.7 and 1.8 are an example of velocity determinations made by Montalbetti (1972) from a plot of correlation coefficients as a function of two-way vertical travel time. The depth to the reflecting layers can be obtained once the velocity is determined. A stratigraphic cross-section can also be made by mapping changes in the interval velocity as a function of location.

Many examples illustrating the effectiveness of various forms of filtering will be discussed in later chapters. For the present, several additional examples of the application of time sequence analysis will be presented from research by the geophysics group at the University of Alberta. They illustrate how statistical techniques may be combined with various physical principles to obtain waves and fields.

The vast quantity of seismic reflection profiling has been mentioned above but there has been very little research carried out on techniques for determining the nature of the reflecting horizons. Clowes and Kanasewich (1970) have taken the power spectra of observed reflected energy from the base of the crust and of synthetic seismograms for various proposed models. Figure 1.9 illustrates sills of alternating high and low velocity material. Other models in which first order discontinuities and linear gradients were used did not match the observed results by at least one order of magnitude.

Two-dimensional cross correlation functions were employed by Agarwal and Kanasewich (1971) to determine automatically the trends of magnetic data using computer analyses. The cross correlations were determined by transforming each matrix of two-dimensional field data into the wave domain with a fast Fourier algorithm, performing the appropriate multiplication, and taking an inverse transform. The trends obtained for an example in the Stony Rapids area, Saskatchewan, were traced by fitting a least squares third degree polynomial to the maximum coefficients in the correlation matrix. The results are illustrated in figure 1.10.

According to an equation originally obtained by Poisson, gravity (U) and magnetic (V) field potentials are related through the density, ρ, and the intensity of magnetization, J, as

$$V = \frac{J}{G\rho} \frac{\partial U}{\partial \nu} \qquad 1.2\text{-}5$$

where G is the gravitational constant and ν is the direction of magnetic polarization. The necessary differential operations are most easily performed in

Applications of time sequence analysis

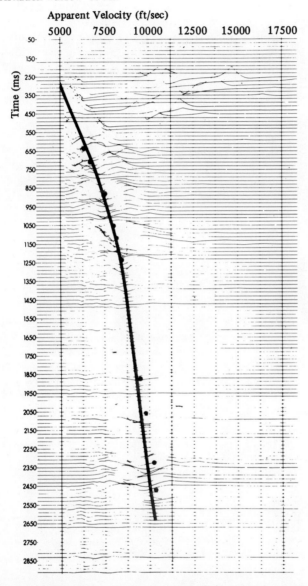

Figure 1.8 Example of statistically normalized cross correlation as a coherence measure. A time gate of 60 ms. with a 20 ms. increment was used. At each time, a sweep of velocities from 5,000 ft/sec. to 20,000 ft/sec. increments was performed. The time-velocity axes are indicated. The input recorded was the same as in the previous example. (From Montalbetti, 1971, courtesy of the Canadian Society of Exploration Geophysicists.)

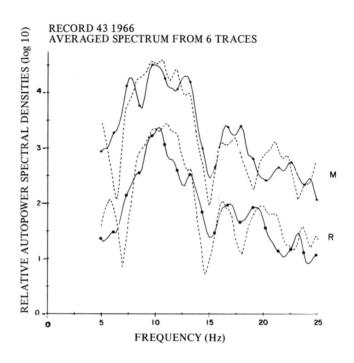
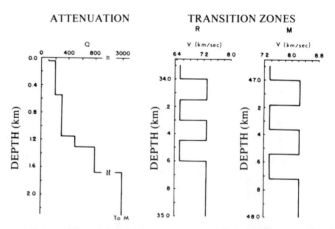

Figure 1.9 Comparison of the auto power spectra for 2-sec. intervals centered about two reflections on field seismograms (solid curves) with spectra computed from reflections on synthetic seismograms (dashed curves). The variation of Q with depth and the characteristics of the transition zone models are shown in the lower part of the figure. (From Clowes and Kanasewich, 1970, courtesy of the J. Geophys. Res.)

Applications of time sequence analysis

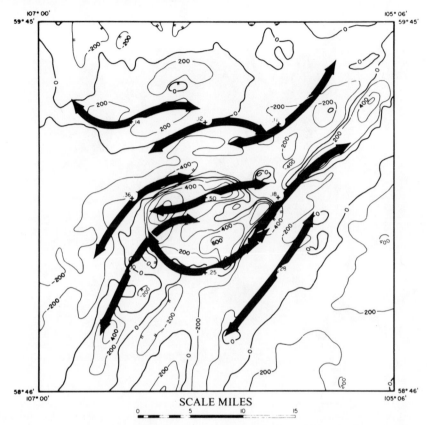

Figure 1.10 Trends in northern Saskatchewan as obtained by using a 12 by 12 mile map. (From Agarwal and Kanasewich, 1971, courtesy of the Society of Exploration Geophysicists.)

the wave number domain. This procedure was suggested by Kanasewich and Agarwal (1970) to generate a theoretical pseudo magnetic field, Z_T, which may be compared to the actual observed magnetic field, Z_0.

A coherence function

$$\mathrm{Coh}^2(\lambda_1,\lambda_2) = \frac{|CP(Z_0,Z_T)|^2}{P(Z_0) \cdot P(Z_T)} \qquad 1.2\text{-}6$$

is used to test the validity of the comparison at each individual wavelength (λ_1,λ_2). In the equation above $CP(Z_0,Z_T)$ is the cross-power spectra of the two data sets while $P(Z_0)$ and $P(Z_T)$ are the auto-power spectra. Gravity and magnetic data from Stony Rapids, Saskatchewan, was used. Figure 1.11 illustrates the high coherence obtained at most wavelengths. It was then

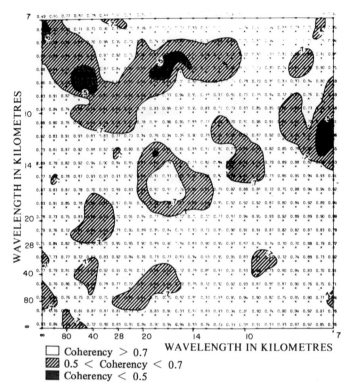

Figure 1.11 Coherence contoured as a function of wavelength for the Stony Rapids anomaly in northern Saskatchewan. (From Kanasewich and Agarwal, 1970, courtesy of the J. Geophys. Res.)

possible to use equation 1.2-5 to compute the ratio of the intensity of magnetization to the density (J/ρ) of the rocks responsible for the anomalous fields. The values obtained are illustrated in figure 1.12 and are in general accord with those measured for basic rocks. Thus it was concluded that both the density contrast and the magnetization are due to norite and associated basic rocks which outcrop in the area.

Geomagnetic studies using an array of 46 Gough magnetic variometers have been carried out in a co-operative program between the University of Alberta and the University of Texas at Dallas. Field measurements by Reitzel, Gough, Porath and Anderson (1970) for one line in the southwestern United States is shown in figure 1.13 together with a contour map of the results after Fourier transformation to obtain spectral amplitudes at a period of 60 minutes. Heat flow values correlate well with the electromagnetic induction anomaly over the Southern Rockies. A model proposed by Porath and Gough (1971) accounts for the correlation. This model with highly conductive slabs in the upper mantle to simulate a rise in the $1500^{0}C$ isotherm is not unique but it

Applications of time sequence analysis

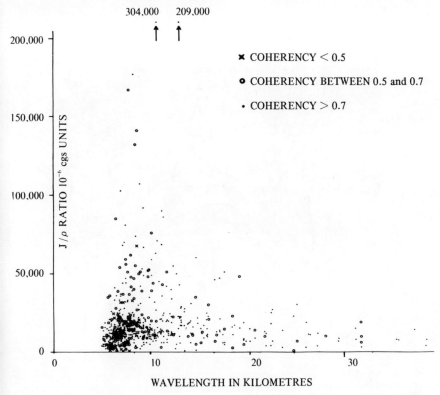

Figure 1.12 Computed values of J/ρ for the Stony Rapids area, Saskatchewan. (From Kanasewich and Agarwal, 1970, courtesy of the J. Geophys. Res.)

seems fairly conclusive evidence that the uplift of the mountains involved the upper mantle and that the process has taken a very long time to reach the present state of equilibrium.

There are many other examples of the use of time sequence analysis to electromagnetic fields. Auto- and cross-spectral estimates are essential in the application of the magnetotelluric method (Cagniard, 1953) to determine the apparent resistivity of the earth. Samson, Jacobs and Rostoker (1971) have also used spectral analysis in space physics research to determine the polarization characteristics of the energy sources and the fields associated with plasmas in the ionosphere.

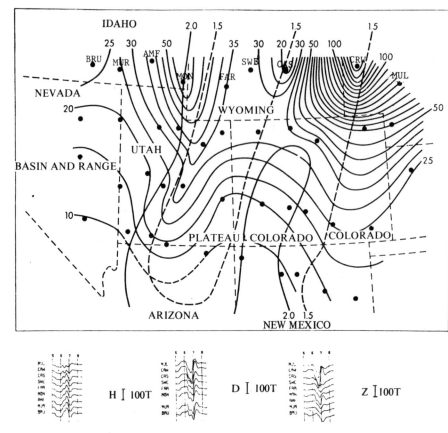

Figure 1.13 Contours of Fourier spectral amplitude at T=60 minutes of the vertical field, Z, of a substorm on 1 September 1967. Unit for Z is 0.1γ. Heat-flow isolines at 1.5 and 2.0 microcal/cm^{2}sec. from a map by Roy, Blackwell, Decker and Decker. Note tracking of high Z along the east flank of each heat-flow high, corresponding to currents under the heat-flow maxima. Dots represent variometers, straight broken lines outline Utah, Colorado and neighbouring states. The insert on the right shows variograms of the northern line of stations for a substorm on 1 September 1967. (From Reitzel et al., 1970, courtesy of the Geophys. J. R. Astr. Soc.)

References

Agarwal,R.G. and Kanasewich,E.R.(1971), Automatic trend analyses and interpretations of potential field data. *Geophysics* **36**, 339-348.

Alsop,L.E., Sutton,G.H. and Ewing,M.(1961), Free oscillation of the earth observed on strain and pendulum seismographs. *J. Geophys. Res.* **66**, 631-641.

References

Alterman,Z., Jarosch,H. and Pekeris,C.L.(1959), Oscillations of the earth, *Proc. Roy. Soc. A*, **252**, 80-95.

Anderson,D.L. and Jordan,T.H.(1972), Interpretation of gross earth inversion models. Abstract. *Ninth International Symposium on Geophysical Theory and Computers, Banff, Canada, University of Alberta*.

Asbel,I.Ja., Keilis-Borok,V.I. and Yanovskaya,T.B.(1966), A technique of a joint interpretation of travel-time and amplitude-distance curves in the upper mantle studies. *Geophysical Journal of Roy. Astr. Soc.* **11**, 25-55.

Backus,G. and Gilbert,J.F.(1961), The rotational splitting of the free oscillations of the earth. *National Academy of Sciences, Proceedings* **47**, 363-371.

Backus,G.E. and Gilbert,J.F.(1967), Numerical applications of a formalism for geophysical inverse problems. *Geophysical Journal of Roy. Astr. Soc.* **13**, 247-276.

Backus,G.E. and Gilbert,J.F.(1968), The resolving power of gross earth data. *Geophysical Journal of Roy. Astr. Soc.* **16**, 169-205.

Backus,G.E. and Gilbert,J.F.(1970), Uniqueness in the inversion of inaccurate gross earth data. *Philosophical Transactions of the Royal Society of London* **266**, 123-192.

Barber,N.F. and Ursell,F.(1948), The generation and propagation of ocean waves and swell. *Phil. Trans. A* **240**, 527-560.

Barber,N.F.(1954), Finding the Direction of Travel of Sea Waves. *Nature* **174**, 1048-1050.

Benioff,H.(1958), Long waves observed in the Kamchatka earthquake of November 4, 1952. *J. Geophys. Res.* **63**, 589-593.

Benioff,H., Press,F. and Smith,S.(1961), Excitation of free oscillations of the earth by earthquakes. *J. Geophys. Res.* **66**, 605-619.

Blackman,R.B. and Tukey,J.W.(1958), The measurement of power spectra. *Dover, New York*.

Bode,H.W.(1945), Network analysis and feedback amplifier design. *Princeton. Van Nostrand, New York*.

Boore,D.M.(1969), PhD thesis at Geophysics Dept., M.I.T.

Boore,D.M.(1972), Finite difference methods for seismic wave propagation in heterogeneous materials. *In methods in computational Physics, Volume 11*, **37**, Academic Press New York.

Burg,J.P.(1967), Maximum entropy spectral analysis. *Paper presented at the 37th Annual Int. SEG meeting, Oklahoma, Oct. 31, 1967. Preprint - Texas Instruments, Dallas*.

Buslenko,P.P. and Schreider,Ju.A.(1961), Method of statistical testing and its realization on digital computers. *(In Russian), Fizmatgiz, Moscow*.

Cagniard,L.(1953), Basic theory of the magnetotelluric method of geophysical prospecting. *Geophysics* **18**, 605-635.

Claerbout,J.F.(1969), Coarse grid calculations of waves in inhomogeneous media with application to delineation of complicated seismic structure. *Geophysics* **35**, 407-418.

Clowes,R.M. and Kanasewich,E.R.(1970), Seismic attenuation and the nature of reflecting horizons within the crust. *J. Geophys. Res.* **75**, 6693-6705.

Cooley,J.W. and Tukey,J.W.(1965), An algorithm for the machine calculation of complex Fourier series. *Mathematics of Computation* **19**, 297-301.

Dziewonski,A.M. and Gilbert,F.(1972), Observations of normal modes from 84 recordings of the Alaskan Earthquakes of March 28, 1964. *Geophys. Journal of Roy. Astr. Soc.* **27**, 393-446.

Gilbert,F., Brune,J.N. and Dziewonski,A.M.(1972), Inversion and resolution for the seismological data set. Abstract. *Ninth International Symposium on Geophysical Theory and Computers. Banff, Canada, University of Alberta.*

Golden,R.M. and Kaiser,J.F.(1964), Design of a wide-band sampled data filter. *The Bell Telephone Technical Journal, 1533-1547.*

Hilterman,F.J.(1970), Three dimensional seismic modeling. *Geophysics* **35**, 1020-1037.

Hurewicz,W.(1947), Filters and servo systems with pulsed data. *Ch. 5, p 231-261 in Theory of Servomechanisms. Editors: James, H. M., Nichols, N. B. and Phillips, R. S., McGraw-Hill, New York.*

Jordan,T.H. and Franklin,J.N.(1971), Optimal solutions to a linear inverse problem in geophysics. *Proc. National Academy of Sciences* **68**, 291-293.

Kanasewich,E.R. and Agarwal,R.G.(1970), Analysis of combined gravity and magnetic fields in wave number domain. *J. Geophys. Res.* **75**, 5702-5712.

Keilis-Borok,V.I. and Yanovskaya,T.B.(1967), Inverse problems of seismology. *Geophysical Journal of the Roy. Astr. Soc.* **13**, 223-234.

Kelvin,Lord(Sir William Thomson), (1864), Dynamical problems regarding elastic spheroidal shells and spheroids of incompressible liquid. *Phil. Trans. Roy. Soc.* **153**, 563-616.

Kolmogorov,A.N.(1939), Interpolation and extrapolation. *Bulletin de l'academie des sciences de USSR. Ser. Math.* **5**, 5-14, 1941. Compt les Rendues, Paris, 1939.

Lamb,H.(1882), On the vibrations of an elastic sphere. *London Math. Soc. Proc.* **13**.

Lamb,H.(1932), Hydrodynamics, p 320, article 209, *Cambridge University Press.*

Lame',M.G.(1853), Memoire sur l'equilibre d'e'lasticite' des envelopes sphe'riques. *Comptes Rendus,* **37**, see also *Journal de mathematiques,* Liouville (1854), **19**, 51-87.

Levshin,A.L., Sabitova,T.M. and Valus,V.P.(1966), Joint interpretation of body and surface waves data for a district in middle Asia. *Geophysical Journal of Roy. Astr. Soc.* **11**, 57-66.

Love,A.E.H.(1911), Some problems of geodynamics. (See Love, A treatise on the mathematical theory of elasticity. Dover Publ. p 286).

Mayne,W.H.(1962), Common reflection point horizontal data stacking techniques. *Geophysics* **27**, 927-938.

Montalbetti,J.F.(1971), Computer determinations of seismic velocities - a review. *Journal of the Canadian Society of Exploration Geophysicists* **7**, 32-45.

Munk,W.H., Miller,G.R., Snodgrass,F.E.and Barber,N.F.(1963), Directional recording of swell from distant storms. *Phil. Trans. Royal Society, London* **A,255**, 505-589.

Ness,N.F., Harrison,J.C. and Slichter,L.B.(1961), Observations of the free oscillations of the earth. *J. Geophys. Res.* **66**, 621-629.

Parker,R.L.(1970), The inverse problem of electrical conductivity in the mantle. *Geophysical Journal of Roy. Astro. Soc.* **22**, 121-138.

Pekeris,C.L.(1960), Talk at the Helsinki meeting of the IUGG. See also National Academy of Sciences, Proceedings **47**, 95-96, 1961.

Porath,H. and Gough,D.I.(1971), Mantle conductive structures in the western United States from magnetometer array studies. *Geophys. J. R. Astr. Soc.* **22**, 261-275.

Press,F.(1968), Earth models obtained by Monte Carlo Inversion. *J. Geophys. Res.* **75**, 5223-5234.

Press,F. and Biehler,S.(1964), Inferences on crustal velocities and densities from P wave delays and gravity anomalies, *J. Geophys. Res.* **69**, 2979-2995.

Reitzel,J.S., Gough,D.I., Porath,H.and Anderson,III,C.W.(1970), Geomagnetic deep sounding and upper mantle structure in the western United States. *Geophys. J. R. Astr. Soc.* **19**, 213-235.

References

Rice,R.B.(1962), Inverse convolution filters. *Geophysics* **27**, 4-18.

Robinson,E.A.(1962), Random wavelets and cybernetic systems. *Hafner Publishing Co., New York*.

Robinson,E.A.(1963), Mathematical development of discrete filters for the detection of nuclear explosions. *J. Geophys. Res.* **68**, 5559-5567.

Samson,J.C., Jacobs,J.A. and Rostoker,G.(1971), Latitude-dependent characteristics of long-period geomagnetic micropulsations. *J. Geophys. Res.* **76**, 3675-3683.

Schneider,W.A. and Backus,M.M.(1968), Dynamic correlation analysis. *Geophysics* **33**, 105-126.

Schuster,A.(1898), Investigation of hidden periodicities. *Terrestrial Magnetism* **3**, 13-41.

Smith,M.L. and Franklin,J.N.(1969), Geophysical application of generalized inverse theory. *J. Geophys. Res.* **74**, 2783-2785.

Taylor,G.I.(1920), Diffusion by continuous movements. *Proc. London Math. Soc., Ser. 2,* **20**, 196-212.

Treitel,S.(1970), Principles of digital multichannel filtering. *Geophysics* **35**, 785-811.

Trorey,A.W.(1970), A simple theory for seismic diffractions. *Geophysics* **35**, 762-784.

Valus,V.P.(1968), Computation of seismic cross-sections conforming to a set of observations: Some direct and inverse seismic problems. *4th issue, ed. Keilis-Borok, V. I., Institute of Physics of the Earth, Schmidt, O. Yu, Moscow, p 3-14*.

Wadsworth,G.P., Robinson,E.A., Bryan,J.G. and Hurley.P.M.(1953), Detection of reflections on seismic records by linear operators. *Geophysics* **18**, 539-586.

Wiener,N.(1930), Generalized harmonic analysis. *Acta Mathematica* **55**, 117-258.

Wiener,N.(1949), Extrapolation, interpolation and smoothing of stationary time series. *J. Wiley & Sons, New York*.

Wiggins,R.W. and Robinson,E.A.(1965), Recursive solution to the multichannel filtering problem. *J. Geophys. Res.* **70**, 1885-1891.

Wiggins,R.A.(1972), The general linear inverse problem: Implications of surface waves and free oscillations for earth structure. *Rev. of Geophysics and Space Physics* **10**, 251-285.

2 Convolution of a time series

2.1 Introduction
Every signal that is recorded in a scientific experiment is modified in some manner by both natural physical processes and the instruments used to measure it. This modification may be thought of as a filtering process and its effect may be described mathematically by a convolution equation. Since the convolution operation is basic to all time sequence analysis it will be described in this section as it applies to digital data.

The physical significance of a convolution is particularly simple in its digital form; however, a more complete analysis with analog signals is also necessary and will be presented later.

The signals which are to be studies in this section will be zero up to some initial time and will approach zero as time increases without limit. In other words we demand that the integral of the square of any signal or its energy be finite.

$$E = \int_{-\infty}^{\infty} [x(t)]^2 \, dt < \infty \qquad 2.1\text{-}1$$

For signals that are sampled at discrete intervals of time

$$E = \sum_{j=0}^{\infty} [x_j]^2 < \infty \qquad 2.1\text{-}2$$

If $x(t)$ is a time varying voltage across a resistor, R, the power dissipated will be

$$P = \frac{E}{R} \qquad 2.1\text{-}3$$

The sampled data, $x = (...x_{-2}, x_{-1}, x_0, x_1, x_2,...)$, may be passed through a linear operator or filter.

Introduction

$$\mathbf{W} = (..., W_{-2}, W_{-1}, W_0, W_1, W_2, ...)$$

$$\mathbf{x} \longrightarrow \boxed{\mathbf{W}} \longrightarrow \mathbf{y}$$

The necessary and sufficient condition for this operator to be stable is that

$$\sum_{t=-\infty}^{\infty} |W_t| < \infty \qquad 2.1\text{-}4$$

The proof (see Appendix 6) is presented by Hurewicz (1947). However, a weaker condition is usually sufficient for a time series.

$$\sum_{t=-\infty}^{\infty} |W_t|^2 < \infty \qquad 2.1\text{-}5$$

This insures that the result from a convolution converges in the mean for a sequence of random variables, **x**.

The output from the filter is given by the convolution of **x** with **W**

$$y_L = \Delta t \sum_{t=-\infty}^{\infty} W_t x_{L-t} \qquad 2.1\text{-}6$$

where Δt will be assumed to be unity from now on. The operation is written in abbreviated notation as

$$\mathbf{y} = \mathbf{W} * \mathbf{x} \qquad 2.1\text{-}7$$

The filter given above is *non-realizable* because it requires all past and future values of the data.

It is more usual to have a finite amount of data. Thus if there are $n+1$ points in the time series, all the values before and after the given sequence are set equal to zero.

$$\mathbf{x} = (...\ 0,\ 0,\ x_0,\ x_1,\ x_2,\ ...\ x_n,\ 0,\ 0,\ ...)$$

Similarly one can design a one-sided linear operator called the impulse response of the filter.

$$\mathbf{W} = (\ldots\ 0,\ 0,\ W_0,\ W_1,\ W_2,\ \ldots\ W_m,\ 0,\ 0,\ \ldots)$$

It is called an impulse response because if one feeds an impulse or a Dirac delta function to the filter, the output from the filter is just the impulse response. In sampled data the equivalent function is a unit impulse which is sometimes called a sigma function (Krutko, 1969).

$$\ldots 0,0,0,0,1,0,0,0,0,\ldots \longrightarrow \boxed{\mathbf{W}} \longrightarrow 0,0,0,0,W_0,W_1 .. W_m,0,0,.$$

The filter is *realizable* because it only operates on past data and does not require future values as in equation 2.1-6.

$$x_0,\ x_1,\ \ldots\ x_n \longrightarrow \boxed{W_0,\ W_1\ \ldots\ W_m} \longrightarrow y_0,\ y_1,\ \ldots\ y_{m+n}$$

The output from this filter is the convolution, $\mathbf{y} = \mathbf{W} * \mathbf{x}$.

$$y_L = \sum_{t=0}^{m} W_t \cdot x_{L-t} \qquad 0 \leqslant L \leqslant m+n \qquad 2.1\text{-}8$$

The convolution may be viewed either as an integration as given in equation 2.1-8, or either of two kinds of geometric operations, or as an algebraic operation. These various interpretations will be illustrated with an example.

2.2 Convolution as a digital integration

Let the input be the wavelet $\mathbf{x} = (2,0,-1)$ and let the impulse response of the filter be $\mathbf{W} = (4,2,1)$. The output is given by equation 2.1-8.

$$y_0 = x_0 W_0 = 8$$

$$y_1 = x_0 W_1 + x_1 W_0 = 4 + 0 = 4$$

$$y_2 = x_0 W_2 + x_1 W_1 + x_2 W_0 = 2 + 0 - 4 = -2$$

$$y_3 = x_1 W_2 + x_2 W_1 = 0 - 2 = -2$$

Convolution as a digital integration

$$y_4 = x_2 W_2 = -1$$

$$y = (8, 4, -2, -2, -1)$$

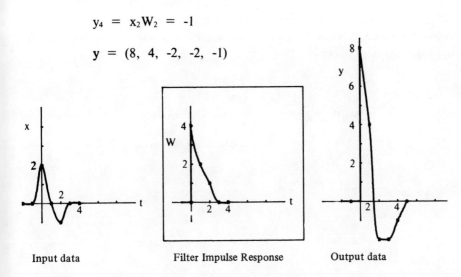

Input data   Filter Impulse Response   Output data

Figure 2.1   Convolution of a wavelet with an impulse response.

2.3 Convolution as a geometric operation of folding

The input data and the impulse response may be written along the x and y axes of a graph and the products formed in the interior (Robinson, 1964). The output from a convolution operation may be obtained by summing the products along diagonal lines as illustrated below in figure 2.2.

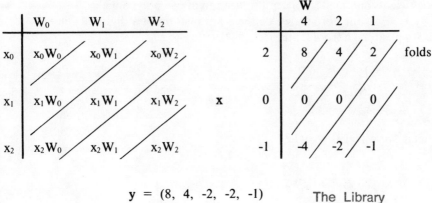

$$y = (8, 4, -2, -2, -1)$$

Figure 2.2   Convolution as a folding operation.

2.4 Convolution as a geometric operation of sliding

Convolution may be regarded as a sliding operation or autocorrelation with the input wavelet reversed as in figure 2.3.

$$y_0 = \boxed{\begin{array}{ccc} x_2 & x_1 & x_0 \end{array}} = x_0 W_0 = 8$$
$$\boxed{\begin{array}{ccc} W_0 & W_1 & W_2 \end{array}}$$

$$y_1 = \boxed{\begin{array}{ccc} x_2 & x_1 & x_0 \end{array}} = x_1 W_0 + x_0 W_1 = 4$$
$$\boxed{\begin{array}{ccc} W_0 & W_1 & W_2 \end{array}}$$

$$y_2 = \boxed{\begin{array}{ccc} x_2 & x_1 & x_0 \end{array}} = x_2 W_0 + x_1 W_1 + x_0 W_2 = -2$$
$$\boxed{\begin{array}{ccc} W_0 & W_1 & W_2 \end{array}}$$

$$y_3 = \boxed{\begin{array}{ccc} x_2 & x_1 & x_0 \end{array}} = x_2 W_1 + x_1 W_2 = -2$$
$$\boxed{\begin{array}{ccc} W_0 & W_1 & W_2 \end{array}}$$

$$y_4 = \boxed{\begin{array}{ccc} x_2 & x_1 & x_0 \end{array}} = W_2 W_2 = -1$$
$$\boxed{\begin{array}{ccc} W_0 & W_1 & W_2 \end{array}}$$

Figure 2.3 Convolution as a sliding operation.

2.5 Convolution as an algebraic operation - z transformation

For digital computation and for analysis of stability an algebraic concept of convolution is preferable. This interpretation considers convolution as the coefficients obtained from a multiplication of two polynomials.

For our purpose we define a z transform of a digitized function with the opposite sign from Hurewicz (1947)

$$W(z) = \sum_{n=0}^{\infty} w_n z^n \qquad \qquad 2.5\text{-}1$$

that is, for wavelets **w** and **x**

$$W(z) = w_0 + w_1 z + w_2 z^2 + \ldots$$

Convolution as an algebraic operation - z transformation

$$X(z) = x_0 + x_1 z + x_2 z^2 + \ldots$$

where z is a complex number. Instead of the symbol W(z) it is sometimes convenient to represent the z transform of a function w by $Z[w]$.

Let U be a unit step such that

$$U_i = \begin{matrix} 1 & i \geq 0 \\ 0 & i < 0 \end{matrix} \qquad 2.5\text{-}2$$

then the z transform of the product of two functions delayed by i units of time is

$$Z[w_{t-i} U_{t-i}] = \sum_{n=0}^{\infty} (w_{t-i} U_{t-i}) z^n \qquad 2.5\text{-}3$$

$$= z^i \sum_{n=0}^{\infty} w_{n-i} U_{n-i} z^{n-i}$$

let $n - i = j$

$$Z[w_{t-i} U_{t-i}] = z^i \sum_{j=-i}^{\infty} w_j U_j z^j$$

But by the definition of the unit step function, U_j is zero for negative i

$$Z[w_{t-i} U_{t-i}] = z^i \sum_{j=0}^{\infty} w_j z^j$$

or if the function w is defined so it is zero before zero time then the unit step can be omitted

$$Z[w_{t-i}] = z^i \sum_{j=0}^{\infty} w_j z^j$$

$$Z[w_{t-i}] = z^i W(z) \qquad 2.5\text{-}4$$

thus a delay of i units of time is easily generated by raising z to the ith power without altering the original z transform of the function.

Now consider the product

$$Y(z) = W(z) X(z) = \sum_{n=0}^{\infty} w_n z^n X(z)$$

or by 2.5-4

$$W(z) X(z) = \sum_{n=0}^{\infty} w_n z^n \sum_{m=0}^{\infty} x_m z^m$$

Rearrange order of summation and replace $z^n x_m$ by its delayed equivalent, equation 2.5-4.

$$W(z) X(z) = \sum_{m=0}^{\infty} \left[\sum_{n=0}^{\infty} w_n x_{m-n} \right] z^m$$

$$\qquad 2.5\text{-}5$$

$$W(z) X(z) = Z\left[\sum_{n=0}^{\infty} w_n x_{m-n} \right]$$

Equation 2.5-5 is the z transform of a convolution equation since the term in brackets is identical to equation 2.1-8. The upper limit is m not ∞ since the function x_{m-n} is zero for (m - n) less than zero.

The z transform is also known as the *generating function* of Laplace (Jordan, 1947, p. 21; or Rice, 1962).

The z transform of the output y is

$$Y(z) = X(z) W(z) = y_0 + y_1 z + y_2 z^2 + \ldots$$

In our example of a convolution of two three-by-three row vectors we obtain

Convolution as an algebraic operation - z transformation

$$w_0 + w_1 z + w_2 z^2$$

$$x_0 + x_1 z + x_2 z^2$$

$$x_0 w_0 + (x_0 w_1 + x_1 w_0)z + (x_0 w_2 + x_1 w_1 + x_2 w_0)z^2$$
$$+ (x_1 w_2 + x_2 w_1)z^3 + x_2 w_2 z^4$$

Notice that the coefficients of the various powers of z give the output data **y**.

In passing one may note that convolutions are commutative, distributive and associative.

i.e.

$\mathbf{x} * \mathbf{w} = \mathbf{w} * \mathbf{x}$ 2.5-6

$\mathbf{x} * (\mathbf{w} + \mathbf{g}) = (\mathbf{x} * \mathbf{w}) + (\mathbf{x} * \mathbf{g})$ 2.5-7

$(\mathbf{x} * \mathbf{w}) * \mathbf{g} = \mathbf{x} * (\mathbf{w} * \mathbf{g})$ 2.5-8

2.6 Convolution of analog signals

It should be noted that convolution need not be restricted to digital data. Smith (1958), in a classic paper, developed a magnetic recording delay-line system in which the coefficients $W_0, W_1 .... W_m$ are simulated by resistors (see figure 2.4).

2.7 Convolution algorithm

For digital calculation, convolution by polynomial multiplication is recommended. Robinson (1966) has given a simple algorithm for performing convolutions.

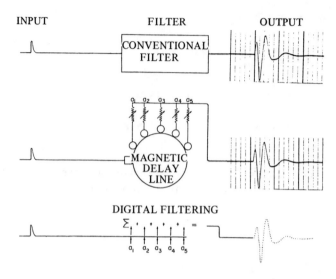

Figure 2.4 A representation of delay-line filtering and digital filtering. (From Smith, 1958, courtesy of the Society of Exploration Geophysicists.)

References

Hurewicz,W.(1947), Filters and servo systems with pulsed data. *Ch. 5, p. 231-261, in Theory of servo mechanisms.* Editors: H. M. James, N. B. Nichols and R. S. Phillips, McGraw-Hill, New York.

Jordan,Charles(1947), Calculus of finite differences. *2nd Edition, New York, Chelsea Publishing Co.*

Krutko,P.D.(1969), Statistical dynamics of sampled data systems. *Chapter 2. Translated from Russian by Scripta Technica Ltd., American Elsevier Inc., New York.*

Laplace,P.S.(1812), Des fonctions génératrices a une variable. *Livre premier, Théorie Analytique des Probabilités, Oeuvres completes* **7**, 1-180. Courcier, Paris.

Rice,R.B.(1962), Inverse convolution filters. *Geophysics* **27**, 4-18.

Robinson,E.A.(1964), Wavelet composition of time series. *Ch. 2 of Econometric Model Building,* Ed. H. S. Wold, NorthHolland Publishing Co., Amsterdam.

Robinson,E.A.(1966), Fortran II programs for filtering and spectral analysis of single channel time series. *Geophys. Prosp. Vol. XIV, p. 8.*

Smith,M.K.(1958), A review of methods of filtering seismic data. *Geophysics* **23**, 44-57.

3 Fast Fourier transforms

3.1 Fourier series

The Fourier series was introduced by Leonhardt Euler and others between 1748 and 1753 when it was proposed that a function, f(x), describing the motion of a string fixed at x = 0 and 1 could be expanded into a sine series with period 2.

$$f(x) = \sum_{n=1}^{\infty} F(n) \cdot \sin n\pi x \qquad 3.1\text{-}1$$

This mathematical description stemmed from a series of refined experiments that Daniel Bernoulli carried out around 1732 regarding the oscillation of strings and other vibrating systems. From these, Euler inferred that a string on a musical instrument such as a violin can vibrate in harmonics, that is, halves, thirds, fourths, etc. Bernoulli (1733) put forward the principle that small oscillations could coexist in the musical vibration of strings so that a solution could be expressed in terms of a trigonometric series. Both Bernoulli and Euler applied the principle to the vibration of a column of air as in an organ pipe. The Fourier coefficients, F(n), were given by Euler in 1777 as

$$F(n) = 2 \int_0^1 f(x) \cdot \sin n\pi x \cdot dx \qquad 3.1\text{-}2$$

The name of the series became associated with Joseph Fourier, a mathematician and engineer, when he presented a talk to the French Academy on 7 December 1807, in which he stated that *any arbitrary function*, defined in a finite interval, can be expressed as a sum of sine and cosine functions. This claim was disputed by many mathematicians including Lagrange. Fourier's work was not published by the French Academy until 1822 and more completely in 1824-26. The validity of the theorem was gradually extended and accepted with the works of Lejeune Dirchlet (1829), Riemann (1854), Fejér (1904) and Lebesgue (1905). The validity of a Fourier series was found to depend only on the integratability of the function f(x).

Many boundary value or initial value problems in physics involve the solution of and equation of the type

$$Ly + \lambda r(t)y = 0 \qquad 3.1\text{-}3$$

where L is a linear homogeneous second order differential operator. An example is the equation

$$\frac{d^2y}{dt^2} + \lambda y = 0 \qquad 3.1\text{-}4$$

where r(t) is 1 and the solution consists of the set of orthogonal functions over some interval (0,T). These may be sine, cosine or exponential functions

$$y_n = \begin{cases} \sin n\pi t/T \\ \cos n\pi t/T \\ e^{\pm i\omega_n t} \end{cases} \qquad 3.1\text{-}5$$

where

$$\omega_n = 2\pi n/T \qquad n = 0, \pm 1, \pm 2, \ldots \qquad 3.1\text{-}6$$

and

$$\lambda_n = 4n^2\pi^2/T^2 \qquad 3.1\text{-}7$$

The λ_n's are called eigenvalues and ω is the circular frequency in radians per second.

The initial conditions on any physical problem are usually expressed as a function, f, of time or position and these are often conveniently transformed into a Fourier series. Any function which is piecewise smooth in the interval $-T/2 \leqslant t \leqslant T/2$ and periodic with period T may be expanded in a complete Fourier series including both sines and cosines

$$f(t) = 0.5\, a_0 + \sum_{n=1}^{\infty} (a_n \cos \omega_n t + b_n \sin \omega_n t) \qquad 3.1\text{-}8$$

where

Fourier series

$$a_n = 2T^{-1} \int_{-T/2}^{T/2} f(t) \cos \omega_n t \, dt \quad n=0,1,2,... \qquad 3.1\text{-}9$$

$$b_n = 2T^{-1} \int_{-T/2}^{T/2} f(t) \sin \omega_n t \, dt \quad n=1,2,... \qquad 3.1\text{-}10$$

The *amplitude spectrum* of any frequency component is

$$|F_n| = (a_n^2 + b_n^2)^{1/2} \qquad 3.1\text{-}11$$

This spectrum is an even function if n is thought to extend to negative values or frequencies. The phase spectrum is

$$\phi_n = \tan^{-1}[+b_n/a_n] \qquad 3.1\text{-}12$$

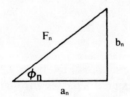

Figure 3.1 Relationship between amplitude, phase spectrum and Fourier coefficients.

The *phase lag* or phase characteristic is used sometimes by engineers and is the negative of phase spectrum (Chen, 1964). The phase spectrum is an antisymmetric or odd function. Therefore if we specify the amplitude and phase for all frequencies, the function is completely determined.

An alternative and equivalent method is to use exponential functions together with positive and negative frequencies (see Appendix 1 for proof)

$$f(t) = \sum_{n=-\infty}^{\infty} F_n \, e^{i\omega_n t} \qquad 3.1\text{-}13$$

where the Fourier coefficients are given by

Fast Fourier transforms

$$F_n = T^{-1} \int_{-T/2}^{T/2} f(t)\, e^{-i\omega_n t}\, dt \qquad 3.1\text{-}14$$

The coefficient F_n is complex. Some proofs and properties of Fourier series and integrals are reviewed in Appendix 1.

3.2 The discrete Fourier transform

If the basic interval of the Fourier series, eq. 3.1-13, is translated by half a period the equation may be written in the form

$$f(t) = T^{-1} \sum_{n=-\infty}^{\infty} F_n\, e^{i\omega_n t} \qquad 3.2\text{-}1$$

$$F_n(\omega) = \int_0^T f(t)\, e^{-i\omega_n t}\, dt \qquad 3.2\text{-}2$$

The normalization factor, $1/T$, may be shifted from 3.2-1 to 3.2-2 if desired. To obtain a set of sampled data the signal is passed through an infinite Dirac comb (see figure 3.2).

$$\nabla(t;\,\Delta t) = \sum_{n=-\infty}^{\infty} \delta(t - n\Delta t) \qquad 3.2\text{-}3$$

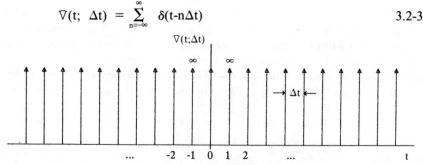

Figure 3.2 An infinite Dirac comb consists of an infinite series of Dirac delta functions spaced Δt apart from each other.

The Fourier transform of this *function* also consists of an infinite series of

The discrete Fourier transform

spectral lines, all with the same weighting function or spectral amplitude (Blackman and Tukey, 1958).

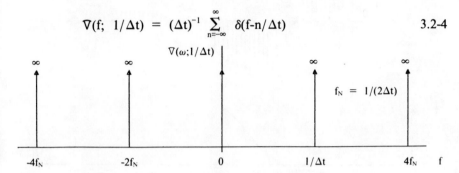

$$\nabla(f;\ 1/\Delta t) = (\Delta t)^{-1} \sum_{n=-\infty}^{\infty} \delta(f-n/\Delta t) \qquad 3.2\text{-}4$$

Figure 3.3 The inverse infinite Dirac comb in the frequency domain.

The equation used for efficient computation is called the discrete Fourier transform. It may be used in place of the Fourier integral or the Fourier series but the relationship between these various transforms is not altogether simple. The development of a discrete Fourier transform will be made with the aid of an example. Figure 3.4 (a) shows the impulse response, W(t), of a seismic system used for digital recording of earthquakes. An impulse about 1 sampling unit in length was applied across a Maxwell bridge to a Willmore Mark II seismometer with a period of 1 second (Kanasewich et al., 1974). The response in the frequency domain is called a transfer function and is given by a Fourier integral of the impulse response. The modulus and phase of the transfer function, Y(f), are shown in part (b) and (c) of the figure. The analog waveform is sampled by an analog to digital converter at a rate of 50 samples per second. Mathematically this is equivalent to a multiplication of the continuous analog signal with an infinitely long Dirac comb.

$$f(n \cdot \Delta t) = W(t) \cdot \nabla(t;\ \Delta t)$$

$$= \sum_{n=-\infty}^{\infty} W(n \cdot \Delta t)\, \delta(t - n \cdot \Delta t) \qquad 3.2\text{-}5$$

The *function*, f, consists of a series of lines which are of infinite height but have a finite area equal to the value of the function W(t) (figure 3.5a). The Fourier transform of **f** is a convolution of Y(f) with a Dirac comb in the frequency domain, $\nabla(f;\ 1/\Delta t)$.

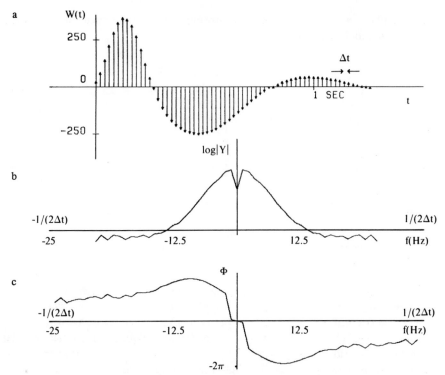

Figure 3.4 (a) Impulse response, W(t), of a broadband seismic recording system, (b) the log of the modulus, |Y(f)|, or amplitude response and (c) the phase response, $\phi(f)$, of the seismic system.

$$F(f) = Y(f) * \nabla(f; 1/\Delta t) \qquad 3.2\text{-}6$$

At this stage, F(f) is a continuous function (figure 3.5b) which is similar to Y(f) in the principal domain, $-1/2\Delta t \leq f \leq 1/2\Delta t$. It repeats as a series with a frequency of $1/\Delta t$ with center on the Dirac pulse shown in figure 3.3. The dashed portion of the curve in figure 3.5 (b) shows areas of concern where aliasing may occur unless the original signal has insignificant power at frequencies above half the sampling rate. If this is not true then the sampling must be carried out at a finer interval or else low pass filtering must be introduced before sampling. It will be shown in Chapter 8 that aliasing introduces leakage of power from higher to lower frequencies and causes a distortion in the spectrum which leads to misinterpretation of spectral peaks.

Computers have finite memories and therefore the sampled data must be truncated by a box-car function, $W_B(t)$ as in figure 3.6. Multiplication in the

The discrete Fourier transform

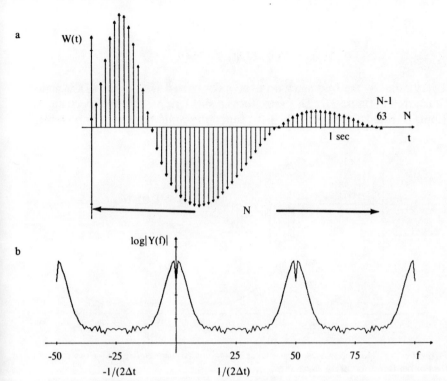

Figure 3.5 (a) The impulse response after sampling with a Dirac comb, $\nabla(t; \Delta t)$. (b) The Fourier transform (log modulus) of the sampled signal. It is continuous and repeats with frequency $1/\Delta t$.

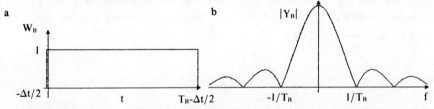

Figure 3.6 (a) The box-car function offset by half a sampling interval to avoid distorting the point at the origin. (b) The modulus of the Fourier transform of the box-car function.

time domain is equivalent to convolution in the frequency domain. The number of samples generated in any interval of duration T_B will be N where $T_B = N\Delta t$.

$$\begin{aligned}
f_B &= W(t) \cdot \nabla(t; \Delta t) \cdot W_B(t) \\
&= \sum_{n=0}^{N-1} W(n \cdot \Delta t) \, \delta(t - n \cdot \Delta t)
\end{aligned} \qquad 3.2\text{-}7$$

$$F_B = Y(f) * \nabla(f; 1/\Delta t) * Y_B(f) \qquad 3.2\text{-}8$$

Every point in the frequency domain is convolved with a sine x/x function as illustrated in figure 3.7. This introduces a slight ripple into the spectrum. If the ripple is to be kept small the box-car function should be as wide as possible.

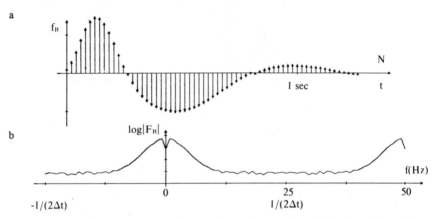

Figure 3.7 (a) The impulse response after sampling and truncation. (b) The log of the modulus of the Fourier transform of the signal in (a).

Although the signal has been sampled at discrete intervals, its Fourier transform is still a continuous function. For digital computation, the transform must be sampled by passing it through a Dirac comb in the frequency domain.

$$\nabla_1(f; 1/T_B) = \sum_{m=-\infty}^{\infty} \delta(f - m/T_B) \qquad 3.2\text{-}9$$

The Fourier transform of this function is

$$\nabla_1(t; T_B) = T_B \sum_{m=-\infty}^{\infty} \delta(t - mT_B) \qquad 3.2\text{-}10$$

For a discrete Fourier transform with N samples the sampling frequency will be taken to be $1/T_B$ where T_B is the length of the box-car function used to truncate our data. In the time domain there is a convolution of $\mathbf{W_B}$ with equation 3.2-10 (figure 3.8).

The discrete Fourier transform

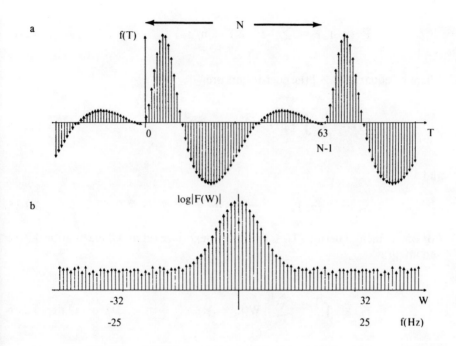

Figure 3.8 (a) The function used to compute the discrete Fourier transform. (b) The log of the modulus of the discrete Fourier transform of **f**.

$$\mathbf{f}(n\Delta t) = W(t) \cdot \nabla(t; \Delta t) \cdot W_B(t) * \nabla(t; T_B) \quad \quad 3.2\text{-}11$$

$$= \left[\sum_{n=0}^{N-1} W(n \cdot \Delta t) \, \delta(t - n \cdot \Delta t) \right] * \left[T_B \sum_{m=-\infty}^{\infty} \delta(t - mT_B) \right]$$

$$= T \sum_{m=-\infty}^{\infty} \sum_{n=0}^{N-1} W(n \cdot \Delta t) \, \delta(t - n \cdot \Delta t - mT_B) \quad \quad 3.2\text{-}12$$

The function in the time domain is now periodic with only N unique values in any interval of length T_B. The Fourier series of a function like **f** in figure 3.8 (a) in the interval $-\Delta t/2$ to $T - \Delta t/2$ can be written compactly as a series of spectral lines

$$F(n/T_B) = \sum_{n=-\infty}^{\infty} F_n \, \delta(f - n/T_B) \qquad 3.2\text{-}13$$

where by equation 3.2-2 the coefficients are

$$F_n = \frac{1}{T_B} \int_{-\Delta t/2}^{T_B - \Delta t/2} f(t) \, e^{-i\omega_n t} \, dt \qquad 3.2\text{-}14$$

and

$$\omega_n = 2\pi n/T_B \qquad n = 0, \pm 1, \pm 2, \ldots \qquad 3.2\text{-}15$$

For one principal period ($T_B = T$) in the integral we set m = 0 in equation 3.2-12 and integrate.

$$F_n = \int_{-\Delta t/2}^{T-\Delta t/2} \sum_{n=0}^{N-1} W(n \cdot \Delta t) \, \delta(t - n \cdot \Delta t) \, e^{-i\omega_n t} \, dt \qquad 3.2\text{-}16$$

Interchanging the sum and integration gives

$$F_n = \sum_{n=0}^{N-1} W(n \cdot \Delta t) \int_{-\Delta t/2}^{T-\Delta t/2} \delta(t - n \cdot \Delta t) \, e^{-i\omega t} \, dt$$

$$F_n = \sum_{n=0}^{N-1} W(n \cdot \Delta t) \, e^{-i\omega_n n \Delta t} \qquad 3.2\text{-}17$$

It is convenient to let the sampling interval be one unit of time ($\Delta t = 1$) in which case equation 3.2-17 can be written with $W(n \cdot \Delta t) = f(T)$.

$$\boxed{F(W) = \sum_{T=0}^{N-1} f(T) \, e^{-2\pi i WT/N} \qquad W = 0, 1, \ldots N\text{-}1} \qquad 3.2\text{-}18$$

This equation is often called the *discrete Fourier transform* (DFT). The function of time may be recovered exactly by the following inverse Fourier transform.

The discrete Fourier transform

$$\boxed{f(T) \;=\; N^{-1} \sum_{W=0}^{N-1} F(W)\; e^{+2\pi i WT/N}} \qquad T = 0, 1, \ldots N\text{-}1 \qquad 3.2\text{-}19$$

N is the total number of points sampled at intervals of Δt seconds and the frequency in radians is

$$\omega = 2W\pi/(N\Delta t) \qquad W = 0, 1, \ldots (N/2) \qquad 3.2\text{-}20$$

If we substitute 3.2-18 into 3.2-19 we should recover the same function back.

$$f(T) = N^{-1} \sum_{W=0}^{N-1} \left[\sum_{V=0}^{N-1} f(V)\, e^{-2\pi i VW/N} \right] e^{2\pi i WT/N} \qquad 3.2\text{-}21$$

This will be true if

$$N^{-1} \sum_{V=0}^{N-1} e^{2\pi i WT/N}\, e^{-2\pi i VW/N} = \delta_N(T - V) \qquad 3.2\text{-}22$$

where $\delta_N(T - V)$ is a Kronecker delta function with the argument considered as modulo N. That is

$$\delta_N[KN] = \begin{cases} 1 & \text{if K is integer} \\ 0 & \text{otherwise} \end{cases} \qquad 3.2\text{-}23$$

The relation in 3.2-21 will hold if the functions, f(T) and F(W), are periodic with period N so that

$$e^{2\pi i WT/N} \equiv e^{2\pi i (W+N)T/N} \equiv e^{2\pi i W(T+N)/N} \qquad 3.2\text{-}24$$

The functions are cyclic as if they were obtained from a circular universe (see figure 3.9).

$$f(T) = \ldots f_{N-2}, f_{N-1}, f_0, f_1, f_2, \ldots f_{N-2}, f_{N-1}, f_0, f_1 \ldots \quad \text{3.2-25}$$

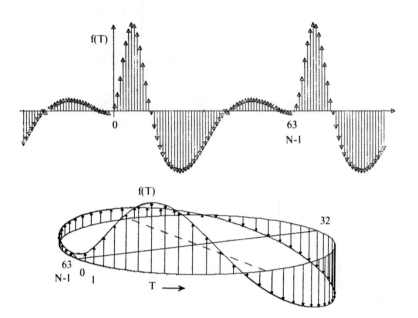

Figure 3.9 Data for fast Fourier series computations may be viewed as cyclically repeating along a time axis (upper diagram) or around a circle (lower diagram).

Both the function and its transform are cyclic as

$$f(T) = f(T + MN) \quad \text{3.2-26}$$

$$F(W) = F(W + MN) \quad M = 0, \pm 1, \pm 2, \ldots \quad \text{3.2-27}$$

and therefore

$$f(-T) = f(N - T)$$
$$\quad \text{3.2-28}$$
$$F(-W) = F(N - W)$$

The inverse Fourier transform, equation 3.2-18, is obtained directly by taking a series of complex conjugates which are indicated by a star.

The discrete Fourier transform

$$f(T) = N^{-1} \left[\sum_{W=0}^{N-1} F^*(W) \, e^{-2\pi i WT/N} \right]^* \qquad 3.2\text{-}29$$

The expression in 3.2-29 is very useful since it has the same form as 3.2-18 and allows us to calculate the Fourier transform and its inverse with the same computer program.

3.3 The principle of the fast Fourier transform

A new and rapid method of computing complex Fourier series has been presented by Cooley and Tukey (1965) following an algorithm of Good (1958). This algorithm has been applied by Gentleman and Sande (1966) to a wide variety of problems including the calculation of auto-and cross-covariances, auto- and cross-power spectra, Laplace transforms, convolution and multi-dimensional Fourier analysis. The fast Fourier method was first presented by Runge (1903, 1905) and later by Runge and König (1924). It seems to have been proposed in one form or another by Stumpff (1939) and Good (1958) but there was only intermittent use of these techniques until the paper by Cooley and Tukey. For a more detailed account see Cooley, Lewis, Welch (1967), Brigham (1974) and Claerbout (1976).

The discrete Fourier transform may be used in deriving an algorithm for digital computation which is efficient and accurate. The first of three different techniques will be described in this section. Central to all techniques is the principle that the number of data points, N, are factorable into k products of 2 or 4. Some algorithms have also been written to accept products of any other factors in any combination (i.e. $N = 168 = N_1 N_2 N_3 N_4 N_5 = 3 \times 7 \times 2 \times 2 \times 2$) but factors of 2 or 4 are most efficiently treated by binary arithmetic and simplify the sines and cosines which must be calculated. The reader is referred to Gentleman and Sande (1966) for a derivation of the general algorithm. The technique is illustrated for the case where N is factored into two integers.

$$N = N_1 N_2 \qquad 3.3\text{-}1$$

Indexes in the time domain will be in Roman letters while those in the frequency domain will be in Greek letters.

a and α are integers from 0 to N_1-1

b and β are integers from 0 to N_2-1

Let
$$W = \alpha + \beta N_1$$
$$T = b + a N_2 \qquad \text{3.3-2}$$

Note that the minimum value of W (or T) is 0 (when $\alpha = 0$, $\beta = 0$) and the maximum value is N - 1 (when $\alpha + \beta N_1 = [N_1 - 1 + (N_2 - 1) N_1]$). Then substituting these into equation 3.2-18

$$F(\alpha + \beta N_1) = \sum_{b=0}^{N_2-1} \sum_{a=0}^{N_1-1} f(b + a N_2)\, e^M \qquad \text{3.3-3}$$

where
$$M = -2\pi i/(N_1 N_2) \left[(\alpha + \beta N_1)(b + a N_2) \right] \qquad \text{3.3-4}$$

or
$$M = -2\pi i \left[\alpha b/(N_1 N_2) + \alpha a/N_1 + \beta b/N_2 + a\beta \right]$$

Now $\exp(-2\pi i a\beta) = 1$ since $a\beta$ is always an integer, therefore

$$M = -2\pi i \alpha b/(N_1 N_2) - 2\pi i \alpha a/N_1 - 2\pi i \beta b/N_2 \qquad \text{3.3-5}$$

Equation 3.3-3 is then written in a form which is independent of a

$$F(\alpha + \beta N_1) = \sum_{b=0}^{N_2-1} e^{-2\pi i b\beta/N_2}\, P(N_1, \alpha, b) \qquad \text{3.3-6}$$

where the N_1 point Fourier transform is $P(N_1,\alpha,b)$ and does not contain β.

$$P(N_1,\alpha,b) = e^{-2\pi i \alpha b/N_1 N_2} \sum_{a=0}^{N_1-1} f(b+a N_2)\, e^{-2\pi i \alpha a/N_1} \qquad \text{3.3-7}$$

It is seen that equation 3.3-6 is really the product of two shorter Fourier transforms. The process may be iterated by repeated factoring of N into smaller radices. The exponential term in front of the summation sign in 3.3-7 is called a *twiddle factor*.

The principle of the fast Fourier transform

Since the number of operations increases as N^2 in any Fourier transform equation, a process of recursive multiplications of shorter (N_1, N_2, N_3, ... N_k) Fourier transformations will reduce the number of operations to $N \log_2 N$. A computer can perform a FFT computation on 2048 points in less than a thousandth of the time required with a fairly efficient Fourier transform program written along traditional lines.

Fast Fourier algorithms are written with the input data vectors being complex.

$$f(T) = x(T) + iy(T) \qquad 3.3\text{-}8$$

Since the data, x(T), is real $x(T) = x^*(T)$, the Fourier transforms are related to their complex conjugates as

$$X(W) = X^*(-W) \qquad 3.3\text{-}9$$

For real data y(T) will be set equal to zero. The output is usually placed into the same storage location in memory as x(T) and y(T) so that the original data is destroyed.

Note that by equation 3.2-17 the coefficients of negative frequencies are stored in the second half of the location of the vector F(W). Thus for N = 9 data points, W = 0 is the zero frequency or DC (direct current) subscript, W = 1, 2, 3, 4 will specify positive frequency coefficients and W = 5, 6, 7, 8 will specify the negative frequency coefficients, -4, -3, -2 and -1 respectively. For real data the real part of X(W) will have even symmetry while the imaginary part will have odd symmetry. For an even number of data points, such as N = 8, positive frequencies are given by W = 1, 2, 3, 4; the negative frequencies are given by W = 5, 6, 7 = (-3 -2 -1). Half the amplitude of W(4) can be considered as contained in W(-4). If one is beginning calculations in the frequency domain, it is important to keep this in mind for the last frequency coefficient, W(N/2). The imaginary part of this last term may have to be set equal to zero before carrying out a fast Fourier transform or else the inverse transform may not give a real output vector. This occurs in making computations of synthetic seismograms. Proper care must be exercised in arranging the vector with correct symmetry and with the required symmetry for the coefficient at W(N/2).

3.4 Fast Fourier transform using matrix notation

The fast Fourier transform algorithm can be modified according to various matrix methods as given by Good (1958, 1960), Robinson (1967), Pease (1968),

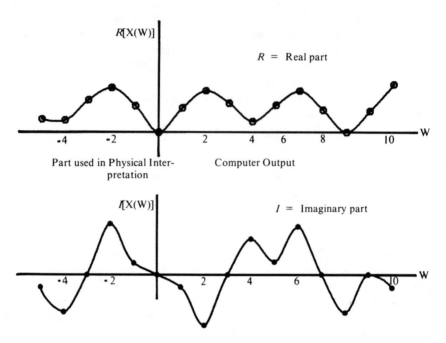

Figure 3.10 Symmetrical relations between positive and negative frequency Fourier coefficients. The data is output by the computer from W = 0 to N - 1 whereas it must be transformed to -N/2 to +N/2 for physical interpretation.

Bailey (1969) and Corinthios (1969). These forms may be carried out on digital computers or on special purpose machines built to perform the analysis.

Let the sampled data set be denoted by a row vector.

$$f = [f_0, f_1, f_2, \ldots f_{n-1}] \qquad 3.4\text{-}1$$

The discrete Fourier transform output is also a row vector.

$$\mathbf{F} = [F_0, F_1, F_2, \ldots F_{n-1}] \qquad 3.4\text{-}2$$

Define an exponential matrix with elements

$$e^{-2\pi i WT/N}, \qquad W, T = 0, 1, 2, \ldots N\text{-}1 \qquad 3.4\text{-}3$$

Fast Fourier transform using matrix notation

or letting

$$\Omega = -2\pi i/N$$

the elements are

$$e^{\Omega WT} = e^{\Omega M} \qquad M = 0, 1, 2, \ldots (N-1)^2$$

That is

$$E = \begin{bmatrix} e^0 & e^0 & e^0 & e^0 & \ldots & e^0 \\ e^0 & e^{\Omega} & e^{2\Omega} & e^{3\Omega} & \ldots & e^{(N-1)\Omega} \\ e^0 & e^{2\Omega} & e^{4\Omega} & e^{6\Omega} & \ldots & e^{2(N-1)\Omega} \\ e^0 & e^{3\Omega} & e^{6\Omega} & e^{9\Omega} & \ldots & e^{3(N-1)\Omega} \\ \vdots & & & & & \vdots \\ e^0 & e^{(N-1)\Omega} & e^{2(N-1)\Omega} & e^{3(N-1)\Omega} & \ldots & e^{(N-1)(N-1)\Omega} \end{bmatrix} \qquad 3.4\text{-}4$$

Note that the matrix **E** is symmetric because of the cyclic nature of the exponential terms.

$$e^{M\Omega} = e^{WT\Omega} = e^{(kN+WT)\Omega} \qquad \begin{array}{l} k, \text{ integer and} \\ 0 \leq M \leq N-1 \end{array} \qquad 3.4\text{-}5$$

Thus when $N = 8$

$$E = \begin{bmatrix} e^0 & e^0 & e^0 & e^0 & e^0 & e^0 & e^0 & e^0 \\ e^0 & e^{1\Omega} & e^{2\Omega} & e^{3\Omega} & e^{4\Omega} & e^{5\Omega} & e^{6\Omega} & e^{7\Omega} \\ e^0 & e^{2\Omega} & e^{4\Omega} & e^{6\Omega} & e^{8\Omega} & e^{10\Omega} & e^{12\Omega} & e^{14\Omega} \\ e^0 & e^{3\Omega} & e^{6\Omega} & e^{9\Omega} & e^{12\Omega} & e^{15\Omega} & e^{18\Omega} & e^{21\Omega} \\ e^0 & e^{4\Omega} & e^{8\Omega} & e^{12\Omega} & e^{16\Omega} & e^{20\Omega} & e^{24\Omega} & e^{28\Omega} \\ e^0 & e^{5\Omega} & e^{10\Omega} & e^{15\Omega} & e^{20\Omega} & e^{25\Omega} & e^{30\Omega} & e^{35\Omega} \\ e^0 & e^{6\Omega} & e^{12\Omega} & e^{18\Omega} & e^{24\Omega} & e^{30\Omega} & e^{36\Omega} & e^{42\Omega} \\ e^0 & e^{7\Omega} & e^{14\Omega} & e^{21\Omega} & e^{28\Omega} & e^{35\Omega} & e^{42\Omega} & e^{49\Omega} \end{bmatrix} \qquad 3.4\text{-}6$$

the matrix can be reduced with equation 3.4-5 to

$$E = \begin{bmatrix} 1 & 1 & 1 & 1 & 1 & 1 & 1 & 1 \\ 1 & e^{\Omega} & e^{2\Omega} & e^{3\Omega} & -1 & e^{5\Omega} & e^{6\Omega} & e^{7\Omega} \\ 1 & e^{2\Omega} & -1 & e^{6\Omega} & 1 & e^{2\Omega} & -1 & e^{6\Omega} \\ 1 & e^{3\Omega} & e^{6\Omega} & e^{\Omega} & -1 & e^{7\Omega} & e^{2\Omega} & e^{5\Omega} \\ 1 & -1 & 1 & -1 & 1 & -1 & 1 & -1 \\ 1 & e^{5\Omega} & e^{2\Omega} & e^{7\Omega} & -1 & e^{\Omega} & e^{6\Omega} & e^{3\Omega} \\ 1 & e^{6\Omega} & -1 & e^{2\Omega} & 1 & e^{6\Omega} & -1 & e^{2\Omega} \\ 1 & e^{7\Omega} & e^{6\Omega} & e^{5\Omega} & -1 & e^{3\Omega} & e^{2\Omega} & e^{\Omega} \end{bmatrix}$$ 3.4-7

In the example above we have substituted 1 for e^0 and -1 for $e^{4\Omega} = e^{N\Omega/2}$. Except for a scale factor, **E** is a unitary matrix and its inverse is given by

$$E^{-1} = E^{*T}/N$$ 3.4-8

The discrete Fourier transform is

$$F = Ef$$ 3.4-9

The inverse Fourier transform is obtained by forming the complex conjugate transform of **E**.

$$f = FE^{*T}/N$$ 3.4-10

For fast computation matrix **E** must be factored into n sparse matrices, S_n, and an ideal shuffle or a permutation matrix, **P**, which reorders the output vector in the desired sequence

$$E = S_1 \, S_2 \, S_3 \, \ldots \, S_n \, P_N$$ 3.4-11

where

$$n = \log_2 N$$ 3.4-12

As in the Cooley-Tukey algorithm for the fast Fourier transform, the output data is computed in a scrambled format to obtain maximum efficiency. The output vector is obtained in permuted binary digit-reverse order when $N = 2^n$. The following table illustrates the case for 8 data points:

Fast Fourier transform using matrix notation

Input Vector	Decimal Index (j)		Binary Index	Reverse Binary		Equivalent Decimal (M_j)	Scrambled Output Vector
f_0	0	=	000	000	=	0	F_0
f_1	1	=	001	100	=	4	F_4
f_2	2	=	010	010	=	2	F_2
f_3	3	=	011	110	=	6	F_6
f_4	4	=	100	001	=	1	F_1
f_5	5	=	101	101	=	5	F_5
f_6	6	=	110	011	=	3	F_3
f_7	7	=	111	111	=	7	F_7

Note that this shuffling separates the even and odd indexed transformed data. The permutation matrix is designed to convert the scrambled vector, F_s, into its desired form, F.

$$F_S \ P_N = F \qquad 3.4\text{-}13$$

For N = 8

$$P_8 = P_8^{-1} = \begin{bmatrix} 1 & 0 & 0 & 0 & 0 & 0 & 0 & 0 \\ 0 & 0 & 0 & 0 & 1 & 0 & 0 & 0 \\ 0 & 0 & 1 & 0 & 0 & 0 & 0 & 0 \\ 0 & 0 & 0 & 0 & 0 & 0 & 1 & 0 \\ 0 & 1 & 0 & 0 & 0 & 0 & 0 & 0 \\ 0 & 0 & 0 & 0 & 0 & 1 & 0 & 0 \\ 0 & 0 & 0 & 1 & 0 & 0 & 0 & 0 \\ 0 & 0 & 0 & 0 & 0 & 0 & 0 & 1 \end{bmatrix} \qquad 3.4\text{-}14$$

Thus $[F_0, F_4, F_2, F_6, F_1, F_5, F_3, F_7,] \ P = [F_0, F_1, F_2, F_3, F_4, F_5, F_6, F_7,]$. The permutation matrix is orthogonal [$P^2 = I$ and $P = P^T$]. If the permutation matrix is omitted, the fast Fourier equation, from 3.4-9, 3.4-11 and 3.4-13, is

$$F_S = S_1 \ S_2 \ ... \ S_n f \qquad 3.4\text{-}15$$

Each sparse matrix is made up of combinations of a basic core matrix

$$\begin{bmatrix} 1 & 1 \\ e^{M_j\Omega} & e^{M_{j+i}\Omega} \end{bmatrix}$$

where M_j is taken in the sequence for integers from a reverse binary order. The number of consecutive 1s and repetitions of the core matrix along the diagonal in S_1, S_2, S_3, ... S_n are 2^{n-1}, ..., 2^2, 2^1, 2^0. As an example, for $N = 8$ the sequence of powers M_j is 0, 4, 2, 6, 1, 5, 3, 7 and there are $n = 3$ sparse matrices as shown below:

$$S_3 = \begin{bmatrix} 1 & 1 & 0 & 0 & 0 & 0 & 0 & 0 \\ e^0 & e^{4\Omega} & 0 & 0 & 0 & 0 & 0 & 0 \\ 0 & 0 & 1 & 1 & 0 & 0 & 0 & 0 \\ 0 & 0 & e^{2\Omega} & e^{6\Omega} & 0 & 0 & 0 & 0 \\ 0 & 0 & 0 & 0 & 1 & 1 & 0 & 0 \\ 0 & 0 & 0 & 0 & e^{1\Omega} & e^{5\Omega} & 0 & 0 \\ 0 & 0 & 0 & 0 & 0 & 0 & 1 & 1 \\ 0 & 0 & 0 & 0 & 0 & 0 & e^{3\Omega} & e^{7\Omega} \end{bmatrix}$$

$$S_2 = \begin{bmatrix} 1 & 0 & 1 & 0 & 0 & 0 & 0 & 0 \\ 0 & 1 & 0 & 1 & 0 & 0 & 0 & 0 \\ e^0 & 0 & e^{4\Omega} & 0 & 0 & 0 & 0 & 0 \\ 0 & e^0 & 0 & e^{4\Omega} & 0 & 0 & 0 & 0 \\ 0 & 0 & 0 & 0 & 1 & 0 & 1 & 0 \\ 0 & 0 & 0 & 0 & 0 & 1 & 0 & 1 \\ 0 & 0 & 0 & 0 & e^{2\Omega} & 0 & e^{6\Omega} & 0 \\ 0 & 0 & 0 & 0 & 0 & e^{2\Omega} & 0 & e^{6\Omega} \end{bmatrix}$$

$$S_1 = \begin{bmatrix} 1 & 0 & 0 & 0 & 1 & 0 & 0 & 0 \\ 0 & 1 & 0 & 0 & 0 & 1 & 0 & 0 \\ 0 & 0 & 1 & 0 & 0 & 0 & 1 & 0 \\ 0 & 0 & 0 & 1 & 0 & 0 & 0 & 1 \\ e^0 & 0 & 0 & 0 & e^{4\Omega} & 0 & 0 & 0 \\ 0 & e^0 & 0 & 0 & 0 & e^{4\Omega} & 0 & 0 \\ 0 & 0 & e^0 & 0 & 0 & 0 & e^{4\Omega} & 0 \\ 0 & 0 & 0 & e^0 & 0 & 0 & 0 & e^{4\Omega} \end{bmatrix}$$

There are only two non-zero elements in each matrix and one of these is always unity so there are N multiply and add operations per matrix pair. The total

Fast Fourier transform using matrix notation

number of multiply and add operations for the computations is $n = N \log_2 N$. The exponential term can be computed once and stored in a table although this is not always done in more general programs.

3.5 Fourier transformation of very long data sets

Runge and König's (1924) original method of fast Fourier transformation is particularly suitable for the analysis of data sets which are longer than ordinary FFT programs are capable of handling because of limitations on the size of the memory in computers. Let f(T) be a vector from $T = 0$ to $2N - 1$. This may be written as two shorter vectors consisting of the even and odd indexed samples. The even set of data is given by

$$f(2T) = \sum_{W=0}^{N-1} F_e(W) \, e^{2\pi i WT/N} \qquad 3.5\text{-}1$$

whereas the odd set is found from

$$f(2T+1) = \sum_{W=0}^{N-1} F_o(W) \, e^{2\pi i WT/N} \qquad 3.5\text{-}2$$

where

$$T = 0, 1, 2, \ldots N-1$$

The inverse Fourier transform of the entire series is given by

$$f(T) = \sum_{W=0}^{2N-1} F(W) \, e^{2\pi i WT/(2N)} \qquad 3.5\text{-}3$$

Equation 3.5-3 can be subdivided into functions of even and odd indices by replacing T by 2T

$$f(2T) = \sum_{W=0}^{2N-1} F(W) \, e^{2\pi i WT/N} \qquad 3.5\text{-}4$$

and T by $2T + 1$

$$f(2T+1) = \sum_{W=0}^{2N-1} F(W) \, e^{2\pi i WT/N} \, e^{2\pi i W/2N} \qquad 3.5\text{-}5$$

where we have let

$$2\pi i \left[2WT/(2N) + W/(2N)\right] = 2\pi i W \left[(2T+1)/(2N)\right]$$

Since $\exp(2\pi i N/N) = 1$ and $\exp(2\pi i N/2N) = -1$, it is possible to separate terms into those from 0 to N - 1 and from N to 2N - 1.

$$\sum_{W=N}^{2N-1} F(W) \, e^{2\pi i WT/N} = \sum_{W=0}^{N-1} F(N+W) \, e^{2\pi i WT/N} \qquad 3.5\text{-}6$$

Also

$$\sum_{W=N}^{2N-1} F(W) e^{2\pi i WT/N} \, e^{2\pi i W/(2N)} = -\sum_{W=0}^{N-1} F(N+W) \, e^{2\pi i WT/N} \, e^{2\pi i W/(2N)} \qquad 3.5\text{-}7$$

Substitute 3.5-6 and 3.5-7 into equations 3.5-4 and 3.5-5 and equate coefficients of like powers in the exponential function with those of 3.5-1 and 3.5-2. From this

$$F_e(W) = F(W) + F(N+W) \qquad 3.5\text{-}8$$

$$F_o(W) = [F(W) - F(N+W)] \, e^{2\pi i W/(2N)} \qquad 3.5\text{-}9$$

Solving for F(W) from the equations above gives

$$F(W) = [F_e(W) + F_o(W) \, e^{-2\pi i W/(2N)}](1/2) \qquad 3.5\text{-}10$$

$$F(W+N) = [F_e(W) - F_o(W) \, e^{2\pi i W/(2N)}](1/2) \qquad 3.5\text{-}11$$

$$W = 0, 1, 2, \ldots N\text{-}1$$

Fourier transformation of very long data sets

Some FFT algorithms are written to compute transforms on a maximum of 2^{13} or fewer data points. However with the pair of equations above it is possible to obtain the Fourier transform on $2^{14} = 16{,}384$ points. Brenner (1969) discusses the computation of long externally stored data in detail.

3.6 Simultaneous computation of two Fourier transforms

It is possible to place a real, non-zero set of data into y(T) in equation 3.3-8 and perform the Fourier computations on x(T) and y(T) simultaneously. Let the Fourier transforms of x(T) and y(T) be X(W) and Y(W) respectively.

$$F[x(T)] = X(W)$$
$$F[y(T)] = Y(W) \quad \quad 3.6\text{-}1$$

Let
$$f(T) = x(T) + iy(T) \quad \quad 3.6\text{-}2$$

where x and y are both real. Since x and y are linearly independent we have

$$F[f(T)] = F(W) = X(W) + iY(W) \quad \quad 3.6\text{-}3$$

where, now, X and Y are both complex and F [f(T)] does not have the symmetry relations discussed in 3.3-9. Taking the complex conjugate of equation 3.6-3 and letting W = N - W

$$F^*(N-W) = X^*(N-W) - iY^*(N-W) \quad \quad 3.6\text{-}4$$

Note that if
$$F[x(T)] = X(W)$$

then
$$F[x^*(T)] = X^*(-W) \quad \quad 3.6\text{-}5$$

and
$$F[x^*(-T)] = X^*(W) \quad \quad 3.6\text{-}6$$

Also if x(T) is real then

$$X(W) = X^*(-W) = X^*(N-W) \qquad 3.6\text{-}7$$

If y(T) is pure imaginary then

$$Y(W) = -Y^*(-W) = -Y^*(N-W) \qquad 3.6\text{-}8$$

From the last two relations it follows that equation 3.6-4 becomes

$$F^*(N-W) = X(W) - iY(W) \qquad 3.6\text{-}9$$

Solving equations 3.6-3 and 3.6-9 simultaneously gives

$$X(W) = [F(W) + F^*(N-W)]/2 \qquad 3.6\text{-}10$$

$$Y(W) = [F(W) - F^*(N-W)]/(2i) \qquad 3.6\text{-}11$$

With these last two relations it is possible to carry out the computation of two Fourier transforms simultaneously. If the data is available the increase in speed and economy makes this worthy of consideration.

3.7 Two-dimensional Fourier transforms

Two-dimensional Fourier transformations are very useful in studying geophysical and geological data in the spatial frequency domain. Tsuboi and Fuchida (1939) have shown how it is possible to transform gravity field data from one elevation to another. The original map of Bouguer gravity anomalies is sampled at discrete intervals to generate a two-dimensional matrix. A double Fourier transform is taken and the original map can be reconstructed at any elevation, z, from the following series:

$$\Delta g(x,y,z) = \sum_{m=0}^{n_0} \sum_{n=0}^{n_0} B_{mn}\, e^{2\pi z(m^2+n^2)^{1/2}/L} \begin{array}{c}\cos\\ \sin\end{array} m2\pi x/L \begin{array}{c}\cos\\ \sin\end{array} n2\pi y/L \qquad 3.7\text{-}1$$

where L is the length of the map in the x or y directions, B_{mn} are the coefficients in the Fourier series expansion and m_0, n_0 are indices for the highest harmonics along the x and y axes. Upward continuation of the field is obtained by taking z negative, and downward continuation, closer to the source, is obtained by taking z positive.

Two-dimensional Fourier transforms

Magnetic anomalies are often simplified if the total magnetic fields are transformed to the magnetic pole following the method of Baranov (1957) and Bhattacharyya (1965). Two-dimensional analysis may be used in this case also. In both gravity and magnetic problems it is easier to use exponential functions instead of sines and cosines and to use a fast Fourier algorithm for computation.

The two-dimensional Fourier transform of a matrix of data, f(X,Y), is given by

$$F(K_1,K_2) = \sum_{X=0}^{N_1-1} \sum_{Y=0}^{N_2-1} f(X,Y)\, e^{-2\pi i[XK_1/N_1 + YK_2/N_2]} \qquad 3.7\text{-}2$$

The inverse Fourier transform is

$$f(X,Y) = (N_1 N_2)^{-1} \sum_{K_1=0}^{N_1-1} \sum_{K_2=0}^{N_2-1} F(K_1,K_2)\, e^{2\pi i[XK_1/N_1 + YK_2/N_2]} \qquad 3.7\text{-}3$$

where

$$X, K_1 = 0, 1, 2, \ldots N_1-1$$

$$Y, K_2 = 0, 1, 2, \ldots N_2-1$$

The wave lengths in the x and y directions are given by

$$\lambda_1 = N_1\, \Delta x / K_1 \qquad K_1 = 0, 1, \ldots N_1/2 \qquad 3.7\text{-}4$$

$$\lambda_2 = N_2\, \Delta y / K_2 \qquad K_2 = 0, 1, \ldots N_2/2 \qquad 3.7\text{-}5$$

Usually one treats a square map where $N_1 = N_2$ and the sampling interval is the same in both directions.

$$\Delta x = \Delta y$$

The data are arranged in the form of a matrix f(X,Y); Naidu (1970), Anderson (1980).

f(0,0) f(1,0) ... f(N_1-1,0)
f(0,1) f(1,1) ... f(N_1-1,1)
f(0,N_2-1) f(1,N_2-1) ... f(N_1-1,N_2-1)

A fast Fourier one-dimensional analysis is carried out on this vector of $N = N_1 N_2$ samples of data. Since equation 3.7-2 is very similar to 3.3-3, a two-dimensional Fourier transform can be represented as a linear super-position of two finite discrete transforms.

$$F(K_1, K_2) = \sum_{X=0}^{N_1-1} g(X, K_2) \, e^{M(X)} \qquad 3.7\text{-}6$$

$$g(X, K_2) = \sum_{Y=0}^{N_2-1} f(X, Y) \, e^{M(Y)} \qquad 3.7\text{-}7$$

where

$$M(X) = -2\pi i X K_1 / N_1 \quad \text{and} \quad M(Y) = -2\pi i Y K_2 / N_2 \qquad 3.7\text{-}8$$

Both discrete Fourier transform in 3.7-6 and 3.7-7 are one-dimensional and can be carried out with a conventional one-dimensional FFT program row by row to yield $y(X, K_2)$ and then column by column to give $F(K_1, K_2)$. The extension to more dimensions is straightforward but computer memory limitations will restrict the application to small N. Some care is necessary in arranging the data with the proper symmetry when it is desired to compare this method of transformation with results obtained by equations such as 3.7-1.

Figure 3.11 illustrates an example of upward continuation on gravity data used originally by Tsuboi (1959) and recomputed with a two dimensional fast Fourier algorithm by Agarwal (1968). In addition the Mth order vertical derivative is easily obtained for an N_1 by N_1 array by multiplying the complex Fourier coefficients by

$$\text{Exp}\left[2\pi/N(K_1^2 + K_2^2) / (N \cdot \Delta X) \right]^M \qquad 3.7\text{-}9$$

where the total number of data points is $N = N_1 N_2$ and ΔX is the sampling interval. The first vertical derivative is shown in figure 3.11 (d). Before the computations are carried out for upward or downward continuation or for derivative studies, the samples can be reorganized to have symmetry about the first and last rows and columns. This will avoid the introduction of sharp discontinuities or short wavelengths with large power at the edge of the map. For instance, in a set of 4 by 4 data points shown in figure 3.12, the matrix before Fourier transformation is arranged in the manner shown in figure 3.13.

Two-dimensional Fourier transforms

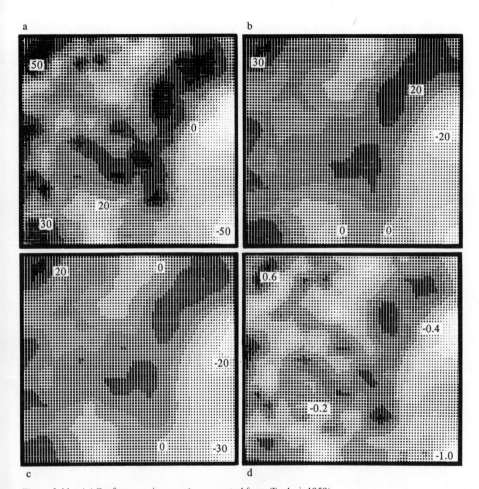

Figure 3.11 (a) Surface gravity map (recomputed from Tsuboi, 1959).
(b) Upward continuation of (a) at a level of 15.9 km above the surface.
(c) Upward continuation of (a) at a level of 15.9 km above the surface obtained by using fast Fourier transform programs.
(d) First vertical derivative map of (c) obtained by using FFT programs.

f_{11}	f_{12}	f_{13}	f_{14}
f_{21}	f_{22}	f_{23}	f_{24}
f_{31}	f_{32}	f_{33}	f_{34}
f_{41}	f_{42}	f_{43}	f_{44}

Figure 3.12   Original map of data.

The technique of arranging two dimensional data with even symmetry as in figure 3.13 can be used for small data sets. It becomes prohibitively expensive when the matrix exceeds a size of about 64 by 64 values (4,096 discrete measurements). Under these circumstances edge discontinuities can be controlled by tapering the border of the matrix in figure 3.12 with a cosine bell (Chapter 9). After any filtering, the data along the edge needs to be disregarded or interpreted with caution. However, the tapering of a two-dimensional function to a zero value is an effective method of analysing large multi-dimensional data sets.

$-f_{11}$	$-f_{12}$	$-f_{13}$	$-f_{14}$	$-f_{13}$	$-f_{12}$	$-$
f_{21}	f_{22}	f_{23}	f_{24}	f_{23}	f_{22}	
f_{31}	f_{32}	f_{33}	f_{34}	f_{33}	f_{32}	
$-f_{41}$	$-f_{42}$	$-f_{43}$	$-f_{44}$	$-f_{43}$	$-f_{42}$	$-$
f_{31}	f_{32}	f_{33}	f_{34}	f_{33}	f_{32}	
f_{21}	f_{22}	f_{23}	f_{24}	f_{23}	f_{22}	

Figure 3.13 Expanded vector of data to obtain even symmetry about the rows and columns indicated by the dashed lines. The symmetry is obtained when the data is repeated by circular repetition outside this fundamental interval.

3.8 Fast Fourier computer programs

There are now a large number of suitable algorithms available for the rapid computation of a discrete Fourier transform. Descriptions and details can be found in Robinson (1967), Brigham (1974), Halpeny (1975) and Claerbout (1976) and in many other references. The Harm Fortran subroutine in the IBM Scientific Subroutine Package (SSP) for one-, two- or three-dimensional cases has been found to be reliable and easy to implement. Note that an engineering sign that is opposite to the one used here occurs in these subroutines. It is particularly important to test any subroutines on simple functions such as sine waves of various frequencies, pulses and square waves. Furthermore the Fourier coefficients must be verified in terms of mathematical predictions from series computations. The cyclic nature of the discrete Fourier transform and the effect of adding trailing zeros should be thoroughly understood before computations and physical interpretation is carried out on Fourier transformed data.

As an example a FORTRAN program that is similar to those given by Cooley, Lewis and Welch (1969) is given below. It avoids complex calling statements which shorten the appearance of the program but do not increase its efficiency. The output Fourier transform replaces the input data; FR and FI. A

Fast Fourier computer programs

recursion relation is used to compute the sines and cosines. If the exponential function is

$$E_N^K = \exp(-2\pi iK/N) = \cos(-2\pi K/N) + i\sin(-2\pi K/N)$$

then the recursion relation is $E_N^{J+K} = E_N^J \cdot E_N^K$, where the initial value is $E_N^0 = 1.0 + 0.0i$. Appendix 7 gives the explicit equations on which this algorithm is based.

```
          SUBROUTINE FASTF(FR,FI,N)
C
C         N IS THE NUMBER OF DATA POINTS = 2**M
C         FR IS THE REAL DATA SET
C         FI IS THE IMAGINARY PART OF DATA SET (= 0.0 IF ONLY REAL)
C         FIRST COMPUTE M
          REAL FR(N), FI(N), GR, GI, ER, EI, EU, EZ
          M = 0
          KD = N
    1     KD = KD/2
          M = M + 1
          IF (KD .GE. 2) GO TO 1
          ND2 = N/2
          NM1 = N-1
          L = 1
C
C         SHUFFLE INPUT DATA IN BINARY DIGIT REVERSE ORDER
          DO 4 K=1, NM1
          IF (K .GE. L) GO TO 2
          GR = FR(L)
          GI = FI(L)
          FR(L) = FR(K)
          FI(L) = FI(K)
          FR(K) = GR
          FI(K) = GI
    2     NND2 = ND2
    3     IF (NND2 .GE. L) GO TO 4
          L = L - NND2
          NND2 = NND2/2
          GO TO 3
    4     L = L + NND2
          PI = 3.14159265
C
C         FIRST ARRANGE ACCOUNTING OF M STAGE
          DO 6 J = 1,M
          NJ = 2**J
          NJD2 = NJ/2
          EU = 1.0
          EZ = 0.0
          ER = COS(-PI/NJD2)
          EI = SIN(-PI/NJD2)
C
```

```
C       COMPUTE FOURIER TRANSFORM IN EACH M STAGE
        DO 6 IT = 1,NJD2
        DO 5 IW = IT,N,NJ
        IWJ = IW + NJD2
        GR = FR(IWJ) * EU - FI(IWJ) * EZ
        GI = FI(IWJ) * EU + FR(IWJ) * EZ
        FR(IWJ) = FR(IW) - GR
        FI(IWJ) = FI(IW) - GI
        FR(IW) = FR(IW) + GR
      5 FI(IW) = FI(IW) + GI
        SEU = EU
        EU = SEU * ER - EZ * EI
      6 EZ = EZ * ER + SEU * EI
        RETURN
        END
```

As a test, SUBROUTINE FASTF was implemented in the following calling program. Several data sets with 8 points and an assumed sampling rate of 1/8 per second were computed. The pure cosine waves had a unit amplitude and a frequency of 1 and 2 Hz. The reader should insure that he or she is able to interpret the output from the discrete Fourier transform. Note that the real and imaginary parts of the Fourier Transform must be divided by N, the number of data points, and the positive and negative frequency components must be summed in order to obtain the Fourier coefficients or the amplitude of the harmonic. For the first (or DC) term and the central one ($N = 4$), the summation is unnecessary. If a phase shift is present the real and imaginary parts of the Fourier transform are non-zero. The calling program for a 2 Hz cosine wave is shown below.

```
            DIMENSION CR(2048), CI(2048)
            DATA CR/1.0,0.0,-1.0,0.0,1.0,0.0,-1.0,0.0/
            DATA CI/0.0,0.0,0.0,0.0,0.0,0.0,0.0,0.0/
            CALL FASTF(CR,CI,8)
            WRITE(6,200) CR
            WRITE(6,200) CI
        200 FORMAT(8F8.4/)
            STOP
            END
```

Input data: Cosine, 2Hz: CR/1.0,0.0,-1.0,0.0,1.0,0.0,-1.0,0.0/
Fourier Transform: FR = [0.0,0.0,4.0,0.0,0.0,0.0,4.0,0.0]
                   FI = [0.0,0.0,0.0,0.0,0.0,0.0,0.0,0.0]
Input data: Cosine, 1Hz: CR/1.0,0.707107,0.0,-0.707107,-1.0,-0.707107,0.0,0.707107/
Fourier Transform: FR = [0.0,4.0,0.0,0.0,0.0,0.0,0.0,4.0]
                   FI = [0.0,0.0,0.0,0.0,0.0,0.0,0.0,0.0]
Input data: Sine, 2Hz: CR/0.0,1.0,0.0,-1.0,0.0,1.0,0.0,-1.0/
Fourier Transform: FR = [0.0,0.0,0.0,0.0,0.0,0.0,0.0,0.0]
                   FI = [0.0,0.0,-4.0,0.0,0.0,0.0,4.0,0.0]
Input data: Unit Step: CR/1.0,1.0,1.0,1.0,1.0,1.0,1.0,1.0/

Fast Fourier computer programs

Fourier Transform: FR = [8.0,0.0,0.0,0.0,0.0,0.0,0.0,0.0]
FI = [0.0,0.0,0.0,0.0,0.0,0.0,0.0,0.0]
Input data: Decaying exponential: CR/1.0,$(0.7)^1$,$(0.7)^2$,$(0.7)^3$,$(0.7)^4$,...,$(0.7)^7$/
Fourier Transform: FR = [3.1412,0.9517,0.6325,0.5681,0.5543,0.5681,0.6325,0.9517]
FI = [0.0,-0.9328,-0.4427,-0.1881,0.0,0.1881,0.4427,0.9328]
Input data: Delayed Unit Spike: CR/0.0,1.0,0.0,0.0,0.0,0.0,0.0,0.0/
Fourier Transform: FR = [1.0,0.7071,0.0,-0.7071,-1.0,-0.7071,0.0,0.7071]
FI = [0.0,-0.7071,-1.0,-0.7071,0.0,0.7071,1.0,0.7071]

References

Agarwal,R.G.(1968), Two dimension harmonic analysis of potential fields, Ph.D. thesis. University of Alberta, Edmonton.
Anderson,G.L(1980), A Stepwise Approach to Computing the Multi-dimensional Fast Fourier Transform. *IEEE Transactions* **ASSP, 28** 280-284.
Bailey,J.S.(1969), A fast Fourier transform without multiplications. *Symposium on Computer Processing in Communications. Vol.* **XIX,** Microwave Research Institute Symposium Series, Jerome Fox, ed. Polytechnic Press. Polytechnic Institute of Brooklyn, Brooklyn, N.Y. 37-46.
Baranov,V.(1957), A new method for interpretation of aeromagnetic maps: Pseudo gravimetric anomalies.*Geophysics***22,**359-383.
Bernoulli,D.(1732-33), Theoremata de oscillationibus corporum filo flexili connexorum et catenae verticaliter suspensae. *L. Euleri, Opera Omnia* **II,11,** 154-173.
Bhattacharyya,B.K.(1965), Two dimensional harmonic analysis as a tool for magnetic interpretation. *Geophysics* **30,** 829-857.
Blackman,R.B. and Tukey,J.W.(1958), The measurement of power spectra. Dover, N.Y.
Brenner,N.M.(1969), Fast Fourier transform of externally stored data. June 1969 - special issue on the fast Fourier transform. *IEEE Transactions Vol.* **AU-17,** 128-132.
Brigham,E.O.(1974), The fast Fourier transform. Prentice-Hall Inc. Englewood Cliffs, New Jersey, 1-252.
Chen,W.H.(1964), Linear Network Design and Synthesis. McGraw-Hill Book Co., N.Y. 31.
Claerbout,J.F.(1976), Fundamentals of Geophysical Data Processing. McGraw-Hill, New York, 1-274.
Cooley,J.W. and Tukey,J.W.(1965), An algorithm for the machine calculation of complex Fourier series. *Mathematics of Computation* **19,** 297-301.
Cooley,J.W., Lewis,P.A.W. and Welch,P.D.(1967), Historical notes on the Fast Fourier Transform. *IEEE Trans.* **AU-15,** (June 1967), 76-79.
Cooley,J.W., Lewis,P.A.W. and Welch,P.D.(1969), The fast Fourier transform and its applications. *IEEE Trans. on Education* **12,** 27-34.
Corinthios,M.J.(1969), A time series analyzer. *Symposium on Computer Processing in Communications.* **XIX,** Microwave Research Institute Symposia Series, Jerome Fox, Ed. Polytechnic Press. Polytechnic Institute of Brooklyn, Brooklyn, N.Y. 47-61.
Euler,L.(1748,1753), De vibratione chordarum exercitatio. *Memoires de l'academie des sciences de Berlin* **4,** 69-85. See Euler, Opera Omnia **I,14** ; 513; **II,10** ; 63-77.
Fejér,L.(1904), Untersuchungun über Fouriersche reihen. *Mathematische Annalen* **58,** 51-69.

Fourier,J.W.(1822), *Theorie Analytiques de la Chaleur, Chapters 3 and 4, 141-303, Gauthier-Villars et fils, Paris.* See the Analytic Theory of Heat, translation by A. Freeman, Cambridge, 1878.

Fourier,J.W.(1824), Théorie du Mouvement del e la Chaleur dans les corps solides. *Memoire de l'Academie Royale des Sciences de l'Institute de France.* See Oeuvres de Fourier **2**, 1-94. Gauthier-Villars et Fils, Paris.

Gentleman,W.M. and Sande,G.(1966), Fast Fourier transforms for fun and profit. Proceedings of the Fall Joint Computer conference, San Francisco. 563-578.

Good,I.J.(1958), An interaction algorithm and practical Fourier Analysis. *J. Roy. Stat. Soc. Series B*, **20**, 361-372. *Addendum* **22**, 372-375 (1960).

Halpeny,O.S.(1975), Appendix 6, Harm, in Digital Filtering and Signal Processing by D. Childers and A. Durling. West Publishing Co, St Paul. 455-456.

I.B.M. Applications Program, (1968). *System 360 Scientific Subroutine Package*, **GH20-0205-4**, 276-283.

Kanasewich,E.R., Siewert,W.P., Burke,M.D., McCloughan,C.H. and Ramsdell,L.(1974), Gain-ranging analog or digital seismic system. *Bulletin Seismological Society of America* **64**, 103-113.

Lejeune Dirichlet,G.(1829), Sur la convergence des séries trigonométriques qui servent a représenter une fonction arbitraire entres des limites données. *Werke* **1**, 117-132. Berlin, Druck und Verlag von Georg Reimer, *Journal für die reine und angewandle Mathematik* **4**, 157-169.

Lebesgue,H.(1905), Recherches sur la convergence des series de Fourier. *Mathematische Annalen* **61**, 251-280.

Naidu,P.S.(1970), Fourier transform of large scale aeromagnetic field using a modified version of the fast Fourier transform. *Pure and Applied Geophysics* **81**, 17-25.

Pease,M.C.(1968), An adaptation of the fast Fourier transform for parallel processing. *Journal of the Association for computing Machinery* **15**, 252-264.

Rabiner,L.R. and Gold,B.(1975), Theory and applications of digital signal processing. Prentice-Hall, Inc. Englewood Cliffs, New Jersey.

Riemann,B.(1854), Veber die Anzahl der Prinzahlen unter einer gegebener Grösse. *Werke (ed.1876)* 136-144. Also see on the possibility of representing a function by means of a trigonometric series. *Collected works of Bernhard Riemann.* Dover Publication, New York, 1953, 227-271.

Robinson,E.A.(1967), Multichannel time series analysis with digital computer programs. Appendix 3, 279-287, Holden Day, San Francisco.

Runge,C.(1903), Zeit für Math. und Physik **48**, 443.

Runge,C.(1905), Zeit für Math. und Physik **53**, 117.

Runge,C. and König,H.(1924), Die Grundlehren der Mathematischen Wissenschaften. *Band XI, Vorlesungen uber Numerisches Rechnen.* Julius Springer, Berlin.

Stumpff,K.(1939), Tafeln und Aufgaben zur Harmonisches Analyse und Periodogrammrechnung. Julius Springer, Berlin.

Tsuboi,C. and Fuchida,F.(1939), Relations between gravity values and corresponding subterranean mass distribution. *Bull. Earthquake Res. Inst. Tokyo* **15**, 636-649.

Tsuboi,C.(1959), Application of double Fourier series to computing gravity anomalies and other gravimetric quantities at higher elevations, starting from given surface gravity anomalies. *Inst. of Geodesy, Photography and Cartography.* Ohio State University Report No. 2.

4 Laplace transforms and complex s plane representation

4.1 Introduction

Heaviside (1899) originated a method of solving ordinary linear differential equations of the type

$$a_0 \frac{d^2y}{dt^2} + a_1 \frac{dy}{dt} + a_2 y = 1 \qquad 4.1\text{-}1$$

for which y(t), dy/dt and d^2y/dt^2 were zero for negative times. A rigorous mathematical foundation was given for this technique by Bromwich (1916), Carson (1926) and Van der Pol (1929). Carson called the integral solution which was obtained a *Laplace transform*. The Laplace transform is related to a complex Fourier transform by a rotation of 90 degrees in the complex frequency plane and it is convenient to introduce the subject by means of a Fourier transform.

Consider the Fourier integral of the following function:

$$f(t) = e^{-\sigma t} \qquad t \geqslant 0$$
$$= 0 \qquad t < 0 \qquad 4.1\text{-}2$$

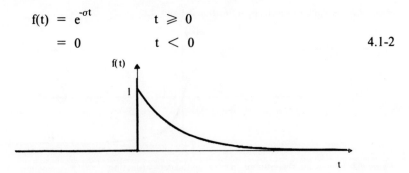

Figure 4.1 The function f(t) = exp(-σt) for t > 0.

The Fourier transform is

$$F(\omega) = \int_0^\infty e^{-\sigma t} e^{-i\omega t} \, dt$$

$$= \left[\frac{e^{-(\sigma t + i\omega t)}}{-(\sigma + i\omega)} \right]_0^\infty \qquad 4.1\text{-}3$$

The upper limit gives a zero and the result is

$$F(\omega) = \frac{1}{\sigma + i\omega}$$

$$= \frac{\sigma}{\sigma^2 + \omega^2} - i\frac{\omega}{\sigma^2 + \omega^2} = R + iI \qquad 4.1\text{-}4$$

The amplitude spectrum and phase lag are then

$$|F(\omega)| = (R^2 + I^2)^{1/2} = \left(\frac{1}{\sigma^2 + \omega^2}\right)^{1/2} \qquad 4.1\text{-}5$$

$$\phi(\omega) = -\tan^{-1} \omega/\sigma \qquad 4.1\text{-}6$$

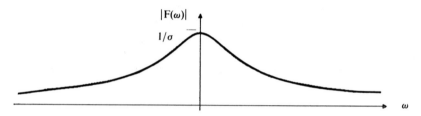

Figure 4.2 Amplitude spectrum of function in figure 4.1.

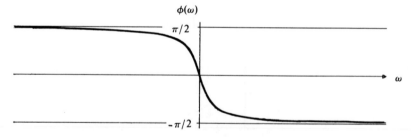

Figure 4.3 Phase lag spectrum of an exponential decay function.

It is convenient to make the following transformation. Let

$$s = \sigma + i\omega \qquad 4.1\text{-}7$$

where ω = angular frequency

σ = convergence factor.

Introduction

Thus the function in 4.1-4 may be rewritten as

$$F(s) = 1/s \qquad 4.1\text{-}8$$

A function F(s) is said to be *analytic* if the function and its derivatives exist in the region. The function in 4.1-8 is analytic everywhere except at s = 0. The Fourier transform is defined only for positive σ. The variable s can be plotted in the complex s plane and the region in which f(s) is analytic is indicated by the shaded area below in figure 4.4.

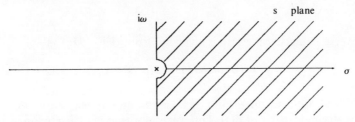

Figure 4.4 Region where f(s) = 1/s is analytic is shown by shading.

4.2 The Laplace transform

The example above suggests a method of handling a function such as a unit step for which the Fourier transform is not defined because it does not satisfy equation 2.1-1. The term exp(-σt) can be used as a convergence factor in which σ is just large enough to insure absolute convergence. The modifications of the amplitude and phase spectrum are very gentle, simple curves as shown in figures 4.2 and 4.3.

$$F(\sigma,\omega) = \int_0^\infty f(t)\, e^{-(\sigma+i\omega)t}\, dt \qquad 4.2\text{-}1$$

or

$$F(s) = \int_0^\infty f(t)\, e^{-st}\, dt \qquad 4.2\text{-}2$$

The function F(s) is called the *Laplace transform* of f(t). The transformation ignores all information in f(t) prior to t = 0. The Laplace transform of a unit step function is given by 4.2-3.

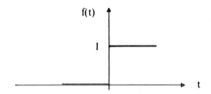

Figure 4.5

$$F(s) = L[f(t)] = \int_0^\infty 1 \, e^{-st} \, dt = 1/s \quad \quad 4.2\text{-}3$$

The inverse Laplace transform is given by Bromwich (1916) as

$$f(t) = L^{-1}[f(s)] = [1/(2\pi i)] \oint F(s) \, e^{st} \, ds \quad \quad 4.2\text{-}4$$

where the contour of integration is taken along the imaginary axis and to the right of all singularities of F(s). This integral was used by Riemann in 1859.

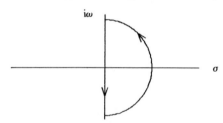

Figure 4.6 Path of integration for the inverse Laplace transform.

4.3 Examples of Laplace transformation
Example 1:

As an example of representation in the complex s plane consider the function

$$f(t) = \begin{cases} A \, e^{-at} \cos \omega_1 t & t > 0 \\ 0 & t < 0 \end{cases} \quad \quad 4.3\text{-}1$$

Examples of Laplace transformation

$$f(t) = A e^{-at}\left[\frac{e^{i\omega_1 t} + e^{-i\omega_1 t}}{2}\right] \quad t > 0$$

$$= \frac{A}{2} e^{(-a+i\omega_1)t} + \frac{A}{2} e^{(-a-i\omega_1)t} \quad \text{4.3-2}$$

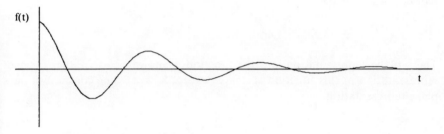

Figure 4.7 Graph of a damped cosine function.

This function can be represented in the s plane by assuming that $f(t) = 0$ for negative time and taking the Laplace transform.

$$F(s) = \mathcal{L}[f(t)] = \int_0^\infty A e^{-at} \cos \omega_1 t \, e^{-st} dt$$

$$= \frac{A}{2} \int_0^\infty e^{(-a+i\omega_1-s)t} dt + \frac{A}{2} \int_0^\infty e^{(-a-i\omega_1-s)t} dt$$

$$= (A/2) \left[\frac{1}{a - i\omega_1 + s} + \frac{1}{a + i\omega_1 + s}\right]$$

$$F(s) = (A/2)\left[\frac{1}{s + a - i\omega_1} + \frac{1}{s + a + i\omega_1}\right] \quad \text{4.3-3}$$

$$F(s) = \frac{A(s + a)}{(s+a)^2 + \omega_1^2} \quad \text{4.3-4}$$

We say that F(s) has a pole of order n at $s = s_1$ if

$$\lim_{s \to s_1} F(s) \to \infty \quad \text{4.3-5}$$

and

$$\left[(s - s_1)^n F(s) \right]_{s=s_1} \qquad 4.3\text{-}6$$

is finite and non-zero. Also F(s) has a zero of order n at $s = s_2$ if $1/F(s)$ has a pole or order n there. From 4.3-4 we see our function has a zero at $s = -a$ or since $s = \sigma + i\omega$, the zero is at $\sigma = -a$; $\omega = 0$. From 4.3-3 it is clear that a pair of poles exist at

$$\sigma = -a, \qquad \omega = \pm \omega_1$$

The s plane representation and the function is shown below. Either form of representation of this function is unique except for a missing constant in the s plane representation.

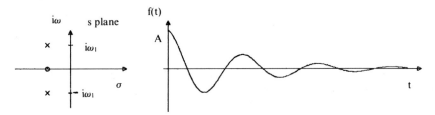

Figure 4.8

$$F(s) = \frac{A(s + a)}{(s + a)^2 + \omega_1^2} \qquad f(t) = A\, e^{-at} \cos \omega_1 t \qquad t > 0 \qquad 4.3\text{-}7$$

Example 2:

If there is no damping, the s plane representation would be as in figure 4.9.

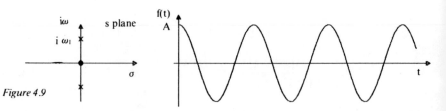

Figure 4.9

$$F(s) = \frac{As}{s^2 + \omega_1^2} \qquad f(t) = A \cos \omega_1 t \qquad 4.3\text{-}8$$

Examples of Laplace transformation

Example 3:

A sine function with damping is shown next. Note that the zero disappears (figure 4.10).

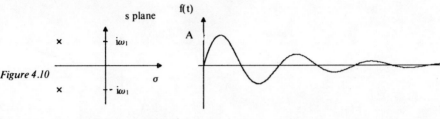

Figure 4.10

$$F(s) = \frac{A\omega_1}{(s+a)^2 + \omega_1^2} \qquad f(t) = A\,e^{-at}\sin\omega_1 t \qquad 4.3\text{-}9$$

Example 4:

A damped sine function (figure 4.11) with a phase shift has a zero at $b = -a - B\cos\theta$.

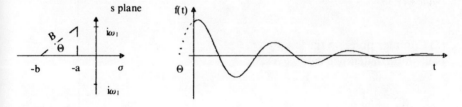

Figure 4.11

$$F(s) = \frac{A(s+b)}{(s+a)^2 + \omega_1^2} \qquad f(t) = \frac{B}{\omega_1}\,e^{-at}\sin(\omega_1 t + \theta) \qquad 4.3\text{-}10$$

Example 5:

Another simple example is the unit step function or DC voltage. This has a single pole at the origin (figure 4.12).

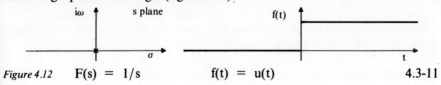

Figure 4.12 $\qquad F(s) = 1/s \qquad\qquad f(t) = u(t) \qquad 4.3\text{-}11$

Example 6: The ramp has two poles at the origin (figure 4.13).

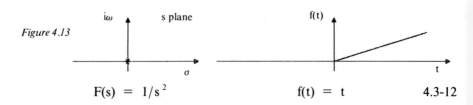

Figure 4.13

$$F(s) = 1/s^2 \qquad\qquad f(t) = t \qquad\qquad 4.3\text{-}12$$

In general,
(1) Poles on the negative real axes contribute terms like
$$e^{-at} \text{ or } te^{-at}$$

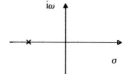

Figure 4.14

(2) Complex poles on the left plane contribute to damped sine waves.

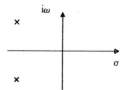

Figure 4.15

(3) Pure imaginary poles contribute sine waves.

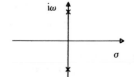

Figure 4.16

(4) Zeros contribute to the amplitude and phase angles but do not contribute to the form of the time function.

(5) Poles with positive real parts contribute to terms that grow exponentially with time and the response is unstable.

4.4 Translation of a function

Translation in the time domain corresponds to multiplication by an exponential function in the s domain. Consider the following two functions shown in figure 4.17.

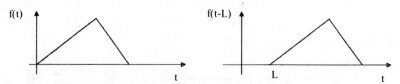

Figure 4.17   A pulse at the origin and the same pulse translated L units.

The Laplace transform of f(t) may be designated

$$L[\ f(t)\ ] = F(s) \qquad 4.4\text{-}1$$

The transform of the translated pulse is given by equation 4.2-2.

$$L[\ f_1(t)\ ] = \int_L^\infty f(t-L)\ e^{-st}\ dt \qquad 4.4\text{-}2$$

Let $t' = t - L$.

$$L[\ f_1(t)\ ] = \int_0^\infty f(t')\ e^{-s(t'+L)}\ dt'$$

$$= e^{-sL} \int_0^\infty f(t')\ e^{-st'}\ dt'$$

$$= e^{-sL}\ F(s) \qquad 4.4\text{-}3$$

Therefore e^{-sL} is an exponential delay factor in the s domain. This fact will be used later in z transformation analysis.

A *square wave pulse* can be built directly as in figure 4.18.

Figure 4.18

$$F(s) = \int_0^L A\, e^{-st}\, dt = \left[-\frac{Ae^{-st}}{s} \right]_0^L$$

$$= A\left[\frac{1 - e^{-sL}}{s}\right] \qquad 4.4\text{-}4$$

Alternatively we can use the unit step function, u(t), and a negative delayed unit step function (figure 4.19).

Figure 4.19

$$F(s) = \frac{A}{s} - \frac{A}{s} e^{-sL} \qquad 4.4\text{-}5$$

For a series of square waves we proceed as follows:

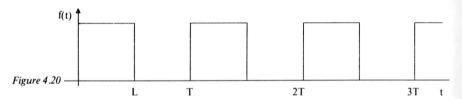

Figure 4.20

Let $f_1(t)$ be one square wave. Then the series of square waves is

$$f(t) = f_1(t) + f_1(t - T) + f_1(t - 2T) + \dots \qquad 4.4\text{-}6$$

$$\mathcal{L}[\,f(t)\,] = F_1(s) + F_1(s)e^{-Ts} + F_1(s)e^{-2Ts} + \dots \qquad 4.4\text{-}7$$

$$= F_1(s)\,[1 + e^{-Ts} + e^{-2Ts} + \dots] \qquad 4.4\text{-}8$$

The part in square brackets is a series of the type

$$1/(1 - e^{-Ts}) = 1 + e^{-Ts} + e^{-2Ts} + \dots \qquad 4.4\text{-}9$$

Translation of a function

therefore

$$\mathcal{L}[f(t)] = F_1(s) / (1 - e^{-Ts}) \qquad 4.4\text{-}10$$

but

$$F_1(s) = [1 - e^{-sT/2}] \cdot A/s \quad \text{where} \quad L = T/2 \qquad 4.4\text{-}11$$

therefore

$$\mathcal{L}[f(t)] = \frac{A}{s} \frac{1 - e^{-sT/2}}{1 - e^{-Ts}} \qquad 4.4\text{-}12$$

$$= \frac{A}{s(1 + e^{-sT/2})} \qquad 4.4\text{-}13$$

References

Aseltine, J.A. (1958), Transform method in linear system analysis. McGraw-Hill Book Company, Inc., New York.

Bromwich, T.J., Ia. (1916), Normal coordinates in dynamical systems. *Proc. London Math. Soc.* **XV**, 401-448.

Carson, J.R. (1926), Electric circuit theory and operational calculus. Chelsea Publishing Company, New York.

Heaviside, O. (1899), Electromagnetic theory, London, **II**, 34. From Vol I, II, III, 1893-1912, D Van Nostrand Co., New York.

Riemann, G.F.B. (1859), 'Ueber die Anzahl der Primzahlen unter einer gegenbener Grøsse'. *Monatsber, Bel. Akad.* **Nov. 1959**, also in *Gesammelte Werke, Druck und Verlag B. G. Teubner, Leipzig* **1876**, 136. Also The Collected Works of Bernhard Riemann, Dover Publ., New York, 1953.

Van der Pol, B. (1929), See Operational calculus based on two-sided Laplace integral. Cambridge University Press, Cambridge, U.K., (with H. Bremmer), 1955.

5 Impulse response, convolution and the transfer function

5.1 The impulse response

All the filters, electronic amplifiers, mechanical recording devices or digital operators will be assumed to be linear systems. A linear system is one such that if the input and output waveforms of one signal are x_1 and y_1

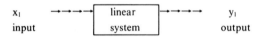

Figure 5.1

and for a second signal they are x_2 and y_2

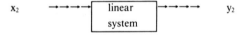

Figure 5.2

then a linear combination of the two is a superposition of the two outputs.

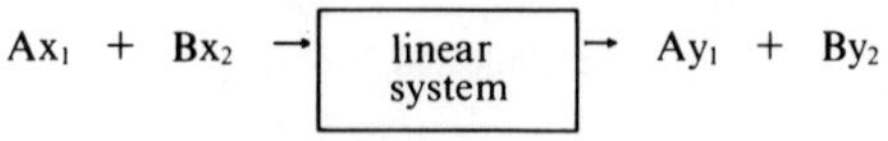

$$Ax_1 + Bx_2 \rightarrow \boxed{\text{linear system}} \rightarrow Ay_1 + By_2$$

Figure 5.3

    Many systems may be described approximately on this basis and the linearity of differential operators makes it possible to superimpose a set of solutions. Under extreme conditions an amplifier may be driven into a non-linear region and may generate harmonics or actually become unstable. Similarly, an electromechanical device such as a seismometer must be considered a non-linear system particularly if it is required to operate at very long periods.

The impulse response

Consider an impulse, similar to a Dirac delta function, applied to a fixed linear network (see Appendix 2 for an introduction to the Dirac delta function).

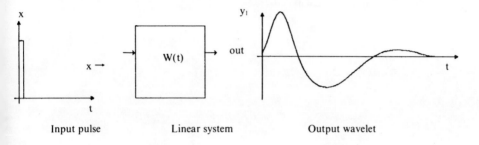

Figure 5.4 Impulse response of a linear system. The impulse response is for a seismic (VASA) system used for recording earthquakes at the University of Alberta.

The impulse response function, W(t), is the response due to an impulse applied at time t = 0. Using 2.2 (Appendix 2) we have

$$y(t) = W(t) = \int_0^\infty W(L) \, \delta(t - L) \, dL \qquad 5.1\text{-}1$$

A true impulse cannot be generated physically but a pulse with a very narrow pulse width forms a good approximation if the pulse width is less than the significant time constants of the system. The output from the system is a smeared out signal because of the limited and distinctive band pass characteristic of the system.

Another method of obtaining the impulse response function is to feed a unit pulse, u(t), to the linear system. The delta Dirac function is the derivative of the unit pulse (see figure 5.5).

$$\delta(t) = \frac{d}{dt} u(t) \qquad 5.1\text{-}2$$

Any other input signal can be expressed as a linear combination of weighted delta functions.

Impulse response, convolution and the transfer function

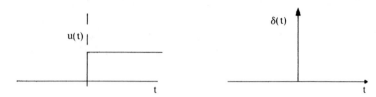

Figure 5.5 The unit step, u, and its derivative.

$$x_i(t) = \int_{-\infty}^{\infty} x_i(L') \, \delta(t - L') \, dL' \qquad 5.1\text{-}3$$

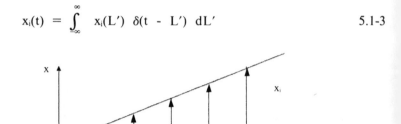

Figure 5.6 Representation of a ramp function by a series of weighted delta Dirac functions.

The output from a linear network must be the same linear combination of delayed impulse responses $W(t - L')$.

$$y(t) = \int_{-\infty}^{\infty} x(L') \, W(t - L') \, dL' \qquad 5.1\text{-}4$$

If the input is a series of discrete spikes, the method of obtaining the output is illustrated below in figure 5.7. Let $t - L' = L$, a lag time in equation 5.1-4.

$$y(t) = \int_{-\infty}^{\infty} W(L) \, x(t - L) \, dL \qquad 5.1\text{-}5$$

Since the filter or circuit cannot respond before the input of the signal, $W(L) = 0$ for negative lag and 5.1-5 becomes

$$y(t) = \int_{0}^{\infty} W(L) \, x(t - L) \, dL \qquad 5.1\text{-}6$$

The impulse response

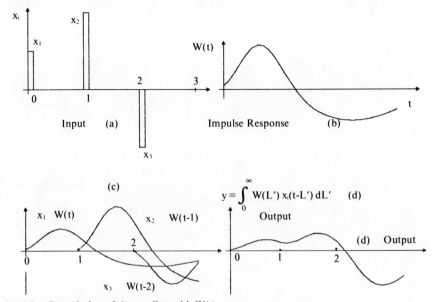

Figure 5.7 Convolution of three spikes with W(t).

The process expressed by the integral in obtaining the output signal is called *convolution in the real domain*. The impulse response function and the input signal are said to be convolved. Usually the integral operation is abbreviated as

$$y(t) = W(t) * x(t) \qquad 5.1\text{-}7$$

A graphical interpretation of the operation involved in a real convolution integral is given in figure 5.8. The first part involves a reflection x(-L) of the input signal about the ordinate axis. The signal is then shifted a time $t = t_0, t_2, \ldots$ etc. and the product of the impulse response function and reflected signal taken at all values of lag, L. The area under the curve is then evaluated to give y(t), the output signal. Thus *convolution* implies a folding, translation, multiplication and an integration. The process of obtaining the convolution of a sampled signal has been discussed in Chapter 2.

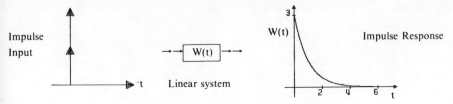

Figure 5.8 a. The response of the system.

Impulse response, convolution and the transfer function

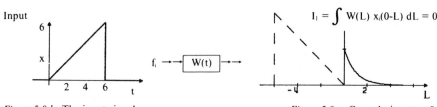

Figure 5.8 b. The input signal.

Figure 5.8 c. Convolution at t=0.

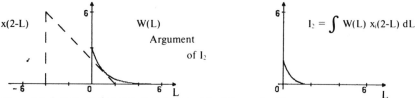

Figure 5.8 d. Convolution at t=2 and area under integrand of I_2.

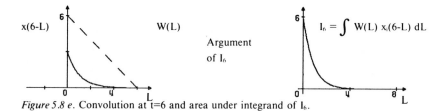

Figure 5.8 e. Convolution at t=6 and area under integrand of I_6.

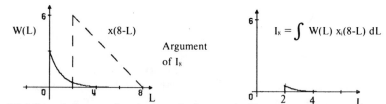

Figure 5.8 f. Convolution at t=8 and area under integrand of I_8.

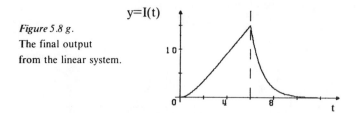

Figure 5.8 g. The final output from the linear system.

Figure 5.8a to g. The steps involved in obtaining the convolution of a signal.

5.2 Convolution in the frequency domain

A useful result is obtained from the Fourier transform of the convolution. Let the output signal $y_0(t)$ from convolution of functions $x_1(t)$ and $x_2(t)$ be an equation similar to 5.1-5.

$$y_0(t) = \int_{-\infty}^{\infty} x_1(L) \, x_2(t - L) \, dL \qquad 5.2\text{-}1$$

Take the inverse Fourier transform, 1.34 (Appendix 1) of each of the signals.

$$x_1(L) = \frac{1}{2\pi} \int_{-\infty}^{\infty} X_1(\omega_1) \, e^{i\omega_1 L} \, d\omega_1$$

$$x_2(t - L) = \frac{1}{2\pi} \int_{-\infty}^{\infty} X_2(\omega_2) \, e^{i\omega_2(t - L)} \, d\omega_2$$

$$y_0(t) = \int_{-\infty}^{\infty}\int_{-\infty}^{\infty}\int_{-\infty}^{\infty} X_1(\omega_1) \, X_2(\omega_2) \, e^{i[(\omega_1 - \omega_2) L + \omega_2 t]} \, dL \, d\omega_1 \, d\omega_2 / (2\pi)^2 \qquad 5.2\text{-}2$$

From the definition of the Dirac delta function, 2.9 (Appendix 2), we can write the Dirac function for a wave with one frequency ($\omega_1 = \omega_2$).

$$\delta(\omega_1 - \omega_2) = (2\pi)^{-1} \int_{-\infty}^{\infty} e^{iL(\omega_1 - \omega_2)} \, dL \qquad 5.2\text{-}3$$

Therefore 5.2-2 becomes

$$y_0(t) = (2\pi)^{-1} \int_{-\infty}^{\infty}\int_{-\infty}^{\infty} X_1(\omega_1) \, X_2(\omega_2) \, e^{i\omega_2 t} \, \delta(\omega_1 - \omega_2) \, d\omega_1 \, d\omega_2$$

Integrating with respect to ω_1 and using the definition of the Dirac function, 2.2 (Appendix 2), the integral has a value only when $\omega = \omega_1 = \omega_2$.

$$y_0(t) = (2\pi)^{-1} \int_{-\infty}^{\infty} X_1(\omega) \, X_2(\omega) \, e^{i\omega t} \, d\omega \qquad 5.2\text{-}4$$

Comparing 5.2-4 to the Fourier transform 1.33 and 1.34 (Appendix 1) we see that

$$X_1(\omega)\ X_2(\omega) = \int_{-\infty}^{\infty} y_0(t)\ e^{-i\omega t}\ dt \qquad 5.2\text{-}5$$

$$\boxed{X_1(\omega)\ X_2(\omega) = Y_0(\omega)} \qquad 5.2\text{-}6$$

That is, the Fourier transform of the convolution of two functions is equal to the product of the frequency spectrum of the functions.

If the signals are zero until time $t = 0$ we can use the Laplace transform

$$X_1(s)\ X_2(s) = Y_0(s) \qquad 5.2\text{-}7$$

or

$$X_1(s)\ X_2(s) = L\ [\ \int_0^t x_1(L)\ x_2(t-L)\ dL\ ] \qquad 5.2\text{-}8$$

The result is very useful because it shows that convolution in the real time domain becomes a multiplication in the complex frequency domain.

Equating 5.2-1 and 5.2-4 and $x_1 = x_2$ and $t = 0$, we obtain Parseval's formula.

$$P = \int_{-\infty}^{\infty} |x(L)|^2\ dL = (2\pi)^{-1} \int_{-\infty}^{\infty} |Y_0(\omega)|^2\ d\omega \qquad 5.2\text{-}9$$

This is related to the total power dissipated as seen in equation 2.1-3.

5.3 The transfer function

Because of the relation between the convolution formula 5.1-7 and 5.2-5 it is convenient to carry out calculations in the frequency domain.

The transfer function, $Y(\omega)$, is defined as the ratio of the Laplace transform of the output variable to the transform of the input variable with all initial conditions zero.

The transfer function

$$Y(\omega) = Y_0(\omega) / X(\omega) = Y_0(s) / X_0(s) \qquad 5.3\text{-}1$$

If the input signal is a Dirac delta function

$$x(t) = \delta(t)$$

$$X(\omega) = 1 \qquad 2.9 \text{ (Appendix 2)}$$

then 5.3-1 defines the transfer function

$$Y(\omega) = Y_0(\omega) \qquad 5.3\text{-}2$$

In other words the Fourier transform of the impulse response gives the transfer function.

$$Y(\omega) = \int_{-\infty}^{\infty} W(t)\, e^{-i\omega t}\, dt \qquad 5.3\text{-}3$$

$$W(t) = (2\pi)^{-1} \int_{-\infty}^{\infty} Y(\omega)\, e^{i\omega t}\, d\omega \qquad 5.3\text{-}4$$

The complex conjugate formulas are useful in calculation.

$$Y(-\omega) = Y^*(\omega) = \int_{-\infty}^{\infty} W(t)\, e^{i\omega t}\, dt \qquad 5.3\text{-}5$$

$$W(t) = (2\pi)^{-1} \int_{-\infty}^{\infty} Y(-\omega)\, e^{-i\omega t}\, d\omega \qquad 5.3\text{-}6$$

According to 2.9 (Appendix 2) and 5.3-2 the transfer function is most simply obtained as the steady state response to a sinusoidal excitation. An oscillator feeds a unit sinusoidal wave of a particular frequency into the system and the output amplitude is measured and plotted as ω changes.

By 5.2-6 and 5.3-3 the output from a system with transfer function, $Y(\omega)$, is

$$Y_0(\omega) = X_1(\omega)\ Y(\omega) \qquad 5.3\text{-}7$$

If $x(t)$ is a single sinusoidal wave

$$X(\omega) = \delta(\omega_1 - \omega)$$

Taking the inverse Fourier transform of 5.3-7

$$y(t) = (2\pi)^{-1} \int_{-\infty}^{\infty} Y(\omega)\ e^{i\omega t}\ \delta(\omega_1 - \omega)\ d\omega = (2\pi)^{-1}\ Y(\omega_1)\ e^{i\omega_1 t} \qquad 5.3\text{-}8$$

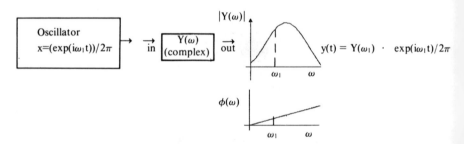

Figure 5.9 The transfer function may be obtained by comparing the input from an oscillator to the output from the linear system being studied. The process must be repeated at many frequencies to obtain the complete phase and amplitude spectrum of $Y(\omega)$.

References

Kuo, B.C. (1964), Automatic control systems. Prentice-Hall, Englewood Cliffs, New Jersey.

6 Correlation and covariance

6.1 Definitions

Let us assume that we have recorded a real function f(t) generated by a random stationary Gaussian process. The signal may also be an ensemble of such functions as f(t) each starting at a different time. The signal must have zero mean or D.C. value. That is, its average over a long time is zero.

$$\mathbf{f} = \lim_{T \to \infty} T^{-1} \int_{-T/2}^{T/2} f(t) \, dt = 0 \qquad 6.1\text{-}1$$

The *variance* is finite and is given by

$$a(0) = \lim_{T \to \infty} T^{-1} \int_{-T/2}^{T/2} [\, f(t) - \mathbf{f} \,]^2 \, dt$$

or by 6.1-2

$$a(0) = \lim_{T \to \infty} T^{-1} \int_{-T/2}^{T/2} f(t)^2 \, dt < \infty \qquad 6.1\text{-}2$$

Suppose we record two signals and sample the data at evenly divided intervals. The data consists of two row vectors and may be complex, if, for example, it consists of Fourier coefficients which are functions of frequency or wavelength.

$$\mathbf{f} = f_{-N}, \ldots f_{-1}, \mathring{f}_0 \, f_1, \ldots f_N$$

$$\mathbf{g} = g_{-N}, \ldots g_{-1}, \mathring{g}_0, g_1, \ldots g_N$$

The *coefficient of correlation* or cross correlation with zero lag between the two sets of data is

$$c_N(0) = \sum_{k=-N}^{N} f_k\, g_k^* \;/\; \left(\sum_{k=-N}^{N} f_k\, f_k^* \sum_{k=-N}^{N} g_k^* g_k \right)^{1/2} \qquad 6.1\text{-}3$$

The star indicates a complex conjugate. Mathematically $c(0)$ is the cosine between two vectors.

$$c_N(0) = \cos \Theta = \mathbf{f} \cdot \mathbf{g} \;/\; (|\mathbf{f}| \cdot |\mathbf{g}|) \qquad 6.1\text{-}4$$

The maximum value of $c_N(0)$ is ± 1 (that is, the function is normalized). This maximum indicates that there is a strong direct or reverse linear correlation between the two signals. If $c_N(0) = 0$, there is no linear correlation. For many purposes the denominator is omitted so that the coefficient is not normalized.

The *autocovariance* of a function with zero mean is

$$a(L) = \lim_{N \to \infty} \frac{1}{2N+1} \sum_{k=-N}^{N} f_{k+L}\, f_k^* \qquad 6.1\text{-}5$$

where L is the lag. For a continuous sequence of data the autocovariance for lag L is

$$a(L) = \lim_{T \to \infty} T^{-1} \int_{-T/2}^{T/2} f(t+L)\, f^*(t)\, dt \qquad 6.1\text{-}6$$

The autocovariance is often called the *autocorrelation* (see Papoulis, 1962, for instance) or the correlation function (Rice, 1944) although strictly speaking this term should refer to the normalized function. Figure 6.1 illustrates several autocovariance functions. Mathematically the autocorrelation is a_N.

$$a_N(L) = a(L) \;/\; a(0) \qquad 6.1\text{-}7$$

The autocorrelation function was introduced by Taylor (1920).

The *cross covariance* of sequence $f(t)$ and $g(t)$ is

$$c(L) = \lim_{N \to \infty} \frac{1}{2N+1} \sum_{k=-N}^{N} f_{k+L}\, g_k^* \qquad 6.1\text{-}8$$

Definitions

For a continous sequence of data

$$c(L) = \lim_{T \to \infty} T^{-1} \int_{-T/2}^{T/2} f(t+L) \, g^*(t) \, dt \qquad 6.1\text{-}9$$

For digital computation it is more convenient to have the initial time index or subscript equal to one.

$$\mathbf{f} = (f_1, f_2, f_3, \ldots f_N)$$

Only a finite amount of real data is available, therefore one only obtains an estimate of the *autocovariance* for digital lag $L = 0, 1, \ldots N-1$.

$$a(L) = N^{-1} \sum_{t=1}^{N-L} f_{t+L} \, f_t \qquad 6.1\text{-}10$$

The estimate of the *cross covariance* is

$$c(L) = N^{-1} \sum_{t=1}^{N-L} f_{t+L} \, g_t \qquad 6.1\text{-}11$$

6.2 Properties of the autocovariance function

Suppose we have the real (R) part of a function with two sinusoidal oscillations.

$$f = R\left[(a_1 + b_1 i) \, e^{i\omega_1 t} + (a_2 + b_2 i) \, e^{i\omega_2 t}\right] \qquad 6.2\text{-}1$$

The first part is an oscillation with frequency ω_1, phase shift ϕ_1, and amplitude r_1.

$$a_1 + b_1 i = r_1 \, e^{i\phi_1} \qquad 6.2\text{-}2$$

The second part has a frequency ω_2 and phase shift ϕ_2. The autocovariance function is

$$a(L) = R \lim_{T \to \infty} T^{-1} \int_{-T/2}^{T/2} [(a_1+b_1 i) e^{i\omega_1(L+t)} + (a_2+b_2 i) e^{i\omega_2(L+t)}] \cdot$$

$$[(a_1-b_1 i) e^{-i\omega_1 t} + (a_2-b_2 i) e^{-i\omega_2 t}] \, dt \qquad 6.2-3$$

The cross terms

$$\int_{-\infty}^{\infty} e^{-i\omega_1 t} e^{+i\omega_2 t} \, dt$$

are zero by the orthogonal properties of sines and cosines. Terms such as

$$\lim_{T \to \infty} T^{-1} \int_{-T/2}^{T/2} e^{i\omega_1 L} \, dt = e^{i\omega_1 L} T/T = e^{i\omega_1 L}$$

Therefore the sum of two sinusoids has an autocovariance

$$a(L) = R \, [a_1^2 + b_1^2) \, e^{i\omega_1 L} + (a_2^2 + b_2^2) \, e^{i\omega_2 L}]$$

which is equal to the sum of the cosines of the two frequencies

$$a(L) = R \, [r_1^2 \, e^{i\omega_1 L} + r_2^2 \, e^{i\omega_2 L}] \qquad 6.2-4$$

The function has a maximum when the lag, L, is zero. Note that all information on the phase of the two frequencies is lost. The information on frequency and amplitude is preserved.

When the lag, L, is zero the autocovariance becomes the variance.

$$a(0) = \lim_{T \to \infty} T^{-1} \int_{-T/2}^{T/2} [f(t)]^2 \, dt \qquad 6.2-5$$

Properties of the autocovariance function

This is equivalent to equation 2.1-1 when f(t) is the voltage or current for a resistive load of 1 ohm. Thus a(0) is the mean power. The peak value of a(L) occurs at L = 0 for a random function.

If the lag, L = ∞, a(∞) = 0 unless there is a direct current (DC) component or a continuous periodic alternating current (AC) component.

The variance as defined by equation 6.1-2 may be written with different limits

$$a(0) = \lim_{T \to \infty} T^{-1} \int_{-T/2-T_1}^{T/2-T_1} [f(t)]^2 \, dt$$

or

$$a(0) = \lim_{T \to \infty} T^{-1} \int_{-T/2}^{T/2} [f(t + T_1)]^2 \, dt \qquad 6.2\text{-}6$$

By Schwarz's inequality (Protter and Morrey, 1964, p. 436)

$$\left| \int_a^b f_1 f_2 \, dt \right|^2 \leq \int_a^b |f_1|^2 \, dt \int_a^b |f_2|^2 \, dt \qquad 6.2\text{-}7$$

For the autocovariance function 6.1-6 this may be written as

$$\left| T^{-1} \int_{-T/2}^{T/2} f(t+L) \cdot f(t) \, dt \right|^2 \leq T^{-1} \int_{-T/2}^{T/2} f^2(t) \, dt \; T^{-1} \int_{-T/2}^{T/2} f^2(t+L) \, dt \qquad 6.2\text{-}8$$

By 6.2-6 the two integrals on the right approach the finite function a(0) as T approaches infinity. Therefore the autocovariance squared $[|a(L)|]^2$ is finite in the limit. By 6.2-8 we obtain the important property that

$$|a(L)| \leq a(0) \qquad 6.2\text{-}9$$

The autocovariance for negative lags is

$$a(-L) = \lim_{T \to \infty} T^{-1} \int_{-T/2}^{T/2} f(t' - L) f(t') dt' \qquad 6.2\text{-}10$$

or letting $t' = t + L$

$$a(-L) = \lim_{T \to \infty} T^{-1} \int_{-T/2}^{T/2} f(t) f(t + L) dt \qquad 6.2\text{-}11$$

The integral on the right is an autocovariance for positive lags.

$$a(-L) = a(L) \qquad 6.2\text{-}12$$

Therefore by 6.2-9 and 6.2-12 the autocovariance and autocorrelation functions are real, finite, even functions. The autocovariance of a transient function is defined to be

$$a(L) = \int_{-\infty}^{\infty} f(t) \cdot f(t + L) \, dt \qquad 6.2\text{-}13$$

In summary autocovariances:
(1) have an absolute maximum at zero lag which is equal to the mean power,
(2) contain no information on the phase relation between frequency components,
(3) approach zero as the lag becomes large for random functions,
(4) are real, finite, even functions. The reader is referred to Lee (1964) for additional discussion. The autocovariance of several simple functions is illustrated in figure 6.1.

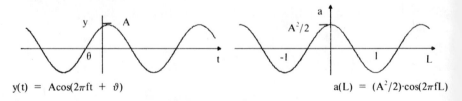

$y(t) = A\cos(2\pi ft + \vartheta)$ $\qquad\qquad a(L) = (A^2/2)\cdot\cos(2\pi fL)$

Figure 6.1a A sinusoid and its autocovariance.

Properties of the autocovariance function

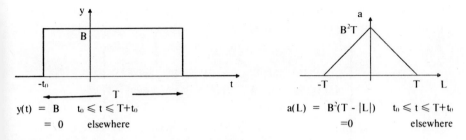

Figure 6.1b   A box-car function and its autocovariance.

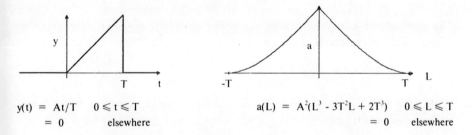

Figure 6.1c   A transient ramp and its autocovariance.

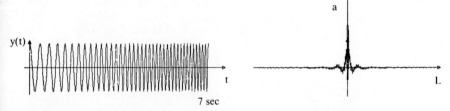

Figure 6.1d   A chirp signal and its autocovariance, a Klauder wavelet. The frequency of the sinusoid increases from 14 to 56 Hertz in 7 seconds. This is a range of 2 octaves.

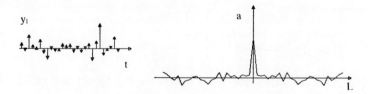

Figure 6.1e   A set of random spikes and their autocovariance, $a(L) \approx \delta(L)$.

Figure 6.1   A set of functions and their autocovariance. From the mathematical definition the mean should be removed from y(t) before the autocovariance is computed. However, this is not done for unipolar transients.

A matrix which is useful in various forms of temporal and spatial filtering is the autocovariance matrix.

$$\mathbf{A} = \begin{bmatrix} a(0) & a(1) & a(2) & \ldots & a(n-1) \\ a(-1) & a(0) & a(1) & \ldots & a(n-2) \\ a(-2) & a(-1) & a(0) & \ldots & a(n-3) \\ \vdots \\ a(-n+1) & a(-n+2) & a(-n+3) & \ldots & a(0) \end{bmatrix} \qquad 6.2\text{-}14$$

The normalized form with ones along the principal diagonal is called the autocorrelation matrix, $\mathbf{A}_N$.

$$\mathbf{A}_N = \mathbf{A}/a(0) \qquad 6.2\text{-}15$$

Both these matrices are symmetric and non-negative (Khintchine, 1934). That is, the determinant and all the principal minors are greater than zero. For $n = 3$

$$\begin{vmatrix} 1 & a_N(1) \\ a_N(-1) & 1 \end{vmatrix} > 0 \qquad 6.2\text{-}16$$

$$\begin{vmatrix} 1 & a_N(2) \\ a_N(-2) & 1 \end{vmatrix} > 0 \qquad 6.2\text{-}17$$

and

$$\begin{vmatrix} 1 & a_N(1) & a_N(2) \\ a_N(-1) & 1 & a_N(1) \\ a_N(-2) & a_N(-1) & 1 \end{vmatrix} > 0 \qquad 6.2\text{-}18$$

If we find that the mean and autocovariance matrix do not change with a change in the origin of time then the system which gives rise to such a data set is said to be a stationary process up to the second order.

6.3 Vibratory sources

Novel methods of generating signals in acoustic, electromagnetic and seismic methods of exploration often make use of the autocovariance or cross covariance functions. Uniquely shaped signals with a long duration in time can be used to transmit a large amount of energy over a broad spectrum. Such signals can then be distinguished more easily in the presence of noise. The autocorrelation or autocovariance technique is used to detect the signal and compress it into a short pulse with the aid of a matched filter. An example is the use of a chirp signal for the shape of an elastic wave source function. The chirp signal (figure 6.1d) was introduced as a uniquely identifiable signal in radar systems (Klauder et al., 1960). Earlier references going into the 1940s are given by Anstey (1964). In seismic exploration a chirp signal is transmitted into the earth by one to four mechanical vibrators in a technique called VIBROSEIS (trademark of Continental Oil Co., Crawford et al., 1960). In equation 6.3-1 $B(t;T)$ is a box-car function which has a value of unity between 0 and T seconds and is zero at other times. Typically T is five to seven seconds.

$$y(t) = B(t;T) \left[\cos[2\pi(f_1 t + kt^2/2) + \phi_0] \right] \quad 0 \leq t \leq T \qquad 6.3\text{-}1$$

The phase of the chirp signal is

$$\phi = 2\pi(f_1 t + kt^2/2) + \phi_0$$

and the instantaneous frequency is

$$f_i = (2\pi)^{-1} \cdot d\phi/dt = f_1 + kt \qquad 6.3\text{-}2$$

Note that the sinusoidal wave increases linearly in frequency from its initial value of f_1 to its terminal value of f_2. For a signal of duration T the rate of change in frequency is

$$k = (f_2 - f_1)/T \qquad 6.3\text{-}3$$

The terminal high frequency is usually specified in terms of a bandwidth ratio, n, in octaves.

$$f_2 = 2^n f_1 \qquad 6.3\text{-}4$$

A seismic recording, $s(t)$, using a VIBROSEIS source may be modelled by convolving the impulse response of the layered earth, $e(t)$, with a chirp signal, $y(t)$.

$$s(t) = y(t) * e(t) \qquad 6.3\text{-}5$$

The elastic response of a sedimentary section is made up primarily of reflected waves from layers with random thicknesses and acoustic impedances. It can be considered an example of white noise (see Appendix 4). A cross covariance is taken of the seismic recording with the chirp signal to give a VIBROSEIS signal, $v(\tau)$.

$$v(\tau) = \sum_t y(t+\tau) \cdot s(t) \qquad 6.3\text{-}6$$

It is usual to assume that the sampling interval, Δt, is unity and that the normalizing factor, $1/N$, is constant throughout and can be omitted. By the convolution equation (2.1-6) it is seen that a cross covariance is equal to a convolution of a signal with the negative lag of the response function. That is

$$v(\tau) = s(\tau) * y(-\tau) \qquad 6.3\text{-}7$$

Using $y(-\tau)$ is equivalent to reversing the chirp signal or reflecting it in a mirror. Making use of the commutative properties of a convolution (equation 2.5-7) and substituting 6.3-5 gives

$$v(\tau) = y(-\tau) * y(\tau) * e(\tau) \qquad 6.3\text{-}8$$

The convolution of a wavelet, $y(\tau)$, and its reverse, $y(-\tau)$, is obviously an autocovariance function if we ignore the normalizing factor. If the wavelet is a chirp signal its autocovariance has been called a Klauder wavelet, $a(\tau)$, as shown in figure 6.1d

$$a(\tau) = y(-\tau) * y(\tau) \qquad 6.3\text{-}9$$

$$a(\tau) = (2\pi k\tau)^{-1} [\cos(2\pi f_0 \tau)] \cdot \sin(\pi k\tau T - \pi k\tau^2)$$

$$+ (4k)^{-0.5} \cos[2\pi k(f_1^2 k^{-2} - \tau^2/4) + 2\phi_0]$$

$$+ 0.25 k^{-0.5} \cos[2\pi k(f_1^2 k^{-2} - \tau^2/4) + 2\phi_0] [C(F_2) - C(F_1)]$$

$$+ 0.25^{-0.5} \sin[2\pi k(f_1^2 k^{-2} - \tau^2/4) + 2\phi_0] [S(F_2) - S(F_1)] \qquad 6.3\text{-}10$$

where $F_1 = 2k^{0.5}[f_1/k + \tau/2]$ and $F_2 = 2k^{0.5}[f_2/k - \tau/2]$

and where ϕ_0 is an initial phase and the central frequency is

Vibratory sources

$$f_0 = (f_1 + f_2)/2 \qquad \text{6.3-11}$$

$C(u)$ and $S(u)$ are the real and imaginary parts of the Fresnel integral, $Z(u)$.

$$Z(u) = C(u) + iS(u) = \int_0^u e^{i\pi a^2/2} \, da \qquad \text{6.3-12}$$

The Fresnel integral is used in evaluating the diffracted pattern produced by diffracted light passing through a slit. It may be computed by subroutines such as found in the IBM Scientific Subroutine Package. The VIBROSEIS signal is then

$$v(\tau) = a(\tau) * e(\tau) \qquad \text{6.3-13}$$

which is a convolution of a Klauder wavelet with the sequence of impulses representing seismic reflections. Note that the autocorrelation in equation 6.3-10 is for a real chirp signal and is different from that given by Klauder et al. (1960) in which they obtain a matched filter for a complex signal. Taking the real part of their equation does not yield the edge contributions of the Fresnel integrals.

A portion of an impulse response with only four reflection pulses, the seismic recording and the VIBROSEIS trace is shown in figure 6.2. The power spectrum of a Klauder wavelet is useful to have on occasion and a typical one is included in the figure. The Fourier transform of a chirp signal is given by

$$S(f) = [T/2(f_2-f_1)]^{1/2} \, e^{-i\pi(f-f_0)^2/k} \, [Z(u_2)-Z(u_1)] \qquad \text{6.3-14}$$

where $Z(u)$ is given by equation 6.3-12

$$u_2 = -2(f-f_0)[T/2\Delta]^{1/2} + [T\Delta/2]^{1/2}$$

$$u_1 = -2(f-f_0)[T/2\Delta]^{1/2} - [T\Delta/2]^{1/2}$$

The bandwidth is $\Delta = f_2-f_1$ and the carrier or mid-frequency is $f_0 = (f_2+f_1)/2$. Approximations to the complex Fresnel integrals are given by Seriff and Kim (1970), Gurbuz (1972) and by Rietsch (1977). For a linear upsweep the power spectrum may be obtained from $S(f) S^*(f)$ or by computing the autocovariance of equation 6.3-10 out to lags of 2T at about 250 samples per second and taking

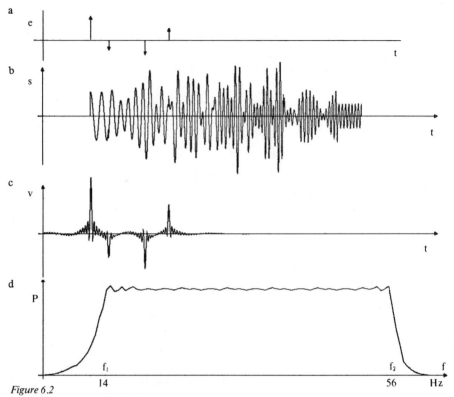

Figure 6.2
(a) A portion of the reflection impulse response for a simple earth model.
(b) A seismic recording using the chirp signal of figure 6.1d and the response of 6.2a.
(c) A VIBROSEIS signal using the Klauder wavelet of figure 6.1d and the response of 6.2a.
(d) The power spectrum of the Klauder wavelet.

a fast Fourier transform. The function B(t,T) may be a box-car or some slowly varying amplitude modulation which tapers the chirp signal. More exotic envelopes have been suggested for the chirp signal but have not found favor because the influence of the coupling in the earth-vibratory system is predominant.

6.4 Multivariate regression

In many problems an experimental observation, y, is described as a function of several independent variables, x_1, x_2, ... x_{n-1}. As an example, the x's may represent the abundance or ratio of certain elements or minerals and y could represent the temperature of mineralization. The independent variables are related to the dependent variable, y, through an unknown coefficients, k_i. If the jth measurement was without error we could write an equation in the form

Multivariate regression

$$y_j = k_0 + k_1 x_{1j} + k_2 x_{2j} + \ldots + k_n x_{nj} \qquad 6.4\text{-}1$$

The problem is not restricted to a linear dependence as the x's can represent complicated functions. For instance, if $x_1 = t$, $x_2 = t^2$, $x_3 = t^3$, etc., then this special case is called polynomial regression. Since each measurement of y_j has a experimental error, e_j, we obtain a series of equations each with this residual term.

$$\begin{aligned} y_1 &= k_0 + k_1 x_{11} + k_2 x_{21} + \ldots + k_n x_{n1} + e_1 \\ y_2 &= k_0 + k_1 x_{12} + k_2 x_{22} + \ldots + k_n x_{n2} + e_2 \qquad 6.4\text{-}2 \\ &\ldots \\ y_m &= k_0 + k_1 x_{1m} + k_2 x_{2m} + \ldots + k_n x_{nm} + e_m \end{aligned}$$

It is usually assumed for the purposes of obtaining confidence limits that y is subject to random error, with a normal distribution, and that the x's are precisely known. Defining the column vectors

$$\begin{aligned} \mathbf{Y} &= [y_1 \ y_2 \ \ldots \ y_m]^t \\ \mathbf{K} &= [k_0 \ k_1 \ k_2 \ \ldots \ k_n]^t \qquad 6.4\text{-}3 \\ \mathbf{E} &= [e_1 \ e_2 \ \ldots \ e_m]^t \end{aligned}$$

and

$$\mathbf{X} = \begin{bmatrix} 1 & x_{11} & x_{12} & \ldots & x_{1n} \\ 1 & x_{21} & x_{22} & \ldots & x_{2n} \\ \cdot & & & & \\ \cdot & & & & \\ 1 & x_{m1} & x_{m2} & \ldots & x_{mn} \end{bmatrix} \qquad 6.4\text{-}4$$

equations 6.4-2 can be written in matrix form as

$$\mathbf{Y} = \mathbf{XK} + \mathbf{E} \qquad 6.4\text{-}5$$

To obtain the best possible values for the coefficients, k_i, a least squares method is used to minimize the sum of the squares of the errors. The errors are given by

$$E = Y - XK \qquad 6.4\text{-}6$$

and the sum of squares is obtained from

$$S = E^t E \qquad 6.4\text{-}7$$

The minimization is achieved by taking the first derivative of the sum of squares with respect to each of the unknown coefficients and setting them equal to zero.

$$\partial S/\partial k_i = 0 \qquad i = 0, 1, 2 .. n \qquad 6.4\text{-}8$$

This yields a set of $n + 1$ equations given by

$$X^T XK = X^T Y \qquad 6.4\text{-}9$$

The solution is found to involve the inverse of $X^T X$.

$$K = [X^T X]^{-1} X^T Y \qquad 6.4\text{-}10$$

An estimate of the variance (Draper and Smith, 1966) is given by

$$\sigma^2 = [Y^t Y - K^t X^t Y]/(m - n - 1) \qquad 6.4\text{-}11$$

The symmetric $n + 1$ by $n + 1$ inverse matrix

$$C = [X^t X]^{-1} \sigma^2 \qquad 6.4\text{-}12$$

is called a *covariance matrix*. The variance and covariance of the model parameters, k_i, are given by the diagonal and off-diagonal elements of C.

$$\text{Variance}(k_j) = c_{jj} \quad ; \quad \text{Covariance}(k_i, k_j) = c_{ij} \qquad 6.4\text{-}13$$

References

References

Anstey,N.A.(1964), Corrleation techniques - a review. *Geophysical Prospecting* **12**, 23-382.

Bartlett,M.S.(1950), Periodogram analyses and continuous spectra. *Biometrika* **37**, 1-16.

Blackman,R.B. and Tukey,J.W.(1958), The measurement of power spectra Dover Publ., New York.

Crawford,J.M., Doty,W.E.N. and Lee,M.R.(1960), Continuous signal seismograph. *Geophysics* **25**, 95-105.

Draper,N.R. and Smith,H.(1966), Applied Regression Analysis. Wiley, New York.

Gurbuz,B.M.(1972), Signal enhancement of vibratory source data in the presence of attenuation. *Geophysical Prospecting* **20**, 421-438.

Khintchine,A.(1934), Korrelationstheorie der stationären stochastischen prozesse. *Math. Annalen* **109**, 608.

Klauder,J.R., Prince,A.C., Darlington,S. and Albersheim,W.J.(1960), The theory and design of chirp radars. *The Bell System Technical Journal* **39**, 745-808.

Lee,Y.W.(1964), Statistical theory of communication. John Wiley & Sons Inc.

Lines,L.R. and Clayton,R.W.(1977), A new approach to VIBROSEIS deconvolution. *Geophysical Prospecting* **25**, 417-433.

Lines,L.R., Clayton,R.W. and Ulrych,T.J.(1980), Impulse response models for noisey VIBROSEIS data. *Geophysical Prospecting* **28**, 49-59.

Papoulis,A.(1962), The Fourier integral and its applications. McGraw-Hill Book Co., New York.

Phillips,R.S.(1947), Statistical properites of time-variable data, Ch. 6, p. 262-307 in *Theory of servo mechanism*. Ed. H. M. James, N. B. Nichols and R. S. Phillips, McGraw-Hill, New York.

Protter,M.H. and Morrey,C.B. Jr.(1964), Modern mathematical analysis. *Addison-Wesley Publ. Co. Inc., Reading, Mass.*

Rice,S.O.(1944-45), Mathematical analysis of random noise. 282, *Bell System Technical Journal* **23, 24**. See also Selected papers on noise and stochastic processes. Ed. N. Wax, 1954, 133-295, Dover Publ., New York.

Rietsch,E.(1977), Computerized analysis of VIBROSEIS signal similarity. *Geophysical Prospecting* **25**, 541-552.

Ristow,D. and Jurczyk,D.(1974), VIBROSEIS deconvolution. *Geophysical Prospecting* **23**, 363-379.

Seriff,A.J. and Kim,W.H.(1970), The effect of harmonic distortion in the use of vibratory surface sources. *Geophysics* **35**, 234-246.

Taylor,G.I.(1920), Diffusion by continuous moment. *Proc. London Math. Soc. Ser. 2*, **20**, A6-212.

7 Wiener-Khintchine theorem and the power spectrum

7.1 The periodogram

Many variables which are measured in geophysics have a cyclical appearance or else they appear to have been modified by a process of reverberation. Therefore, it is natural to use harmonic analysis to study these effects. The Fourier transform of a function of time or space can be used on transient phenomena if there is an absence of noise. When the variables are quantities like the electric and magnetic field in magnetotelluric studies or ground displacement in microseisms it is impossible to assign a beginning or end to the signal. Since the energy as given by (2.1-1) is not finite a Fourier transform calculation is impossible and it is necessary to obtain an energy density estimate. An often used approach was Schuster's (1898) periodogram which is defined as

$$\text{Periodogram Energy Density Estimate} = \lim_{T \to \infty} T^{-1} \left| \int_{-T/2}^{T/2} f(t)\, e^{i\omega t}\, dt \right|^2 \qquad 7.1\text{-}1$$

In computing this for a finite length of digitized data the periodogram is

$$|F(\omega)|^2 = |F(2\pi J/N)|^2 = N^{-1} \left| \sum_{K=0}^{N-1} f(K)\, e^{-i2\pi JK/N} \right|^2 \qquad 7.1\text{-}2$$

Unfortunately the variance of this estimate does not decrease as the length of the record is increased but stays constant (Fisher, 1929; Hannan, p. 53, 1960). As the record becomes longer and longer a greater frequency resolution is obtained but the statistical reliability of the result is not increased (Bartlett, 1948). The periodogram calculated for a record from a white noise generator or from a series of numbers from a table of random numbers has a very spiked appearance. These fluctuations do not decrease as the sample length is decreased. A time series of a section of stationary white noise may be regarded as an observation of N real numbers and N imaginary numbers, the imaginary ones being identically zero. In carrying out a Fourier transform these data become N amplitudes and N phases involving N/2 positive frequencies and N/2 negative frequencies. The spectral estimate given by a periodogram yields N/2 independent terms each of which is distributed as a chi-squared variable with

The periodogram

two degrees of freedom. This is equivalent to drawing up a histogram with such a small group interval that it only includes two observations per interval.

Methods are now available to decrease the irregular appearance and erratic behavior of the periodogram as the record length changes following a suggestion by Daniell (1946) and these have now come into favor (Jones, 1965; Tukey, 1965). They will be discussed in Chapter 9 together with spectral windows.

7.2 Wiener-Khintchine theorem

An alternative mathematical approach was initiated by Wiener (1930) through the autocovariance or autocorrelation function. A similar method was suggested by Khintchine (1934) although in a more abstract presentation. The development of these ideas was carried out by many mathematicians but as it applied to Gaussian stochastic processes it is due mainly to Tukey (1949) as presented in Blackman and Tukey (1958). The autocorrelation is a convenient function because it is easily modified in desirable ways so that the power spectrum obtained from it is a smooth function and the estimate of the variance is smaller than for a periodogram.

The condition for the estimate is that the energy density or power be finite.

$$E = \lim_{T \to \infty} T^{-1} \int_{-T/2}^{T/2} |f(t)|^2 \, dt < \infty \qquad 7.2\text{-}1$$

The division by the length of data, T, is to be noted when comparing to equation 2.1-1. The autocovariance (equation 6.1-6) is

$$a(L) = \lim_{T \to \infty} T^{-1} \int_{-T/2}^{T/2} f(t) \, f(t + L) \, dt \qquad 7.2\text{-}2$$

The signal f(t) extends over the range -T/2 to T/2 but f(t + L) extends from -T/2 + L to -T/2 + L.

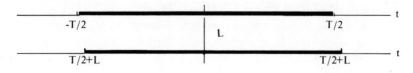

Figure 7.1

To avoid complications we require that $f(t) = 0$ if t exceeds the range $|t| > T/2$. The Fourier transform of $f(t)$ is

$$F(\omega) = \int_{-\infty}^{\infty} f(t) \, e^{-i\omega t} \, dt \qquad 7.2\text{-}3$$

The inverse Fourier transform is

$$f(t) = (2\pi)^{-1} \int_{-\infty}^{\infty} F(\omega) \, e^{i\omega t} \, d\omega \qquad 7.2\text{-}4$$

Then the autocovariance is

$$a(L) = \lim_{T \to \infty} (2\pi T)^{-1} \int_{-T/2}^{T/2} f(t) \int_{-\infty}^{\infty} F(\omega) \, e^{i\omega(t+L)} \, d\omega \, dt \qquad 7.2\text{-}5$$

or changing the order of integration

$$a(L) = \lim_{T \to \infty} (2\pi T)^{-1} \int_{-\infty}^{\infty} F(\omega) \, e^{i\omega t} \int_{-T/2}^{T/2} f(t) \, e^{i\omega t} \, dt \, d\omega$$

substituting (7.2-3) with negative frequencies

$$a(L) = \lim_{T \to \infty} (2\pi T)^{-1} \int_{-\infty}^{\infty} F(\omega) \, e^{i\omega t} \, F(-\omega) \, d\omega \qquad 7.2\text{-}6$$

Defining the power spectrum as

$$P(\omega) = \lim_{T \to \infty} T^{-1} [F(\omega) \cdot F(-\omega)] \qquad 7.2\text{-}7$$

or equivalently as

Wiener-Khintchine theorem

$$P(\omega) = \lim_{T \to \infty} T^{-1} |F(\omega)|^2 \qquad 7.2\text{-}8$$

(7.2-6) becomes

$$a(L) = (2\pi)^{-1} \int_{-\infty}^{\infty} P(\omega) \, e^{i\omega t} \, d\omega \qquad 7.2\text{-}9$$

The power spectrum, $P(\omega)$, is the inverse Fourier transform of the autocovariance function.

$$P(\omega) = \int_{-\infty}^{\infty} a(L) \, e^{-i\omega L} \, dL \qquad 7.2\text{-}10$$

The truncated autocovariance was shown in 6.2-9 and 6.2-12 to be an even function which tends to finite limit. The Fourier transform of the autocovariance is also an even function and approaches a finite limit if the initial condition 7.2-1 is true. Let

$$P(\omega) = d\Lambda / d\omega \qquad 7.2\text{-}11$$

where

$$\Lambda(\omega) = \int_{-\infty}^{\omega} P(\omega') \, d\omega' \qquad 7.2\text{-}12$$

is called the *cumulative spectral distribution* function or integrated spectrum. Then equation 7.2-9 can be written with 7.2-11 as

$$a(L) = (2\pi)^{-1} \int_{-\infty}^{\infty} e^{i\omega L} \, d\Lambda(\omega) \qquad 7.2\text{-}13$$

This relation is known as the Wiener-Khintchine theorem as it was first discovered by Wiener (1930) and independently rediscovered by Khintchine

(1934). Its form, as a Fourier-Stieltjes transform, is a direct consequence of the autocovariance being a non-negative definite function. For sampled data equation 7.2-13 becomes

$$a(L) = (2\pi)^{-1} \int_{\omega=-\pi}^{\pi} e^{i\omega L} \, d\Lambda(\omega) \qquad 7.2\text{-}14$$

The circular frequency, ω, varies only from $-\pi$ to π because the lag, L, is integer and so the exponential function is periodic. For discrete data the integrated spectrum, Λ, is a monotonically increasing function over the range $-\pi \leqslant \omega \leqslant \pi$.

$$\Lambda(\omega) = \int_{-\pi}^{\omega} P(\omega') \, d\omega' \qquad 7.2\text{-}15$$

The important conclusion that follows from the Wiener-Khintchine theorem is that the power spectrum and the autocovariance are Fourier transforms of each other. If f(t) is the voltage across a one ohm resistor, $P(\omega) \, d\omega$ is the average power dissipated from frequencies ω to $\omega + d\omega$.

The integrated spectrum, Λ, varies from zero at $\omega = -\pi$ to a_0, the variance or autocovariance with zero lag, at $\omega = \pi$. The function Λ consists of two parts, Λ_1 and Λ_2, for physically realizable processes. The first part, Λ_1, is an absolutely continuous part which arises from signals generated either by a random process or by a non-random (deterministic) process. The second part, Λ_2, is only approximated in nature and consists of step functions. This part is from deterministic signals such as sinusoidal oscillations at one frequency to produce sharp spectral lines. Some examples would be signals registering the length of the day, the oscillation of a crystal, or the emission of light in a laser beam although even these are not absolutely sharp spectral lines. Figure 7.2 illustrates the monotonically increasing nature of the cumulative spectral distribution and its relation to the power spectrum as obtained from equation 7.2-15.

7.3 Cross-power spectrum

If $f_1(t)$ and $f_2(t)$ are two periodic functions they can be expanded by a Fourier series as in equation 3.1-13 and 3.1-14. The cross covariance of these two series is obtained from 6.1-9

$$c(L) = T^{-1} \int_{-T/2}^{T/2} f_1^*(t) \, f_2(t + L) \, dt \qquad 7.3\text{-}1$$

Cross-power spectrum

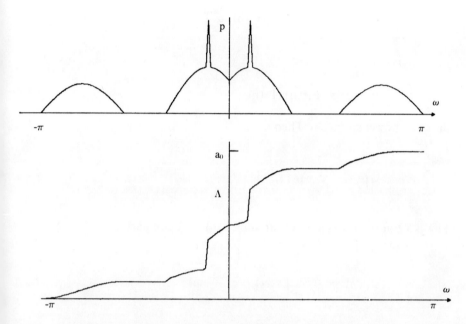

Figure 7.2 The power spectrum and the cumulative spectral distribution for signals generated by a deterministic process. Note that a random (stochastic) process would not have regions where the power is zero or sharp spectral lines. Physically realizable processes also do not have zero power over any frequency range.

or substituting 3.1-13

$$c(L) = T^{-1} \int_{-T/2}^{T/2} f_1^*(t) \sum_{n=-\infty}^{\infty} F_2(n) \, e^{+i\omega_n(t+L)} \, dt \qquad 7.3\text{-}2$$

Changing the order of summation and integration gives

$$c(L) = \sum_{n=-\infty}^{\infty} F_2(n) \, e^{+i\omega_n L} \, T^{-1} \int_{-T/2}^{T/2} f_1^*(t) \, e^{+i\omega_n t} \, dt \qquad 7.3\text{-}3$$

Comparing the integral to the complex conjugate of 3.1-14 it is seen that

$$c(L) = \sum_{n=-\infty}^{\infty} F_1^*(n) \ F_2(n) \ e^{i\omega_n L} \qquad 7.3\text{-}4$$

Let us call

$$P_{12}(n) = F_1^*(n) \ F_2(n) \qquad 7.3\text{-}5$$

the cross-power spectrum. Then

$$c(L) = \sum_{n=-\infty}^{\infty} P_{12} \ e^{i\omega_n L} \qquad 7.3\text{-}6$$

This is a Fourier series and we can solve for P_{12} by comparing 3.1-12

$$P_{12}(n) = T^{-1} \int_{-T/2}^{T/2} c(L) \ e^{-i\omega_n L} \ dL \qquad 7.3\text{-}7$$

The cross covariance is an unsymmetric function in general and the order in which the correlation is made in 6.1-9 is important since

$$C_{12}(-L) = C_{21}^*(L) \qquad 7.3\text{-}8$$

and

$$P_{12}(n) = P_{21}^*(n) \qquad 7.3\text{-}9$$

If $f_1(t) = f_2(t)$ we obtain *autocovariance* for a periodic function.

$$a(L) = T^{-1} \int_{-T/2}^{T/2} f_1(t) \ f_1^*(t + L) \ dt \qquad 7.3\text{-}10$$

From 7.3-4

$$a(L) = \sum_{n=-\infty}^{\infty} |F_1(n)|^2 \ e^{i\omega_n L} \ ; \quad \omega_n = 2\pi n/T \qquad 7.3\text{-}11$$

Cross-power spectrum

the *power spectrum* is obtained as in 7.3-5 or 7.3-7

$$P(\omega_n) = |F_1(n)|^2 = T^{-1} \int_{-T/2}^{T/2} a(L) \, e^{-i\omega_n L} \, dL \qquad 7.3\text{-}12$$

The power spectrum and autocovariance are Fourier transforms of each other.

If the lag, L, is zero we obtain Parseval's theorem from 7.3-10 and 7.3-11

$$a(0) = T^{-1} \int_{-T/2}^{T/2} |f_1(t)|^2 \, dt = \sum_{n=-\infty}^{\infty} |F(n)|^2 \qquad 7.3\text{-}13$$

or from 7.3-12 and 7.3-11.

$$a(0) = \sum_{n=-\infty}^{\infty} P(n) \qquad 7.3\text{-}14$$

As for a continuous signal it is seen from 7.3-13 and 2.1-1 that if $f_1(t)$ is the voltage across a load of 1 ohm, the autocovariance, $a(0)$, is the *mean power obtained from all the harmonics*.

Since the autocovariance function is an even function it can be written as a cosine Fourier series with zero phase shift. From equation 7.3-11 and 7.3-12

$$a(L) = p(0) + 2 \sum_{n=1}^{\infty} P(n) \cos \omega_n L \qquad 7.3\text{-}15$$

The Fourier coefficients, a_n and b_n, are given by 3.1-9 and 3.1-10 and are related to $F(n)$ by 1.21 (Appendix 1).

$$F(n) = 0.5(a_n - ib_n) \qquad n = 0, \pm 1, \pm 2, \ldots \qquad 7.3\text{-}16$$

From 7.3-12 the power spectrum of a periodic function is

$$P(n) = |F_1(n)|^2 = (a_n^2 + b_n^2) / 4 \qquad 7.3\text{-}17$$

The autocovariance function of a periodic function is obtained from 7.3-15.

$$a(L) = a_0^2/4 + 0.5 \sum_{n=1}^{\infty} (a_n^2 + b_n^2) \cos \omega_n L \qquad 7.3\text{-}18$$

7.4 Compensation for instrument response

Modification in the recorded signal may be compensated by an inverse convolution in the time domain. It will be shown in this section that if one is only interested in the power spectrum it is easier to correct the distortion produced by the recording instrument in the frequency domain. The input signal will be $x(t)$, the impulse response of the instrument will be $W(t)$, and the output as finally digitized will be $y(t)$.

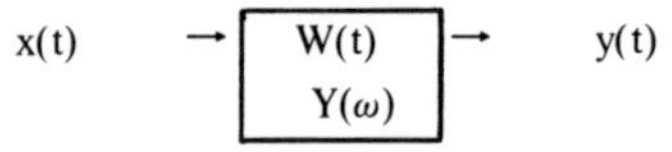

```
x(t)  →  | W(t)  |  →  y(t)
         | Y(ω)  |
```

Figure 7.3 Symbols used in convolution of a signal by a recording instrument.

The output is given by the convolution integral, 5.1-5.

$$y(t) = \int_{-\infty}^{\infty} W(L_1) \, x(t - L_1) \, dL_1 \qquad 5.1\text{-}5$$

The autocovariance coefficient of the output function is given by 6.1-6.

$$a(L) = \lim_{T \to \infty} T^{-1} \int_{-T/2}^{T/2} y(t) \, y(t + L) \, dt \qquad 6.1\text{-}6$$

Compensation for instrument response

Substituting 5.1-5 for the output functions

$$a(L) = \lim_{T \to \infty} T^{-1} \int_{-T/2}^{T/2} \int_{-\infty}^{\infty} \int_{-\infty}^{\infty} W(L_1) \, W(L_2) \, x(t-L_1) \, x(t+L-L_2) \, dL_1 \, dL_2 \, dt \quad 7.4\text{-}1$$

Let the autocovariance of the original function be

$$a_x(L+L_1-L_2) = \lim_{T \to \infty} T^{-1} \int_{-T/2}^{T/2} x(t-L_1) \, x[(t-L_1) + (L+L_1-L_2)] \, dt \quad 7.4\text{-}2$$

then 7.4-1 becomes

$$a(L) = \int_{-\infty}^{\infty} \int_{-\infty}^{\infty} W(L_1) \, W(L_2) \, a_x(L+L_1-L_2) \, dL_1 \, dL_2 \quad 7.4\text{-}3$$

The Fourier transform of $a_x(L + L_1 - L_2)$ is the power spectrum $P(\omega)$ as in 7.3-12 so

$$a(L) = (2\pi)^{-1} \int_{-\infty}^{\infty} \int_{-\infty}^{\infty} \int_{-\infty}^{\infty} W(L_1) \, W(L_2) \, P(\omega) \, e^{i\omega(L+L_1-L_2)} \, dL_1 \, dL_2 \, d\omega \quad 7.4\text{-}4$$

Rearranging terms in 7.4-4

$$a(L) = (2\pi)^{-1} \int_{-\infty}^{\infty} [\int_{-\infty}^{\infty} W(L_1) \, e^{i\omega L_1} \, dL_1] [\int_{-\infty}^{\infty} W(L_2) \, e^{-i\omega L_2} \, dL_2] \, P(\omega) \, e^{i\omega L} \, d\omega \quad 7.4\text{-}5$$

The integrals inside the brackets are the transfer functions, $Y^*(\omega)$ and $Y(\omega)$, from equation 5.3-3.

$$a(L) = (2\pi)^{-1} \int_{-\infty}^{\infty} |Y(\omega)|^2 \, P(\omega) \, e^{i\omega L} \, d\omega \quad 7.4\text{-}6$$

This is a Fourier transform of $P_y(\omega) = |Y(\omega)|^2 P(\omega)$ and making use of the theorem that the Fourier transform of the autocovariance is a power spectrum

$$P_y(\omega) = \int_{-\infty}^{\infty} a(L) \, e^{-i\omega L} \, dL \qquad 7.4\text{-}7$$

where $P_y(\omega)$ is the power spectrum of the output from the instrument. Thus by 7.4-6

$$P_y(\omega) = |Y(\omega)|^2 \, P(\omega) \qquad 7.4\text{-}8$$

If the transfer function is known then it is easy to correct the power spectrum for the modifications produced in recording the signal. The true power spectrum is

$$P(\omega) = P_y(\omega) \,/\, |Y(\omega)|^2 \qquad 7.4\text{-}9$$

The amplitude response, $|Y(\omega)|$, of a typical sensing device used for earthquake detection was shown in figure 3.4.

References

Bartlett,M.S.(1948), Smoothing periodograms from time-series with continuous spectra. *Nature* **161**, 686-687.
Blackman,R.B. and Tukey,J.W.(1958), The measurement of power spectra. Dover Publications, New York.
Daniell,P.J.(1946), Discussion on symposium on autocorrelation in time series. *Supplement to the J. Roy. Statist. Soc.* **8**, 88-90.
Fisher,R.A.(1929), Tests of significance in harmonic analysis. *Proc. Roy. Soc. London, Ser. A* **125**, 54-59.
Hannan,E.J.(1960), Time series analysis. John Wiley & Sons, New York.
Jones,R.H.(1965), A reappraisal of the periodogram in spectral analysis. *Technometrics* **7**, 531-542.
Khintchine,A.(1934), Korrelationstheorie der stationären stochastischen Prozzese. *Math Annalen* **109**, 604-615.
Schuster,A.(1898), Investigation of hidden periodicities. *Terrestrial Magnetism* **3**, 13-41. Also *Trans. Camb. Phil. Soc.* **18**, 107-135 (1900) and *Proc. Roy. Soc. London, A*, **77**, 136-140 (1906).
Tukey,J.W.(1949), The sampling theory of power spectrum estimates. Symposium on application of autocorrelation analysis to physical problems, 47-67, Woods Hole, Massachusetts, Office of Naval Research, Washington, D.C.
Tukey,J.W.(1965), Uses of numerical spectrum analysis in geophysics. Presented at International Association of Statistics in the Physical Sciences and the International Statistical Institute, Beograd, Yugoslavia, Sept. 14-22, 1965, Princeton University.
Wiener,N.(1930), Generalized harmonic analysis. *Acta Math.* **55**, 117-258. See also Harmonic analysis of irregular motion. *Jour. Math. and Phys.* **5**, 99-189 (1926). More accessible is a pocket book on Time Series: *Extrapolation, Interpolation, and Smoothing of Stationary Time Series.* The M.I.T. Press, Cambridge, Mass. (1966).

8 Aliasing

8.1 The Nyquist frequency

The first step in obtaining a power spectrum of some recorded data is the conversion to digital form by an analog to digital converter. This process is equivalent to a multiplication of the continuous functions by an infinite Dirac comb, $\nabla(t; \Delta t)$.

$$\nabla(t; \Delta t) = \sum_{n=-\infty}^{\infty} \delta(t-n\Delta t) \qquad 3.2\text{-}3$$

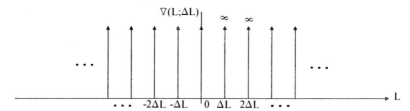

Figure 8.1 The infinite Dirac comb in the time domain.

The Fourier transform of the Dirac comb is a frequency Dirac comb (see example 12.6; Papoulis, 1962).

$$\nabla(f; 1/\Delta t) = (\Delta t)^{-1} \sum_{n=-\infty}^{\infty} \delta(f-n/\Delta t) \qquad 3.2\text{-}4$$

Note that although a single Dirac delta function has a continuous function for its Fourier transform, the infinite Dirac comb has a discontinuous one. The harmonic functions interfere destructively with one another everywhere except at places where the frequency is

$$f = \omega/2\pi = n/\Delta t \qquad 8.1\text{-}1$$

Suppose a sinusoidal wave is sampled at a uniform interval of time, Δt. The frequencies within the interval $0 \leq f \leq f_N$ are said to be in the *principal interval*. The highest frequency, f_N, in the interval is called the Nyquist (1928) or *folding frequency*.

The Nyquist frequency

$$f_N = 1/2\Delta t \qquad 8.1\text{-}2$$

The relation exists because it is necessary to have at least two samples to detect any frequency component (figure 8.2). The Fourier transform of an infinite Dirac comb may be plotted in terms of twice the Nyquist frequency (figure 8.3).

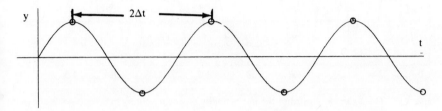

Figure 8.2 The highest frequency detectable after digitization has a period of $2\Delta t$.

In the frequency domain the process of digitization may be described as a convolution of the frequency comb with the Fourier transform of the observed data, y(t). This produces a continuous function in which all the information is contained in the principal interval (see figure 8.3).

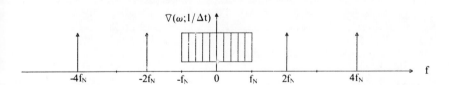

Figure 8.3 The infinite Dirac frequency comb. The observed data will be in the principal interval shown by the shaded area.

8.2 Aliasing

Because of the discrete sampling interval it is impossible to distinguish frequencies above the Nyquist frequencies from those in the principal interval. This is illustrated in figure 8.4. A similar situation is met in the propagation of waves through periodic structures such as crystals (see Brillouin, 1946). Thus if any signals are present with frequencies higher than f_N, their power will be reflected back or *aliased* into the power spectrum over the principal range. It is therefore important to filter out frequencies higher than f_N.

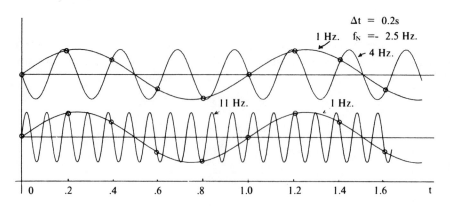

Figure 8.4 This example illustrates the impossibility of distinguishing between signals above the Nyquist frequency after digitization.

Suppose we are given a continuous time series and have calculated its autocovariance and power spectrum (figure 8.5). Suppose the power spectrum, $P(\omega)$, is zero for $|f| > f_N$.

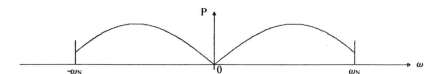

Figure 8.5 Power spectrum of a hypothetical signal.

If the time series is sampled at uniform intervals

$$t = 0, \pm \Delta t, \pm 2\Delta t, \ldots$$

the *aliased power spectrum* is

$$P_a(\omega) = \int_{-\infty}^{\infty} [\nabla(t; \Delta t) \cdot a(t)] e^{-i\omega t} dt \qquad 8.2\text{-}1$$

where the quantity in brackets is the digitized form of the autocovariance.

Making use of the convolution, the aliased power spectrum can also be obtained directly in the frequency domain (Blackman and Tukey, 1958).

Aliasing

$$P_a(\omega) = \nabla(\omega;\ 1/\Delta t) * P(\omega) \qquad 8.2\text{-}2$$

or

$$P_a(\omega) = \int_{-\infty}^{\infty} \sum_{n=-\infty}^{\infty} \delta(\omega'/2\pi - n/\Delta t)\ P(\omega/2\pi - \omega'/2\pi)\ d\omega'/2\pi \qquad 8.2\text{-}3$$

The integral has a value only where $\omega'/2\pi = n/\Delta t$ or

$$P_a(\omega) = \sum_{n=-\infty}^{\infty} P(\omega/2\pi - n/\Delta t) \qquad 8.2\text{-}4$$

This equation states that the aliased power spectrum obtained from a digitized function is a periodic function (see figure 8.6). The section outside the principal aliases is discarded when making use of the power spectrum.

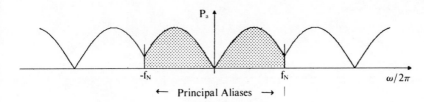

Figure 8.6 The aliased power spectrum P_a. The true power spectrum of the continuous (undigitized) signal is given by the heavy dark curve, $P(\omega)$.

If the time series had contained power in frequencies above the Nyquist frequency the power spectrum as calculated would not be a true representation within the principal alias. Harmonics with frequencies above f_N must be attenuated by a high cut filter *before* calculating the autocovariance so that the higher frequencies are not aliased into the principal section.

If inadvertently we do not filter properly to begin with it is useful to know what frequencies are contributing power to the principal alias. Let us suppose the true power spectrum of the continuous function is P(f). Let the aliased power spectrum of the digitized data have the appearance shown in figure 8.6. Then since conservation of energy must be maintained, the power (energy/time) in the principal alias of P_a must equal the total power in the true power spectrum P.

$$\int_{-f_N}^{f_N} P_a(f) \, df = \int_{-\infty}^{\infty} P(f) \, df \qquad 8.2\text{-}5$$

In the principal alias the power density is given by 8.2-4 or in terms of f.

$$P_a(f) = \sum_{n=-\infty}^{\infty} P(f - 2nf_N) \qquad 8.2\text{-}6$$

For positive n the function P is even and therefore $P(-x) = P(x)$ or

$$P(f - 2nf_N) = P(2nf_N - f)$$

Following Blackman and Tukey (1958) equation 8.2-6 may be written with only

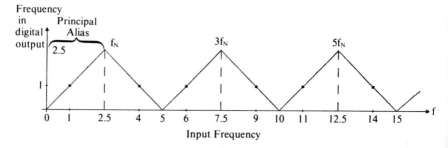

Figure 8.7 Aliasing produced when insufficient high cut filtering is carried out prior to digitization. The input frequency to the analog to digital converter is shown along the abscissa. The diagram is for a sampling rate of $\Delta t = 0.2$ conversions per second.

positive n as the sum of three terms, the first being with $n = 0$.

$$P_a(f) = P(f) + \sum_{n=1}^{\infty} [P(2nf_N - f) + P(2nf_N + f)] \qquad 8.2\text{-}7$$

In equation 8.2-7 the frequency is only allowed to vary over the range 0 to f_N by equation 8.2-5. In expanded form the aliased power is

Aliasing

$$P_a(f) = P(f) + P(2f_N-f) + P(2f_N+f) + P(4f_N-f) + P(4f_N+f) + \ldots \qquad 8.2\text{-}8$$

For example if the sampling interval is 0.2 seconds, the Nyquist frequency is 2.5 cycles per second. By equation 8.2-8 the aliased frequencies contributing power at $f = 1$ c.p.s. are $(2f_N-f)$, $(2f_N+f)$, $(4f_N-f)$, etc., or at 4, 6, 9, 11, etc., c.p.s. since

$$P_a(1) = P(1) + P(4) + P(6) + P(9) + P(11) + \ldots \qquad 8.2\text{-}9$$

If $n = 12$ in equation 8.2-7, $2nf_N = 60$ so any 60 c.p.s. pickup will be reflected into the DC power ($f = 0$). Aliasing is illustrated in figure 8.4 or on a frequency graph in figure 8.7.

Aliasing is called folding because the frequency spectrum can be folded back about the Nyquist frequency as shown below in figure 8.8. The effect of

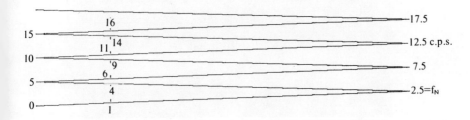

Figure 8.8 Aliasing as a folding operation.

sampling a chirp signal (figure 6.1) as it rises above the Nyquist frequency is shown in figure 8.9. The analog input signal is

$$y(t) = A \sin \phi \qquad 8.2\text{-}10$$

where the phase is

$$\phi = 2\pi(f_0 t + kt^2/2) \qquad 8.2\text{-}11$$

and the instantaneous frequency, f_i, is given by

$$\frac{1}{2\pi} \frac{d\phi}{dt} = f_i = f_0 + kt \qquad 8.2\text{-}12$$

Aliasing

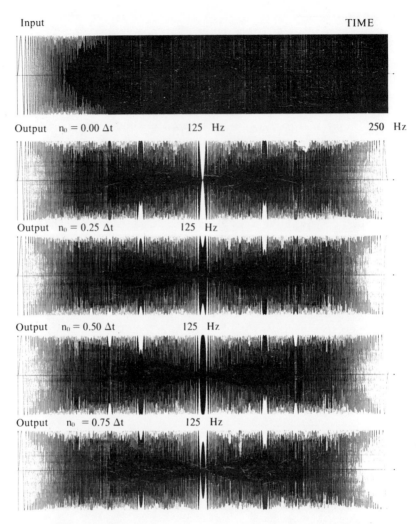

Figure 8.9 A chirp analog input signal and its output from an analog to digital converter. The sampling interval is Δt = 0.004s and the Nyquist frequency is 125 Hertz. The digitized output is shown with various initial phases, n_0.

When the analog chirp signal is sampled by an analog-to-digital converter the output is

$$y[(n + n_0) \Delta t] = A \sin 2\pi \left[f_0(n + n_0) \Delta t + [(n + n_0) \Delta t]^2 k/2 \right] \qquad 8.2\text{-}13$$

where Δt is the sampling interval in seconds, n is an integer, n_0 gives the initial phase and k is the rate of increase of the frequency in cycles per

Aliasing

second-squared. Note that after sampling the apparent maximum amplitude of the digitized signal is variable between half the Nyquist frequency, $f_N/2$, and f_N. This is caused by interference of the input signal with the Dirac comb produced by the a-to-d converter. The shape of this beating effect depends critically upon the initial phase. Above the Nyquist frequency the chirp signal is folded down between frequencies 0 and f_N in the digitized output.

If all the useful information on the recorded data is contained between D.C. (zero frequency) and half the Nyquist frequency, $f_N/2$, then digitization of the continuous signal produces *no* loss of information. This is a highly significant factor because it gives us a method of storing data in a very compact format as digital information. Note that between half the Nyquist frequency and the Nyquist frequency there is loss of phase information as we do not know where the sampling occurs if there are less than 4 samples per period. If one requires information on periods greater than 1 second and has a dynamic range of 15 bits (typically it will be 14 bits amplitude information and 1 bits for time and channel identification), then a single 1600 b.p.i., 9 track magnetic digital tape will store one month of continuous data.

References

Blackman,R.B. and Tukey,J.W.(1958), The measurement of power spectra. Dover Publ. Inc., New York.
Brillouin,L.(1946), Wave propagation in periodic structure. Dover Publ. Inc., New York.
Nyquist,H.(1928), Certain topics in telegraph transmission theory. *A.I.E.E. Transactions,* 617,644.
Papoulis,A.(1962), The Fourier integral and its applications. McGraw-Hill Book Co., New York.
Peterson,R.A. and Dobrin,M.B.(1966), A pictorial digital atlas. United Geophysical Corporation, Pasadena, California.

9 Power spectral estimates

9.1 Introduction

There are almost as many methods of computing a power spectral estimate as there are programmers writing algorithms for this purpose. Many of these make use of some combination of weighting factors which will be discussed here and the discrete Fourier transform. Because of the availability of the fast Fourier transform the newer algorithms for estimating the power spectrum make use of a periodogram (Jones, 1965) in modified form rather than the Fourier transform of a smoothed and truncated autocovariance function as originally proposed by Blackman and Tukey (1959). The latest techniques known as the maximum likelihood method (MLM) of Capon (1969) and Lacoss (1971) and the maximum entropy method (MEM) of Burg (1967) arise from the predictive aspects of optimum filter theory (see Chapters 11 and 12).

9.2 Effects of data truncation

The power spectrum which is obtained from a finite length of data is distorted in some way and considerable effort must be expended to minimize this effect. The truncation of the data may be viewed as a *multiplication* of an infinitely long record by a time window having a rectangular or box-car shape (figure 9.1).

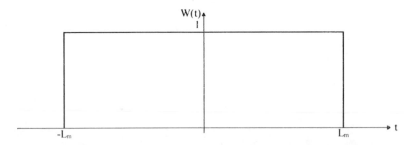

Figure 9.1 Rectangular window or a box-car function as applied to data in the time domain.

In the frequency domain the truncation is equivalent to a *convolution* of the Fourier transform of the data, $F(\omega)$, and the Fourier transform or *kernel* of the rectangular window, $W(\omega)$ (see figure 9.2).

Effects of data truncation

$$W(\omega) = 2L_m \frac{\sin \omega L_m}{\omega L_m} \qquad 9.2\text{-}1$$

The periodogram is obtained from the square of the convolution, $[F(\omega)*W(\omega)] [F(\omega)*W(\omega)]^*$.

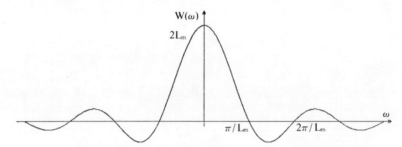

Figure 9.2 Kernel or Fourier transform of the rectangular window.

When the Fourier transform of the rectangular window is convolved with the transform of the signal, the resulting function of frequency involves a modification similar to that shown in figure 9.2 at *every* calculated frequency. This is illustrated for the simple case of the Fourier transform of single 60 cycle sinusoidal wave (see figure 9.3).

Figure 9.3 An infinitely long sinusoid and its Fourier transform.

If the signal in figure 9.3 is truncated the spectral lines acquire a certain breadth and have side lobes similar to the transform of the rectangular window (figure 9.4).

The finite breadth of the spectral *lines* is a common feature of all physical processes which exhibit periodicity, the most striking example of this being contained in Heisenberg's Uncertainty Principle. However, if the autocovariance function is truncated (as opposed to the signal, which was

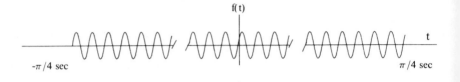

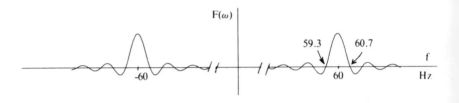

Figure 9.4 A truncated 60 cycle signal and its Fourier transform. The signal was switched on for $\pi/2$ seconds.

truncated in figure 9.4) then its Fourier transform, which is the power spectrum, may be negative at some frequencies. Since a negative power is physically meaningless this is an undesirable feature and efforts are made to remove or reduce its effects by various spectral windows. Special estimates using the autocovariance function are seldom used now and the subject is discussed in Appendix 6.

9.3 Tapering the data

A large discontinuity may occur between the value of the function at the beginning and the end of the vector of data. It is useful to taper the recorded information so that the first and last points approach the mean value which is usually zero. An excellent taper consists of a pair of *cosine bells*. For N' data points the nth component is multiplied by

$$W_n = \begin{cases} (1 + \cos \frac{\pi(n+L)}{M})/2 & -(L+M) < n < -L \\ 1 & -L \leqslant n \leqslant L \\ (1 + \cos \frac{\pi(n-L)}{M})/2 & L < n < L+M \end{cases} \quad 9.3\text{-}1$$

The Fourier transform of this window is

Tapering the data

$$W(\omega) = 2L \frac{\sin \omega L}{\omega L} + 2M \frac{\sin \omega M}{\omega M} \cos \omega(L+M)$$

$$+ M\left[\cos[(\omega+\omega_0)M+\omega L] \frac{\sin(\omega+\omega_0)M}{(\omega+\omega_0)M}\right.$$

$$\left.+ \cos[(\omega-\omega_0)M+\omega L] \frac{\sin(\omega-\omega_0)M}{(\omega-\omega_0)M}\right]$$

where $\omega_0 = \pi/2M$ and $\omega = 2\pi f$. Here M determines the period of the cosine bell. When using the fast Fourier transform algorithm one can add zeros before and/or after the vector of sampled data until the total number of points equals

$$N = 2^k$$

where k is any integer. Figure 9.5 illustrates a pair of cosine bells and a data window extended to form $N = 2^k$ points including zeros. Tapering is not too important for very long vectors but for the study of individual phases in seismology it is particularly essential so as to minimize the effects of other phases and to avoid individual correction of large numbers beginning and ending a set of data.

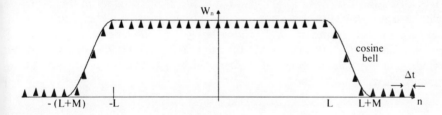

Figure 9.5   A data window used to taper a discrete function.

9.4 Daniell power spectral estimate

The length of data necessary to resolve two closely spaced peaks and the reliability of the power spectral estimate are interrelated quantities. To overcome the difficulties inherent in the periodogram, particularly when studying noisy data, the periodogram may be weighted over a small range of frequencies. This may reduce the variance of the power spectral estimate so that the estimate will be replicated if the experiment is repeated. The variance of the estimate, P_e, is given by

$$\text{Variance } [P_e(\omega)] = T^{-1} \int_{-\infty}^{\infty} P^2(\omega_1) \ W^2(\omega-\omega_1) \ d\omega_1 \qquad 9.4\text{-}1$$

where $P(\omega_1)$ is the true power spectral density and $W(\omega)$ is the response of a smoothing filter or spectral window. If P^2 is approximately constant over the principal lobe of the spectral window then it may be taken outside the integration sign. Defining the asymptotic variance (A.V.) as

$$\text{A.V.}(\omega) = T^{-1} \int_{-\infty}^{\infty} W^2(\omega) \ d\omega \qquad 9.4\text{-}2$$

the variance of the estimate is approximated by

$$\text{Variance } [P_e(\omega)] \simeq P^2(\omega) \ [\text{A.V.}(\omega)] \qquad 9.4\text{-}3$$

If the signal is not a stochastic process, with approximately constant power over all frequencies, then the variance, and therefore the error in the estimate, will be larger. A Fourier transform of the spectral window will give its shape in the lag or time domain, $W(L)$. This will show the degree of smoothing introduced into the autocovariance function. The principal lobe or width of the window in the time domain will be denoted by L_m. The frequency resolution obtainable using L_m will be

$$\Delta f = 1/L_m \qquad 9.4\text{-}4$$

Daniell (1946) suggested that one might use a spectral window whose response is a rectangular (box-car) shape in the frequency domain (figure 9.6). This function may be approximated by taking the average over $2m + 1$ adjoining frequencies. The Daniell spectral estimate, using a fast Fourier transform (Cooley, Lewis and Welch, 1967), is given by

$$P_D(W) = \frac{N'}{N} \frac{1}{2m+1} \sum_{j=-m}^{m} F(W-j) \ F^*(W-j) \qquad W \ 0,1,...,N'/2 \qquad 9.4\text{-}5$$

The star indicates a complex conjugate. A discrete Fourier transform, using a

Daniell power spectral estimate

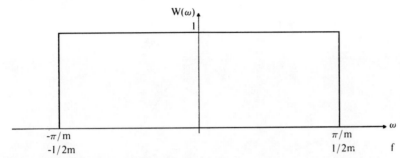

Figure 9.6   The Daniell or rectangular spectral window.

fast Fourier transform algorithm, is used in the computation of F(W).

$$F(W) = \sum_{T=0}^{N'-1} f(T) \, e^{-2\pi i WT/N'} \qquad 9.4\text{-}6$$

The N samples of data, $T = 0,..N-1$, are augmented by the addition of trailing zeros at $T = N,..., N'-1$ to make N' a power of 2. As will be seen, the ratio N/N' should be about 0.7 for a smooth spectral window. The angular frequency, ω, in radians per second is given by

$$\omega = 2\pi W/(N' \cdot \Delta t) \qquad W = 0,1,...N'/2 \qquad 9.4\text{-}7$$

The width of the window (figure 9.6) in the frequency domain is given by

$$\Delta \omega = 2\pi / (m \cdot \Delta t) \qquad 9.4\text{-}8$$

The parameter m determines the resolution and the choice must be made by the experimenter. A value between 3 and 8 is usually taken.

In the time domain (or lag domain if we are calculating power spectra with the autocovariance function), the inverse Fourier transform of the rectangular window is shown in figure 9.7 and equation 9.4-10.

$$W(L) = m/(2\pi) \int_{-\pi/m}^{\pi/m} 1 \cdot e^{i\omega L} \, d\omega \qquad 9.4\text{-}9$$

$$= m/(2\pi i L) \, [e^{i\omega L}]_{-\pi/m}^{\pi/m} = \frac{e^{i\pi L/m} - e^{-i\pi L/m}}{2i\pi L/m}$$

$$= \frac{\sin(\pi L/m)}{\pi L/m} \qquad 9.4\text{-}10$$

This is shown in figure 9.7.

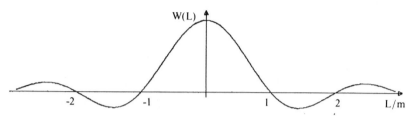

Figure 9.7 The inverse Fourier transform of the ideal Daniell window shown in figure 9.6.

The response of the spectral window is modulated by the discrete Fourier transform and the addition of trailing zeros to the data.

$$H(\omega) = \frac{1}{2m+1} \sum_{j=-m}^{m} \frac{\sin^2[N\omega - 2\pi Nj/N']}{2\pi N \sin^2[\omega - 2\pi j/N']} \qquad 9.4\text{-}11$$

The shape of the window for $N/N' = 0.7$ and 1.0, $m = 7$ is shown in figure 9.8. Note that the ripple has decreased when the augmented data set is used. The variance of the estimate is proportional to

$$\text{Var}[P_D(\omega)] \propto P^2(\omega) / N \cdot \Delta\omega \qquad 9.4\text{-}12$$

where the half power width of the window is

$$\Delta f = 1/(N \cdot \Delta t) + 2m/(N' \cdot \Delta t) \qquad 9.4\text{-}13$$

where Δt is the sampling interval. The equivalent degrees of freedom of a chi-square distribution is $N/(2 \cdot \Delta f)$. The Daniell spectral estimate is recommended for most applications where the data set has a small (100 - 500) or medium (500 - 4,000) number of samples.

9.5 Bartlett's power spectral estimate

For very long data sets (greater than 2,000) it is recommended that the average of several periodogram estimates be used to obtain a power spectrum. Bartlett (1948) suggested that the variance would be reduced if the time series made up of N points is divided into I subsets, each with L points (figure 9.8). The periodogram equation (7.1-2) is obtained for each subset and the average energy density obtained for each frequency. That is

Bartlett's power spectral estimate

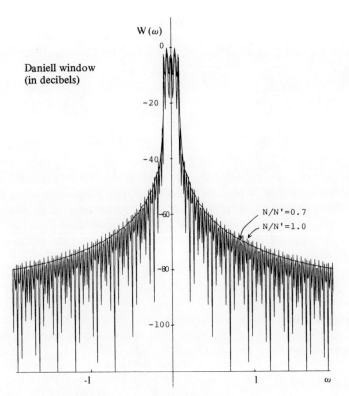

Figure 9.8 Daniell window for $N/N' = 0.7$ and 1.0 for $m = 7$.

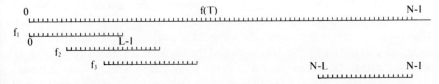

Figure 9.9 Subdivision of a data set with N points in I subsets for a Bartlett spectral estimate.

$$P_e(W) = \frac{L'}{L} \frac{1}{I} \sum_{i=1}^{I} F_i(W) \, F_i(W)^* \qquad 9.5\text{-}1$$

where the star indicates the complex conjugate and the frequency in cycles per second or Hertz is

$$f = W / (L' \cdot \Delta t) \qquad W = 0,1,...L'/2 \qquad 9.5\text{-}2$$

The discrete Fourier transform is used with a fast Fourier algorithm to obtain $F_i(W)$.

$$F_i(W) = \sum_{T=0}^{L'-1} f_i(T) \, e^{-2\pi i W T / L'} \qquad 9.5\text{-}3$$

The L samples in f_i are augmented to L' by the addition of sufficient trailing zeros to make L' a power of 2 so that a fast Fourier algorithm may be used most efficiently. The spectral estimate, P_e, is weighted by L'/L to compensate the power properly as a result of the increased resolution obtained by the addition of the zeros. Cooley et al. (1967) have shown that the variance of the estimate decreases as $1/I$.

$$\text{Var}\,[P_e(\omega)] = P^2(\omega) / I \qquad 9.5\text{-}4$$

For this estimate of variance it is assumed that the true power spectrum, P, is approximately constant at all frequencies in the principal interval. They find the equivalent degrees of a chi-square distribution is $2/I$.

9.6 Prewhitening

Power spectral estimates are most precise when the power is distributed evenly over all frequencies. In many cases it is found that the power has one or more broad peaks. The average value of the power at any particular frequency, f, may be greatly distorted during computation since the effect of a spectral window is to spread the power from the large peaks into adjacent frequencies. To avoid this difficulty the data is passed through a filter which compensates or pre-emphasizes the frequencies with lower amplitudes. This process of bringing the resultant spectrum close to that of white noise is called *prewhitening*. In this manner the deleterious effects of the side lobes in the spectral windows (kernels) are reduced. After the window has been applied and the spectrum obtained, an inverse filter is applied which removes the effect of the prewhitening filter. That is, the prewhitened power spectrum, $P(\omega)$, is divided by the square of the modulus (gain) of the transfer function, Y, of the whitening filter.

$$P_e(\omega) = P(\omega) / |Y(\omega)|^2 \qquad 9.6\text{-}1$$

It is debatable if prewhitening is a very useful procedure in practice. However,

Prewhitening

it is essential that the sample mean be reduced to zero before taking the power estimate. Otherwise the DC power or zero frequency term may dominate at low frequencies. A further practice sometimes carried out is to remove any linear trend since this also distorts the estimate at all frequencies, but particularly at long periods (Jenkins, 1961).

9.7 Matrix formulation of the power spectrum

For some theoretical purposes it is useful to be able to express the spectral density estimate in matrix notation, as carried out by Lacoss (1971). The autocovariance function, a_n, will be given by

$$a_n = \frac{1}{N-n} \sum_{k=1}^{N-n} x_k \, x_{k+n} \qquad n = 0,1,...N-1$$

$$a_n = a_{-n} \qquad\qquad 9.7\text{-}1$$

If the sampling interval is Δt and the time domain spectral window is W_n, then the estimate of the power spectrum at frequency, f, is given by

$$P(f) = \sum_{n=-(N-1)}^{N-1} W_n \, a_n \, e^{-2\pi i f n \Delta t} \qquad 9.7\text{-}2$$

A Bartlett window (see Appendix 5) is defined as follows:

$$W_n = \begin{cases} \dfrac{N - |n|}{N} & |n| < N \\ \\ 0 & |n| \geq N \end{cases} \qquad 9.7\text{-}3$$

Equation 9.7-2 with the Bartlett window may be written in matrix notation as

$$P_B = N^{-1}(1, e^{i2\pi f \Delta t}, ... e^{i2\pi f(N-1)\Delta t}) \begin{bmatrix} a_0 & a_1 & a_2 & ... & a_{N-1} \\ a_{-1} & a_0 & a_1 & ... & a_{N-2} \\ \vdots & & & & \\ a_{-(N-1)} & a_{-(N-2)} & & ... & a_0 \end{bmatrix} \begin{bmatrix} 1 \\ e^{-i2\pi f \Delta t} \\ \\ e^{-i2\pi f(N-1)\Delta t} \end{bmatrix}$$

or more compactly as

$$P_B = N^{-1} \, E^T A E^* \qquad 9.7\text{-}4$$

Where E is a column vector, T indicates a transpose, the * indicates a complex conjugation, and A is an N by N correlation matrix.

To establish 9.7-4 we first multiply out AE^*.

$$AE^* = \begin{array}{l} a_0 + a_1 e^{-i2\pi f \Delta t} + \ldots + a_{N-1} e^{-2\pi i f(N-1)\Delta t} \\ a_{-1} + a_0 e^{-2\pi i f \Delta t} + \ldots + a_{N-2} e^{-2\pi i f(N-1)\Delta t} \\ \vdots \\ a_{-(N-1)} + a_{-(N-2)} e^{-2\pi i f \Delta t} + \ldots + a_0 e^{-2\pi i f(N-1)\Delta t} \end{array} \qquad 9.7\text{-}5$$

Next form the terms above in a sum and evaluate $E^T A E^*$.

$$E^T A E^* = (1, e^{2\pi i f \Delta t}, \ldots e^{2\pi i f(N-1)\Delta t}) \begin{array}{l} \sum_{n=0}^{N-1} a_n e^{-2\pi i n f \Delta t} \\ \sum_{n=-1}^{N-2} a_n e^{-2\pi i (n+1) f \Delta t} \\ \vdots \\ \sum_{n=-(N-1)}^{0} a_n e^{-2\pi i (n+N-1) f \Delta t} \end{array}$$

$$= \sum_{n=0}^{N-1} a_n e^{-2\pi i n f \Delta t} + \sum_{n=-1}^{N-2} a_n e^{-2\pi i n f \Delta t} + \ldots \sum_{n=-(N-1)}^{0} a_n e^{-2\pi i n f \Delta t}$$

$$9.7\text{-}6$$

The sums above can be arranged graphically as in figure 9.10 for the case where $N = 5$. The E_n represent

$$a_n e^{-i2\pi f n \Delta t}$$

and the lines are drawn through the summed points.

Matrix formulation of the power spectrum

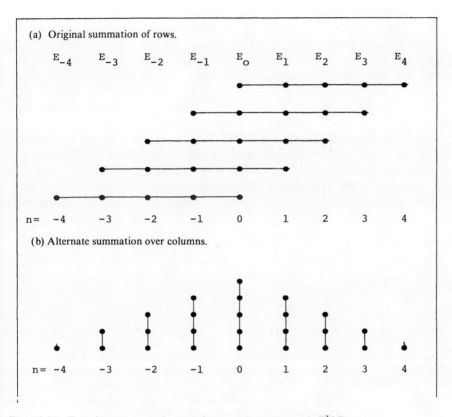

Figure 9.10 Two alternate ways of summation to evaluate the matrix $\mathbf{E^TAE^*}$.

In reformulating the summation over the columns as figure 9.10(b) the matrix becomes

$$\mathbf{E^TAE^*} = \sum_{n=-(N-1)}^{N-1} (N - |n|) \, a_n e^{-2\pi ifn\Delta t} \qquad 9.7\text{-}7$$

Therefore the power spectral estimate with a Bartlett window is given by

$$P_B = N^{-1}\mathbf{E^TAE^*} = \sum_{n=-(N-1)}^{N-1} \frac{(N - |n|)}{N} \, a_n e^{-2\pi ifn\Delta t} \qquad 9.7\text{-}8$$

9.8 Spectral estimates using the autocovariance

Before the general availability of the fast Fourier transform, a truncated autocovariance function, a(L), was multiplied by a time domain window, W(L), to produce a modified autocovariance function, $a_m(L)$. The spectral estimate was obtained from a Fourier transform of a_m.

$$a_m(L) = W(L) \ a(L) \qquad 9.8\text{-}1$$

This procedure came to be known as the Blackman and Tukey (1958) method. The technique is now obsolete but the various empirical windows used in previous scientific literature should be noted. These are discussed in Appendix 6. A large amount of literature exists on this subject and reference is made to Hannan, Chapters 5 and 8 (1960); Jenkins (1961); Parzen (1961); and Papoulis (1973).

References

Bartlett,M.S.(1948), Smoothing periodograms from time series with continuous spectra. *Nature* **161**, 666-668.

Bartlett,M.S.(1950), Periodogram analyses and continuous spectra. *Biometrika* **37**, 1-16.

Blackman,R.B. and Tukey,J.W. (1958), The measurement of power spectra. Dover Publication, New York.

Burg,J.P.(1967), Maximum entropy spectral analysis. Abstract and preprint: 37th Annual International SEG Meeting, Oklahoma City, Oklahoma, 1967.

Capon,J.(1969), High resolution frequency-wave number spectrum analysis. *Proc. IEEE* **57**, 1408-1418.

Cooley,J.W., Lewis,P.A.W. and Welch,P.D. (1967), The fast Fourier transform algorithm and its applications. Research Paper RC-1743, IBM Watson Research Center, Yorktown Heights, N.Y.

Daniell,P.J.(1946), Discussion on Symposium on autocorrelation in time series. *Supplement to the J. Royal Statist. Soc.* **8**, 88-90.

Hannan,E.J.(1960), Time series analysis, Chapter 8. John Wiley and Sons, New York.

Jenkins,G.M. and Priestly,M.B.(1957), The spectral analysis of time series. *J. Royal Statis. Soc.* **8**, 19, 1-12.

Jenkins,G.M.(1961), General considerations in the analysis of spectrum. *Technometrics* **3**, 133-166.

Jones,R.H.(1965), A reappraisal of the periodogram in spectral analysis. *Technometrics* **7**, 531-542.

Lacoss,R.T.(1971), Data adaptive spectral analysis methods. *Geophysics* **36**, 661-675.

Papoulis,A.(1973), Minimum-bias windows for high resolution. *IEEE Transactions on Informations Theory* **19**, 9-12.

Parzen,E.(1961), Mathematical considerations in the estimations of spectra. *Technometrics* **3**, 167-189.

10 Joint spectral analysis on two data sets

10.1 Cross covariance with the fast Fourier transform

The cross covariance will give a measure of the linear dependence between two series, **x** and **y**, at some lag L. Because the discrete Fourier transform can be carried out so rapidly with a fast Fourier algorithm, Sande (1965) has suggested the autocovariance and cross covariance be carried out entirely in the frequency domain. The autopower and cross-power spectrum may then be obtained by an inverse Fourier transform after the covariances are properly weighted with a lag window. Sufficient zeros must be added to the time series so that difficulties are not encountered with the circular recursion of the data. Following Gentleman and Sande (1966), the relevant equations for obtaining autopower or cross-power spectra are given below.

The two time series are sampled at interval of Δt and the mean removed.

$$\mathbf{x} = (x_0, x_1, x_2, \ldots x_{N-1})$$

$$\mathbf{y} = (y_0, y_1, y_2, \ldots y_{N-1})$$

The required cross covariance for lag L is

$$C(L) = N^{-1} \sum_{K=0}^{N-L-1} x_K \, y_{K+L} \qquad L = 0, 1, \ldots N\text{-}1 \qquad 10.1\text{-}1$$

The autocovariance is obtained by setting $\mathbf{x} = \mathbf{y}$. The Fourier transforms of **x** and **y** are

$$X(W) = \sum_{T=0}^{N'-1} x(T) \, e^{-2\pi i W T/N'} \qquad 10.1\text{-}2$$

and

$$Y(W) = \sum_{T=0}^{N'-1} y(T) \, e^{-2\pi i W T/N'} \qquad 10.1\text{-}3$$

where the two time series x(T) and y(T) have been tapered with cosine bells and zeros added so that N' may be factored in the radix 2 or 4. If the Daniell window is used N' should be equal to or greater than 2N. That is, enough zeros must be added at the beginning, end or both sides of the time series to avoid aliasing the periodogram with the side lobes of the window. This will occur since X(W) and Y(W) are cyclic functions which repeat with period N'. For a Parzen or Hanning window N' may only have to be 120 percent of N and so fewer zeros need be added. Calculating the power spectrum from the auto- or cross-correlation function, only 10 or 20 percent of the lags are used with these windows.

The cross covariance is obtained by a Fourier transform of the product X*(W) Y(W) as in equation 3.2-29.

$$C(L) = (N')^{-1} \left[\sum_{W=0}^{N'-1} (X^*(W) \ Y(W))^* \ e^{-2\pi i WT/N'} \right]^* \qquad 10.1\text{-}4$$

Negative lags need not be calculated as they are given by the equation above through the cyclic repetition of the series. That is, $C(-1) = C(N'-1)$, etc.

10.2 Cross-spectral analysis and coherence

From cross-spectral analysis it is possible to obtain not only the coherence which is a measure of the correlation between two processes at each frequency but the phase spectrum which measures the phase difference at each frequency. The complex cross-power spectrum is obtained with a Fourier transform of

$$C_m = C(L) \ W(L) \qquad 10.2\text{-}1$$

where W(L) is the desired window in the lag or time domain. The apparent modified cross covariance is reduced to M points, where M is equal to 2^k, k being an integer whose value will depend on the type and shape of the window used.

$$P(W) = \sum_{L=0}^{M-1} C_m(L) \ e^{-2\pi i WL/M} \qquad 10.2\text{-}2$$

The cospectra is given by the real part of P(W).

Cross-spectral analysis and coherence

$$CSP(W) = R\,[P(W)] \qquad 10.2\text{-}3$$

The quadrature spectra is given by the imaginary part of P(W).

$$QSP(W) = I\,[P(W)] \qquad 10.2\text{-}4$$

The phase spectra is obtained by

$$\Theta(W) = \tan^{-1}\,[QSP(W) / CSP(W)] \qquad 10.2\text{-}5$$

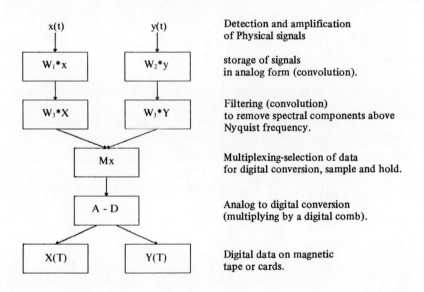

Figure 10.1 Electronic processing necessary prior to digital computation. The star indicates convolution.

Following Tukey (1965) the coherency is defined as

$$C_y(W) = \frac{\text{cospectra} + i\ \text{Quadrature spectra}}{[(\text{autopower of } X)(\text{autopower of } Y)]^{1/2}} \qquad 10.2\text{-}6$$

The coherence is defined as

$$C_e(W) = \frac{(CSP)^2 + (QSP)^2}{(\text{autopower of } X)(\text{autopower of } Y)} \qquad 10.2\text{-}7$$

These quantities will be discussed in the next section. A block diagram outlining the steps necessary to obtain a power estimate are given in figures 10.1 and 10.2. The procedure outlined above was for calculating the cross spectrum with smoothing being done in the lag domain. It is also possible to carry out the computation of a cross periodogram with smoothing being done in the frequency domain to obtain a cross-spectral estimate. Thus following section 9.5, the equation 9.5-1, the estimate would be given by

$$P_{XY}(W) = \frac{L'}{LI} \sum_{i=1}^{I} X_i(W)\, Y_i^*(W) \qquad 10.2\text{-}8$$

where $X(W)$ and $Y(W)$ are obtained by a fast Fourier transform of data vectors subdivided in the way illustrated in figure 9.9.

Most often the data set will not be long enough and some form of the Daniell weighting function, as in equation 9.4-5, will be used.

$$P_{XY}(W) = \frac{N'}{N} (2m+1)^{-1} \sum_{j=-m}^{m} X(W-j)\, Y^*(W-j) \qquad 10.2\text{-}9$$

where

$$\omega = 2\pi W/N' \qquad W = 0, 1, \ldots N'/2$$

$$\omega = -2\pi(N'-W)/N' \qquad W = N'/2 + 1, \ldots N'-1$$

The phase spectrum and coherence are obtained as in 10.2-5 and 10.2-7. Note that the cospectrum is an even function of frequency while the quadrature spectrum is an odd function.

10.3 Coherency

Cross-spectral analysis arises most naturally in physical processes involving electromagnetic phenomena. In the study of monochromatic polarized light it is found that the x and y components of the electric field may be in phase in which case the light is said to be linearly polarized or plane polarized (figure 10.3). Alternatively the two components may be out of phase in which case the

Coherency

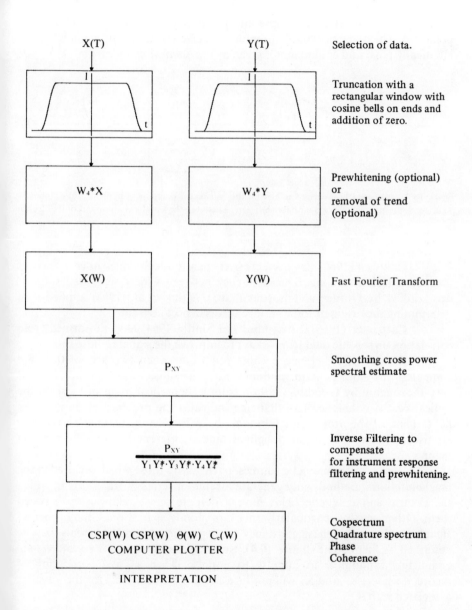

Figure 10.2 Outline of steps necessary to obtain a cross-power estimate.

electric vector moves in an ellipse and the radiation is said to be elliptically polarized. The plane polarized component is given by the *cospectrum* while the elliptically polarized component is given by the *quadrature* spectrum.

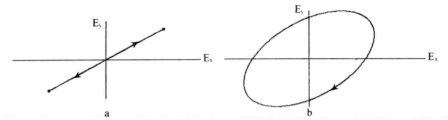

Figure 10.3 (a) Linearly polarized light with the two vector component in phase. (b) Elliptically polarized light in which there is a phase shift of $\pi/4$ between the x and y components of the electric field.

Cantwell (1960) has used cross correlation techniques in carrying out magnetotelluric sounding. The coherency between widely separated stations was studied by Orange and Bostick (1965). Munk et al. (1963) applied it in determining the origin of some surface waves in oceanography.

Carpenter (1965), Shimshoni and Smith (1964) have performed cross correlation on seismic data from arrays to improve the signal to raise ratio.

Studies of coherence become important when two sets of data are compared and both contain random noise. The subject has been investigated very thoroughly by Goodman (1957) from a theoretical point of view. In the optical case a coherence less that one indicates the presence of unpolarized light. That is, the atoms emit photons in an incoherent manner so that the electric field of each quantum of light assumes a different direction in a random manner.

Some electromagnetic sources, particularly those which are used in the magnetotelluric method, may generate a random series of transients in which the electric and magnetic fields have a definite phase relationship between them. Other sources, particularly small, randomly spaced ones, may generate signals in which the phase constancy is not preserved and the signals may be described as incoherent (figure 10.4). Suppose the noise inputs ϵ_{n1} and ϵ_{n2} are completely random from impulse to impulse. Then a cross correlation of recorded signals X_H and X_E will produce a non-zero result only for the wavelets associated with ϵ_S.

If no smoothing is carried out, the coherence, regardless of the nature of the processes, is

$$C_e(W) = |X(W) \ Y^*(W)|^2 / [|X(W)|^2 \ |Y(W)|^2] \equiv 1 \qquad 10.3\text{-}1$$

Therefore, the coherence will always be unity unless some band-pass filtering or

Coherency

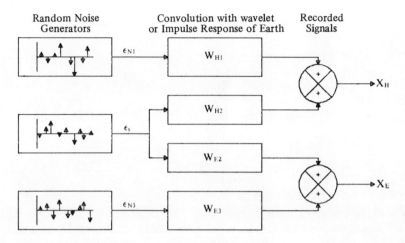

Figure 10.4 Random inputs which produced signals whose coherence depends on the relative power of ϵ_s to ϵ_{N1} and ϵ_{N2}. Two coherent signals will have originated from one source.

smoothing with a spectral window is included as in equation 10.2-8 or 10.2-9. Alternatively, this is equivalent to applying a lag window to the cross covariance. By averaging neighboring spectral estimates the noise which can be regarded as a random vector tends to cancel out because the phase is unrelated in the two signals being cross correlated (Madden, 1964). The data must be treated very carefully on digitization since the introduction of random phase between the two signals when multiplexing can decrease the coherence. In summary one may say that the coherence $C_e(W)$ measures the amount of association between two signals in the band of frequencies centered on W.

10.4 Bispectral analysis

An autocovariance is a statistical comparison of a time series as the lag, L, is varied. Higher order correlations can be made by comparing the function simultaneously at two or more lags. Wiener (1950) has defined a second order autocovariance function (also called a third moment or third cumulent) as

$$b(L_1,L_2) = \int_{-\infty}^{\infty} f(t) \, f(t + L_1) \, f(t + L_2) \, dt \qquad 10.4\text{-}1$$

Its Fourier transform is a complex quantity called the bispectrum.

$$B(\omega_1,\omega_2) = [2\pi]^{-2} \int_{-\infty}^{\infty} \int_{-\infty}^{\infty} b(L_1,L_2) \, e^{-i(\omega_1 L_1 + \omega_2 L_2)} \, dL_1 \, dL_2 \qquad 10.4\text{-}2$$

This is equivalent to the product of three Fourier transforms of the function, f(t).

$$B(\omega_1,\omega_2) = F(\omega_1) \ F(\omega_2) \ F^*(\omega_1 + \omega_2) \qquad 10.4\text{-}3$$

Haubrich (1965) has defined a normalized equivalent called the *bicoherence*.

$$R^2(\omega_1,\omega_2) = \frac{|B(\omega_1,\omega_2)|^2}{P(\omega_1) \ P(\omega_2) \ P(\omega_1+\omega_2)} \qquad 10.4\text{-}4$$

Each of the Fourier transforms in 10.4-3 and the power spectra, $P(\omega)$, in 10.4-4 must be weighted with a suitable window before the calculation for bicoherence is made.

Hasselmann, Munk and MacDonald (1963) show that the symmetry properties require that $B(\omega_1,\omega_2)$ be plotted as a parameter in only one octant on a graph of ω_1 versus ω_2. The bispectra will be zero if the signal is a sample of Gaussian noise.

Bispectral studies appear to be useful in studying non-linear operations in physical processes. One such operator may take the square of all signals. Consider a signal consisting of two cosine waves.

$$x = A \ \cos(\omega_1 t+\phi_1) + B \ \cos(\omega_2 t+\phi_2) \rightarrow \boxed{x^2} \rightarrow f(t)$$
$$10.4\text{-}5$$
$$f(t) = A^2\cos^2(\omega_1 t+\phi_1)+B^2\cos^2(\omega_2 t+\phi_2)+2AB\cos(\omega_1 t+\phi_1) \cdot \cos(\omega_2 t+\phi_2)$$

The cross term may be expanded into a sum and difference of frequencies.

$$2AB \ \cos(\omega_1 t+\phi_1) \ \cos(\omega_2 t+\phi_2) = AB \ \cos[(\omega_1+\omega_2)t + (\phi_1+\phi_2)]$$
$$+ \ AB \ \cos[(\omega_1-\omega_2)t + (\phi_1-\phi_2)] \qquad 10.4\text{-}6$$

Therefore the non-linear device passes the original frequencies ω_1 and ω_2 but it is also a heterodyning mechanism which generates $(\omega_1+\omega_2)$ and $(\omega_1-\omega_2)$. Such effects could be detected with a bispectral analysis. In oceanography surf beats and energy transfer between gravity waves are non-linear processes. The latter has been investigated successfully by Hasselmann et al. (1963). Preliminary investigations have been made for non-linear interactions in generations of microseisms (Haubrich, 1965) and pearl oscillations in electromagnetic data (Madden, 1964) but the results are not encouraging. Zadro and Caputo (1968) have proposed that some of the peaks in spectra from the great Chilean and

Bispectral analysis

Alaskan earthquakes arise from the interaction of different modes of free oscillation as a consequence of non-linear phenomena.

10.5 Spectral estimation of a moving average process

Power spectral estimation will only be reliable if the method of computation is suitable for the assumed physical model (Treitel et al., 1977 and Gutowski et al., 1978). The techniques described in Chapter 9 and 10 involve a *smoothed periodogram* and are applicable to a *moving average* (MA) process (Box and Jenkins, 1976) which are described by a simple convolution equation such as 2.1-8. A moving average process is one in which the current signal is a linear combination of present and past values of an input white noise signal. The output or measured process, y_L, is obtained by passing an input white noise signal, **s**, through a filter **W**, with coefficients 1, $-W_1$, $-W_2$, ... $-W_j$. For convenience, the first term of the filter is normalized to unity and a finite length filter has been assumed. The convolution equation at time t is by equation 2.1-8.

$$y_t = \mathbf{s} * \mathbf{W} = s_t - W_1 s_{t-1} - W_2 s_{t-2} - \ldots - W_j s_{t-j} \qquad 10.5\text{-}1$$

A second signal, **x**, may be obtained by convolution with an input, **n**.

$$x_t = \mathbf{n} * \mathbf{W} = n_t - W_1 n_{t-1} - W_2 n_{t-2} - \ldots - W_j n_{t-j} \qquad 10.5\text{-}2$$

In terms of z transforms the two equations become polynomials of order j.

$$y_t = (1 - W_1 z - W_2 z^2 - \ldots - W_j z^j) \, s_t \qquad 10.5\text{-}3$$

$$x_t = (1 - W_1 z - W_2 z^2 - \ldots - W_j z^j) \, n_t \qquad 10.5\text{-}4$$

Note that $W_1(zs_t) = W_1 s_{t-1}$; $W_2(z^2 s_t) = W_2 s_{t-2}$, etc., because of the delay properties of a z transform variable. Both equations may be factored by finding the roots of the polynomial equations (Loewenthal, 1977).

$$y_t = (\alpha_1 + \beta_1 z)(\alpha_2 + \beta_2 z) \ldots (\alpha_j + \beta_j z) \, s_t \qquad 10.5\text{-}5$$

A similar equation may be written for x_t. If equation 10.5-5 is set equal to zero we see that the only singularities are zeros. This *all zero* process is like an engineering feed forward system and we shall see in Chapter 13 that if the filter is stable all the zeros in the complex plane lie outside the unit circle, $|z| = 1$. In terms of z transforms, the MA process is

$$X(z) = W(z) \ N(z) \qquad 10.5\text{-}6$$

The z transform of the autocovariance function of x_t is

$$A(z) = X(z) \ X(z^{-1}) \qquad 10.5\text{-}7$$

Letting $z = \exp(-2\pi i f)$ is equivalent to taking the Fourier transform of the autocovariance function to obtain the power spectrum or periodogram of x_t. That is, $A(z)$ is evaluated on the unit circle, $|z| = 1$.

$$P(f) = |X(f)|^2 = X[\exp(-i2\pi f)] \ X[\exp(+i2\pi f)] \qquad 10.5\text{-}8$$

The power spectral estimate of a moving average process is

$$P(f) = |W(f)|^2 \ |N(f)|^2 \qquad 10.5\text{-}9$$

The cross-power spectral estimate of two moving average processes is

$$P_{XY}(f) = |W(f)|^2 \ S[\exp(-i2\pi f)] \ N[\exp(+i2\pi f)] \qquad 10.5\text{-}10$$

In using real data the estimates should be made using a Daniell or a Bartlett weighting scheme as described previously. Note that the physical role of W and NN* or SN* in 10.5-9 and 10.5-10 may be interchanged. That is, W can be considered to originate from an input signal while S and N would involve two different filters at different spatial locations.

The moving average process is a very general model which will describe any stable linear system. However, on occasion, the number of terms, j, in the filter may be rather large. In the next chapter a technique is described for determining the power spectrum which may be applicable to short data sets. However it is *only* valid if the process involves *negative feedback*. It should not be used without *a priori* knowledge of the physical model just to obtain spectral estimates with sharp peaks or high resolution. In many geophysical cases involving an impulsive or time limited source (such as an elastic wave from an explosion or an earthquake) the signal from one or more ray paths can be assumed to decay rapidly below the noise level because of attenuation and multiple scattering. In these cases the signal may be isolated and tapered to zero to eliminate the influence of other rays. The shortened signal may be lengthened by the addition of trailing zeros to obtain additional resolution prior to the computation of a smoothed periodogram. This technique assumes that the autocorrelation function decays smoothly to an insignificant level but in many physical situations this is precisely what happens.

References

References

Box,G.E.P. and Jenkins,G.M.(1976), *Time Series Analysis: Forcasting and Control* Chapter 3, Holden-Day Inc., San Francisco, 46-84.

Cantwell,T.(1960), Detection and analysis of low frequency magnetotelluric signals. Ph.D. thesis, M.I.T..

Carpenter,E.W.(1965), Explosion seismology. *Science* **147**, 363-373.

Gentleman,W.M. and Sande,G.(1966), Fast Fourier transforms for fun and profit. Proceedings of the Fall Joint Computer Conference, San Francisco, 563-578.

Goodman,N.R.(1957), On the joint estimation of the spectra, cospectrum and quadrature spectrum of a two-dimentional stationary Gaussian process. *Scientific Paper No. 10* Engineering Statistics Laboratory, New York University.

Gutowski,P.R., Robinson, E.A. and Treitel,S.(1978), Spectral estimation: Fact or Fiction. *IEEE Transaction on Geoscience Engineering* **GE-16**, 80-94.

Hamon,B.V. and Hannan,E.J.(1963), Estimating relations between time series. *Journal of Geophysical Research* **68**, 6033-6041.

Hasselmann,K., Munk,W. and MacDonald,G.(1963), Bispectra of ocean waves. *Ch. 8 of Time Series Analysis*, Ed. M. Rosenblatt J. Wiley & Sons, New York, 125-139.

Haubrich,R.A.(1965), Earth Noise, 5 to 500 millicycles per second. *Journal of Geophysical Research* **70**, 1415-1427.

Loewenthal,D.(1977), Numerical computation of the roots of polynomials by spectral factorization. *Topics in Numerical Analysis* **3**, Editor: J.J.H. Miller. Academic Press, London, 237-255.

Madden,T.(1964), Spectral, cross spectral and bispectral analysis of low frequency electromagnetic phenomena, 429-450 in Natural electromagnetic phenomena below 30 $Kc_{/s}$. Ed. D.F. Bliel, Plenum Press, New York.

Munk,W.H., Miller,G.R., Snodgrass,F.E. and Barber,N.F.(1963), Directional recording of swell from distant storms. *Phil. Trans. Royal Society, London* **A255**, 505-589.

Orange,A.S. and Bostick,F.X.(1965), Magnetotelluric micropulsations at widely separated stations. *Journal of Geophysical Research* **70**, 1407-1413.

Sande,G.(1965), On an alternative method of calculating covariance functions. Unpublished manuscript, Princeton University.

Shimshoni,M. and Smith,S.(1964), Seismic signal enhancement with 3-component detectors. *Geophysics* **29**, 664-671.

Treitel,S., Gutowski,P.R. and Robinson,E.A.(1977), Empirical spectral analysis revisited. *Topics in Numerical Analysis* **3**, Editor: J.J.H. Miller. Academic Press, London, 429-446.

Tukey,J.W.(1965), Uses of Numerical Spectrum Analysis in Geophysics. *Preprint presented at Int. Assoc. for Statistics in the Physical Sciences* Beograd, Yugoslavia, Sept. 14-22.

Wiener,N.(1958), Non-linear problems in random theory. John Wiley & Sons, New York.

Zadro,M.B. and Caputo,N.(1968), Spectral, bispectral analysis and Q of the free oscillations of the earth. *Supplemento al Nuovo Cimento* **Vol 6**, Series 1, 67-68.

11 Maximum entropy spectral analysis

11.1 Data adaptive spectral analysis methods

Traditional methods of obtaining a power spectral density using a smoothed periodogram based on a Fourier transform of the input data have been given in Chapters 9 and 10. This estimate of power is the one that should be used when the input data has been generated by a moving average (MA) process. This process is described by a convolution equation in which the z transform of the filter operator is described by the zeros of the polynomial. To obtain a reliable estimate in the presence of noise there is a smoothing of the autocovariance function by a time domain window or a smoothing of the squared magnitude of the Fourier transform. The smoothing implies that the autocovariance (and the impulse response of the filters) decay to zero and that outside a certain time window these functions are negligible. The smoothing filter or window is seldom ideal and there may be some leakage of power through side lobes in the transfer function. Perhaps more serious is the limitations on the resolution in the frequency domain.

Several new methods have been devised for estimating the spectra with increased resolution by making assumptions about the physical process generating the signal. The advantage of these techniques is that they usually avoid making any assumption about the autocovariance function outside the sample interval. The disadvantage is that they are *only* applicable to physical models that have a feedback system and one must deduce the order of the polynomial describing the system.

The first method was devised by Burg (1967) and is most useful for short lengths of data sampled at equal intervals. It is called the maximum entropy method (MEM) or the maximum entropy spectral analysis (MESA). It is identical to the autoregressive (AR) spectral estimator which was devised by Akaike (1969,a,b) and is also mentioned by Parzen (1969). The second technique is the maximum likelihood method (MLM) and was proposed by Capon (1969) for wave-number analysis using arrays of data. Lacoss (1971) has shown how to use the same techniques for an estimate of the power spectral density of one channel of information. In the MLM technique a window function is computed at every frequency in such a way that it adapts itself to the noise under analysis. The MEM estimator retains all the estimated lags, without smoothing, and uses Wiener optimum filter theory to design a prediction filter which will whiten the input time series. From the whitened output power and the response of the prediction filter it is possible to compute the input power spectrum.

Data adaptive spectral analysis methods

A third method was devised by Treitel, Gutowski and Robinson (1977) and is the least squares solution for spectra of a signal generated by an autoregressive moving average (ARMA) process as described by Box and Jenkins (1970). The z transform of the impulse response of an ARMA process has both poles and zeros and is described by engineers as a feedback-feedforward system. All three methods including the additional new ones of Pisarenko (1972,1973) based on the autocovariance function are able to achieve a very high resolution on short segments of data through an analysis of the data itself. As a result Lacoss calls these techniques data adaptive methods. All of these models assume that the process is stationary and that the system is triggered by a linear combination of random impulses.

An autoregressive (AR) model, which was discovered by Yule (1927), can be described as a stochastic process in which the current value, x_t, of a time series is obtained from a convolution of a wavelet with past values of the signal, $\mathbf{x}$, and a new input, $\mathbf{n}$, which is usually a sample of random (white) noise. The wavelet is the impulse response, $\mathbf{D}$, from a stable filter and will be described in Chapter 13 as a minimum delay or minimum phase wavelet. The white noise input, n_t, at any time t can be thought of as a series of random impulses or numbers with zero mean and constant variance as in equations 6.1-1 and 6.1-2. Following Akaike (1969) these physical systems can be modelled closely by a finite number of terms, m, and the autoregressive process is said to be or order m.

$$x_t = n_t + d_1 x_{t-1} + d_2 x_{t-2} + \ldots + d_m x_{t-m} \qquad 11.1\text{-}1$$

From 2.5-1 the z transform of 11.1-1 is obtained by setting $X(z) = x_0 + x_1 z + x_2 z^2 + \ldots x_t z^t$ and $N(z) = n_0 + n_1 z + n_2 z^2 + \ldots + n_t z^t$. Note that in accordance with the delay property of the z variable $d_1 z x_t = d_1 x_{t-1}, d_2 z^2 x_t = d_2 x_{t-2}$, etc.

$$X(z) - X(z)[d_1 z + d_2 z^2 + \ldots + d_m z^m] = N(z) \qquad 11.1\text{-}2$$

or

$$X(z) = N(z) / [1 - d_1 z - d_2 z^2 - \ldots - d_m z^m] \qquad 11.1\text{-}3$$

The solution for any value of x is obtained by equating like powers on both sides of the equation. The wavelet or impulse response of the filter $[1, -d_1, -d_2, \ldots -d_m]$ appears in the denominator as in an engineering feedback circuit. Its z transform can be set equal to zero and factored to find the roots of the equation. In the complex z plane these are the poles of the function $X(z)$. The z transform of an autoregressive process can be written compactly as

$$X(z) = N(z) / D(z) \qquad 11.1\text{-}4$$

It should be compared to a moving average process in equation 10.5-6.

As a matter of interest the first order autoregressive system (m = 1) is also called a first order Markov process. The value of x_t, at any time t depends only upon the immediately preceeding value of x and a sample of random noise. It is a filter with a very short memory. An example of a second order autoregressive model suggested by Yule (Jenkins and Watts, 1969; Bartlett, 1946) is a simple pendulum with the damping proportional to its velocity due to the resistance of the air. The pendulum is excited by random sized impulses at regular discrete time intervals (figure 11.1a). Each impulse causes the pendulum to perform a decaying harmonic oscillation as in figure 4.10. Note that this second order autoregressive model has two poles on the left side of the s plane or equivalently it will have two poles outside the unit circle in the complex z plane. The random impulses will cause the pendulum to oscillate with periodic motion in an erratic manner. A higher order example of an autoregressive model is a layered elastic system in which the waves from a source are recorded at depth by a seismometer placed in a basal semi-infinite layer (figure 11.1b) (Hubral et al.1980). The reflection coefficients at the interfaces are assumed to have a random distribution so that impulses arrive at the seismometer with random amplitudes due to an infinite number of possible ray paths. A mathematical derivation of the model is given in Chapter 16.

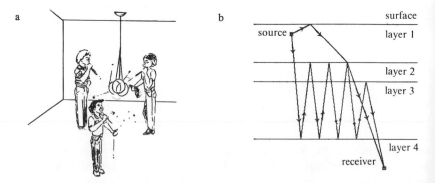

Figure 11.1 (a) A second order autoregressive process is illustrated by a damped pendulum which is excited by random impulses. This model was suggested by Yule in 1921 when he put forward the idea of a pendulum subjected to bombardment by boys equipped with peashooters. (b) A higher order autoregressive process is modelled by a source and receiver in a layered medium over a half space as in a sedimentary basin. Two ray paths out of an infinite number of possible ones are shown.

Data adaptive spectral analysis methods

There are only a few realistic physical systems that can be described by an autoregressive process. Most physical devices are more complex than simple pendulums and the exciting function is more complex than a series of random impulses at equal discrete intervals. In the geophysical example of a layered medium the receiver is most likely to be placed on the surface in which case the process is no longer autoregressive. Furthermore half spaces do not exist in nature, even as approximations, so that there will be waves reflected from interfaces below the seismometer in which case the process is also not autoregressive. A more general type combines equations 11.1-1 and 10.5-2 for the autoregressive and moving average functions to produce an ARMA process.

$$x_t = n_t + d_1 x_{t-1} + d_2 x_{t-2} + ... + d_m x_{t-m} - W_1 n_{t-1} - W_2 n_{t-2} - ... - W_j n_{t-j} \qquad 11.1\text{-}5$$

The z transform of an autoregressive moving average process is

$$X(z) - X(z)[d_1 z + d_2 z^2 + ... + d_m z^m] = N(z)[1 - W_1 z - W_2 z^2 - ... - W_j z^j] \qquad 11.1\text{-}6$$

or

$$X(z) = N(z) W(z) / D(z) \qquad 11.1\text{-}7$$

where the mth order polynomial D(z) and the jth order polynomial W(z) are

$$D(z) = 1 - d_1 z - d_2 z^2 - ... - d_m z^m \qquad 11.1\text{-}8$$

$$W(z) = 1 - W_1 z - W_2 z^2 - ... - W_j z^j \qquad 11.1\text{-}9$$

This system of equations describes many physical processes and includes the important one of a receiver on the surface of a layered half space as a model of exploration by the seismic reflection method. Unfortunately the two orders, m and j, are seldom if ever known and an efficient ARMA spectral computing algorithm has not been developed to this time.

Despite its very limited physical suitability, the maximum entropy spectral method has been widely used as a spectral estimator. The most interesting examples in which ARMA or more complex processes are modelled by an autoregressive spectral estimator are given by Ulrych and Bishop (1975) and Ulrych and Clayton (1976). Before applying an autoregressive spectral estimate using a maximum entropy algorithm there should be a physical or a theoretical justification of the actual process that is in operation. Gutowski, Robinson and Treitel (1976) point out the dangers of an incorrect assumption about the process in spectral estimation. Just because one obtains sharp peaks with a maximum entropy spectral estimate does not mean that the true spectrum has been realized. Bearing in mind that a cautious approach must be followed, the details of the maximum entropy or autoregressive spectral estimate will be explored in the following sections.

11.2 The concept of entropy

Entropy as a concept in communication systems was introduced by Shannon (1948). It relates the amount of randomness or uncertainty in a system that is transmitting information. If we have a set of M symbols, x_i, they can be combined together, in the manner of an alphabet, to make up a set of messages. The probability of occurrence of any symbol, x_i, is given by $p(x_i)$ where the total probability is unity.

$$\sum_{i=1}^{M} p(x_i) = 1 \qquad \qquad 11.2\text{-}1$$

A measure of the information gained from a message is obtained by finding the minimum number of *yes* and *no* questions required to identify the message. Thus if the message is a number less than 64, and each number has an equal probability of 1/64 of occuring, we would first ask if the number is less than 32. If no, we could ask if it is less than 16, etc. Thus the number of questions with two possible answers is six. Note that each question adds one unit of information and a convenient measure in a binary system which is additive is the logarithm to base 2 (1 unit = $\log_2 2$). The addition of 6 such units is

$$6 \, \log_2 2 = \log_2 2^6 = \log_2 64 = 6$$

From this let us define a quantity called *self information*, S, which is the reciprocal of the probability of occurrence.

$$S(x_i) = \log 1/p(x_i) = -\log p(x_i) \qquad \qquad 11.2\text{-}2$$

Note that the self information decreases if a symbol occurs with greater probability whereas a rare event is thought to contain a more significant message. If the probability of x occurring is one then the self information is zero. Thus if a *message* consists only of an infinite string of yeses, there is no information or significance to this.

Over a long period of time, T, the number of times that x occurs is $p(x_i)T$ and so the total information obtained about a system is $-p(x_1)T \cdot \log_2 p(x_1) - p(x_2)T \cdot \log_2 p(x_2) - \ldots$.

The average information per unit time, also called the expectation value, *E*, of the self information is then

The concept of entropy

$$E[S(x_i)] = \sum_{i=1}^{M} p(x_i) S(x_i) \qquad 11.2\text{-}3$$

and this is taken to be the *entropy*.

$$H = - \sum_{i=1}^{M} p(x_i) \log p(x_i) \qquad 11.2\text{-}4$$

This definition of entropy is similar to that found in thermodynamics where it measures the degree of disorder in a physical process. Here it is a measure of the uncertainty that the system will be in a particular state. The base of the logarithm can be any convenient number but is logically 2 if a binary arithmetic is used in the system. In our example, yes = 1 and no = 0. Brillouin (1956) notes that entropy is a measure of our ignorance of a system.

One can show (see Thomas, 1968, section 8.4, or Shannon, 1948) that the entropy of a discrete distribution is a maximum when the probabilities are all equal. Thus for a binary system that only transmits a 0 and 1, the probabilities are p and 1 - p.

$$H = -p \log_2 p - (1-p) \log_2 (1-p) \qquad 11.2\text{-}5$$

The entropy is a maximum if $p = 1/2$ in which case H is 1.

If the variance of x is constrained to be a constant, σ^2, then one can show that the entropy is a maximum if the probability has a normal or Gaussian distribution.

$$p(x) = [2\pi\sigma^2]^{-1/2} e^{-x^2/2\sigma^2} \qquad 11.2\text{-}6$$

The entropy (Thomas, 1968) is given by

$$H = [1/2] \log_e 2\pi e \sigma^2 \qquad 11.2\text{-}7$$

11.3 Maximum entropy spectral analysis

Burg (1967, 1970) has shown how it is possible to obtain the power spectrum of your data by generating a new data set with a special filter so that the new set shows the most randomness and has the maximum entropy. This leads to an estimate with a very high resolution since the method uses the available lags in the autocovariance function without modification and, in principle, can make a non-zero estimate or prediction of the autocovariance function beyond those which can be calculated directly from the data. It is a method of maximizing the amount of information from the available time series.

Consider first the design of a $N + 1$ point prediction error filter following the technique of Peacock and Treitel (1969). From optimum filter theory of Wiener (equation 14.5-20) the impulse response, $\mathbf{D}^{-1}$, which is to be used as a prediction operator is given by

$$\mathbf{AD'} = \mathbf{C} \qquad 11.3\text{-}1$$

where we write D' instead of $\mathbf{D}^{-1}$ for the inverse or deconvolution operator derived in Chapter 14. $\mathbf{A}$ is the N by N autocovariance matrix and $\mathbf{C}$ is a column vector giving the cross covariance between the input signal, x, and the desired output. For a prediction operator the desired signal is the input signal at some time $t + T$ in the future. In matrix form equation 11.3-1 is

$$\begin{bmatrix} a_0 & a_1 & \cdots & a_{N-1} \\ a_1 & a_0 & & a_{N-2} \\ \cdot & \cdot & & \cdot \\ \cdot & \cdot & & \cdot \\ \cdot & \cdot & & \cdot \\ a_{N-1} & a_{N-2} & & a_0 \end{bmatrix} \begin{bmatrix} d_1 \\ d_2 \\ \cdot \\ \cdot \\ \cdot \\ d_N \end{bmatrix} = \begin{bmatrix} c_0 \\ c_1 \\ \cdot \\ \cdot \\ \cdot \\ c_{N-1} \end{bmatrix} \qquad 11.3\text{-}2$$

The output signal, obtained by applying the deconvolution operator, will be y_L and it should be an estimate of the input at some future time $t + T = L$. The symbol T is the prediction distance in units of time.

$$y_L = \sum_{t=0}^{N-1} d_{t+1} \, x_{L-t} \qquad 11.3\text{-}3$$

The error for the series is the difference between the true signal at time $t + T$ and the predicted signal, y_L

$$\epsilon_{t+T} = x_{t+T} - y_{t+T} \qquad 11.3\text{-}4$$

Maximum entropy spectral analysis

or

$$\epsilon_{t+T} = x_{t+T} - \sum_{t=0}^{N-1} d_{t+1}\ x_{L-t} \qquad 11.3\text{-}5$$

Taking the z transform of both sides, then we have

$$z^{-T} E(z) = z^{-T} X(z) - X(z)\ D'(z) \qquad 11.3\text{-}6$$

Multiply both sides by z^T

$$E(z) = X(z)\ [1 - z^T D'(z)] \qquad 11.3\text{-}7$$

Notice that the error series can be computed directly from a convolution of the input signal with the quantity

$$[1 - z^T D'(z)]$$

which is called the prediction error operator.

The unnormalized cross covariance in equation 11.3-2, apart from a factor of 1/N, for a prediction operator is given by

$$c_L = \sum_{t=0}^{N-1} x_t\ y_{t+L} = \sum_{t=0}^{N-1} x_t\ x_{t+T+L} = a_{T+L} \qquad 11.3\text{-}8$$

and is related to the autocovariance function at lag $T + L$. Let us set the prediction distance, T, at one unit of sampled time. Then the matrix equation 11.3-2 becomes

$$\begin{bmatrix} a_0 & a_1 & \cdots & a_{N-1} \\ a_1 & a_0 & & a_{N-2} \\ \cdot & \cdot & & \cdot \\ \cdot & \cdot & \cdot\cdot & \cdot \\ \cdot & \cdot & & \cdot \\ a_{N-1} & a_{N-2} & & a_0 \end{bmatrix} \begin{bmatrix} d_1 \\ d_2 \\ \cdot \\ \cdot \\ \cdot \\ d_N \end{bmatrix} = \begin{bmatrix} a_1 \\ a_2 \\ \cdot \\ \cdot \\ \cdot \\ a_N \end{bmatrix} \qquad 11.3\text{-}9$$

This system of N equation may be written out and an autocovariance coefficient, a_i, added and subtracted to each equation so that the right-hand side vanishes. The matrix is then augmented by the addition of a new equation to regain the symmetry. The rectangle encloses the original set of equations in 11.3-9.

$$
\begin{array}{rcrcrcrcl}
-a_0 & + & a_1 d_1 & + & a_2 d_2 & + \ldots + & a_N d_N & = & -P_m \\
-\hat{a}_1 & + & a_0 d_1 & + & a_1 d_2 & + \ldots + & a_{N-1} d_N & = & a_1 - a_1 \\
-a_2 & + & a_1 d_1 & + & a_0 d_2 & + \ldots + & a_{N-2} d_N & = & a_2 - a_2 \\
\vdots & & & & & & & & \\
-\hat{a}_N & + & a_{N-1} d_1 & + & a_{N-2} d_2 & + \ldots + & a_0 d_N & = & a_N - a_N \\
\end{array}
$$

11.3-10

Multiplying through by -1 the matrix form is then

$$
\begin{bmatrix}
a_0 & a_1 & \ldots & a_N \\
a_1 & a_0 & \ldots & a_{N-1} \\
\vdots & \vdots & & \vdots \\
a_N & a_{N-1} & \ldots & a_0
\end{bmatrix}
\begin{bmatrix}
1 \\
-d_1 \\
\vdots \\
-d_N
\end{bmatrix}
=
\begin{bmatrix}
P_m \\
0 \\
\vdots \\
0
\end{bmatrix}
$$

11.3-11

Define a new set of filter operators by a column vector called Γ by Burg and **D** in this chapter.

$$
\mathbf{D} = \begin{bmatrix} 1 \\ D_1 \\ \vdots \\ D_N \end{bmatrix} \equiv \begin{bmatrix} 1 \\ -d_1 \\ \vdots \\ -d_N \end{bmatrix}
$$

11.3-12

Maximum entropy spectral analysis

To obtain the filter coefficients and the factor P_m, one must solve the equation

$$\begin{bmatrix} a_0 & a_1 & \ldots & a_N \\ a_1 & a_0 & \ldots & a_{N-1} \\ \cdot & \cdot & & \cdot \\ \cdot & \cdot & & \cdot \\ \cdot & \cdot & & \cdot \\ a_N & a_{N-1} & & a_0 \end{bmatrix} \cdot \begin{bmatrix} 1 \\ D_1 \\ \cdot \\ \cdot \\ \cdot \\ D_N \end{bmatrix} = \begin{bmatrix} P_m \\ 0 \\ \cdot \\ \cdot \\ \cdot \\ 0 \end{bmatrix} \qquad 11.3\text{-}13$$

which is Burg's (1967) equation for the prediction error filter at a distance of one time unit. In matrix notation

$$\mathbf{AD} = \tilde{\mathbf{P}} \qquad 11.3\text{-}14$$

Since the input wavelet consists of N terms and the cross correlation contains only the non-zero term this implies that the prediction operator shortens the output wavelet to a spike in which P_m is the *mean square error* or the *mean output power*.

    D may be considered a set of prediction filter weights which, when convolved with the input data, will generate a white noise series. The elements of the noise series will be uncorrelated with each other and the filter will have created the greatest destruction of entropy which is possible (figure 11.2).

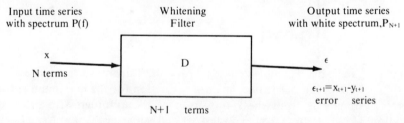

Figure 11.2 Convolution with a prediction error operator. In the frequency domain the output power spectrum is the product of the input power spectrum and the power response of the filter.

The input power spectrum may be obtained by correcting the output power for the response of the filter. In the frequency domain this is equivalent to equation 7.4-9.

$$\text{Input power spectrum} = \frac{\text{Output power spectrum}}{\text{Power response of filter}}$$

The input power spectrum of the time series is designated by Burg as the maximum entropy estimate of power, P_E.

$$P_E(f) = \frac{P_m/f_N}{2\left|1 + \sum_{n=1}^{N} D_n\, e^{-2\pi i f n \Delta t}\right|^2} \qquad 11.3\text{-}15$$

where f_N is the Nyquist frequency and specifies the bandwidth for a sampling interval of Δt.

$$f_N = 1/2\Delta t \qquad 11.3\text{-}16$$

In matrix notation equation 11.3-15 is given by

$$P_E(f) = (P_m/f_N) / (2\mathbf{E}^T\mathbf{D}^*\mathbf{D}^T\mathbf{E}^*) \qquad 11.3\text{-}17$$

where $\mathbf{E}$ is the following $N + 1$ column vector:

$$\mathbf{E} = \begin{bmatrix} 1 \\ e^{2\pi f \Delta t i} \\ \vdots \\ e^{2\pi f n \Delta t i} \\ \vdots \\ e^{2\pi f N \Delta t i} \end{bmatrix} \qquad 11.3\text{-}18$$

The star, *, indicates a complex conjugate and the T denotes a transpose.

The simplest derivation of Burg's equation for the maximum entropy spectral estimate is from the z transform of the autoregressive process in equation 11.1-3. Letting $z = \exp(-2\pi i f \Delta t)$ and squaring $X(z) = N(z)/D(z)$ to obtain the power spectral density function one has

$$P_E(f) = \frac{\sigma^2}{\left|1 - \sum_{j=1}^{m} d_m\, e^{-2\pi i f j \Delta t}\right|^2} \qquad 11.3\text{-}15a$$

where σ^2 is the variance of a zero-mean white noise sequence, n or ϵ. Note that this equation is similar to Burg's equation for the maximum entropy method.

Maximum entropy spectral analysis

Burg's equation, 11.3-15, may be derived by means of Lagrange's method of undetermined multipliers as shown by Smylie, Clarke and Ulrych (1973) and by Chen and Stegan (1974). We wish to find the stationary value for entropy

$$H = \int_{-f_N}^{f_N} \log P_E(f) \, df \qquad 11.3\text{-}19$$

provided that the autocovariance, a_n, for lag $n\Delta t$ is the inverse Fourier transform of the power spectrum, $P_E(f)$.

$$a_n = \int_{-f_N}^{f_N} P_E(f) \, e^{2\pi i f n \Delta t} \, df \qquad 11.3\text{-}20$$

The accessory conditions may be written as

$$\int_{-f_N}^{f_N} [P_E(f) \, e^{2\pi i f n \Delta t} - a_n/2f_N] \, df = 0 \qquad 11.3\text{-}21$$

for each value of n. The maximization of the integral in 11.3-19 is carried out by coupling the $2N + 1$ accessory conditions through the undetermined Lagrange multipliers, λ_n.

$$\delta \int [-\log P_E - \sum \lambda_n (P_E \exp(2\pi i f n \Delta t) - a_n/2f_N)] = 0 \qquad 11.3\text{-}22$$

The maximum must satisfy the Euler-Lagrange relation (equation 3.13, Appendix 3) in which we identify L as the integrand, y as P_E and x as f. The solution is easily found to be

$$P_E(f) = \frac{1}{\sum_{n=-N}^{N} \lambda_n \, e^{2\pi i f n \Delta t}} \qquad 11.3\text{-}23$$

Because the power spectrum is real and positive or zero, this equation can be written as

$$P_E(f) = \frac{P_m \Delta t}{\mathbf{E}^T \mathbf{D} * \mathbf{D}^T \mathbf{E}*} \qquad 11.3\text{-}17$$

or as Burg's equation 11.3-15

$$P_E(f) = \frac{P_m/f_N}{2|1 + \sum_{n=1}^{N} D_n e^{-2\pi i f n \Delta t}|^2} \qquad 11.3\text{-}15$$

in which P_m is constant and the denominator is minimum phase. The Fourier transform of the autocovariance function in 11.3-20 is

$$P_E(f) = (2f_N)^{-1} \sum_{n=-N}^{N} a_n e^{-2\pi i f n \Delta t} \qquad 11.3\text{-}24$$

Note that $a_n = a_{-n}$ and 11.3.24 equals 11.3-15. Equating coefficients of equal power in the exponential term yields equation 11.3-13. This provides a solution for the unknown filter coefficients, D_n, or the Lagrange multipliers, λ_n, in terms of a prediction error filter with a mean square error P_m.

The filter coefficients of **D** are obtained from equation 11.3-14 in which **A** is an $N + 1$ Toeplitz matrix that must be semi-positive definite and its determinant non-negative. A Toeplitz matrix is one in which the elements on symmetric diagonals are equal. This assures that the power spectra will remain positive at all points. Specifying the autocovariance function is a major problem in any method of spectral estimation and Burg (1967, 1970) devotes considerable attention to this. The computation of the filter coefficients as recommended by Burg is examined in detail by Anderson (1974). Beginning with equation 11.3-11 let the i th filter coefficient for an $m + 1$ long prediction error filter be d_{mj}. That is, equation 11.3-11 is rewritten as

$$\begin{bmatrix} a_0 & a_1 & \ldots & a_m \\ a_1 & a_0 & \ldots & a_{m-1} \\ \cdot & \cdot & & \cdot \\ \cdot & \cdot & & \cdot \\ \cdot & \cdot & & \cdot \\ a_m & a_{m-1} & \ldots & a_0 \end{bmatrix} \begin{bmatrix} 1 \\ -d_{m1} \\ \cdot \\ \cdot \\ \cdot \\ -d_{mm} \end{bmatrix} = \begin{bmatrix} P_m \\ 0 \\ \cdot \\ \cdot \\ \cdot \\ 0 \end{bmatrix} \qquad 11.3\text{-}25$$

Maximum entropy spectral analysis

A recursion equation is developed for m = 0, 1, 2, ... M where the chosen maximum value of M < N. The value of M to be used is obtained by determining the final prediction error (FPE) as given by Akaike (1970). The relation between the various filter symbols used in this chapter and by Burg (1970), who uses Γ, is

$$\mathbf{D} = \begin{bmatrix} 1 \\ \Gamma_2 \\ \cdot \\ \cdot \\ \cdot \\ \Gamma_{m+1} \end{bmatrix} = \begin{bmatrix} 1 \\ D_1 \\ \cdot \\ \cdot \\ \cdot \\ D_m \end{bmatrix} = \begin{bmatrix} 1 \\ -d_{m1} \\ \cdot \\ \cdot \\ \cdot \\ -d_{mm} \end{bmatrix} \qquad 11.3\text{-}26$$

The filter computation is initiated by setting m = 0. Thus equation 11.3-25 becomes

$$[a_0] \; [1] \; = \; [P_0] \qquad 11.3\text{-}27$$

From a definition of the autocorrelation for zero lag the average power of a one point prediction filter is

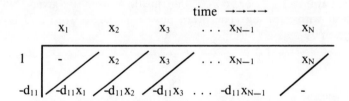

Figure 11.3   Convolution as a geometric operation (see figure 2.2) for a dipole (1, -d_{11}).

$$P_0 = N^{-1} \sum_{i=1}^{N} x_i^2 \qquad *11.3\text{-}28$$

Equation numbers marked with an asterisk in this chapter are required for computer programming of a maximum entropy spectral estimate.

In proceeding with the solution of a two point prediction filter, (1, -d_{11}), we first perform a convolution on the input data, x_1, x_2 . . . . x_N (see figure 11.3). We avoid running off the end of the data so as not to make assumptions about the presence of trailing or leading zeros. The prediction error is

$$\epsilon_i = x_{i+1} - d_{11} x_i \qquad i = 1, \ldots N-1 \qquad 11.3\text{-}29$$

The prediction error filter may also be reversed and run backwards to yield an equally valid error series (figure 11.4).

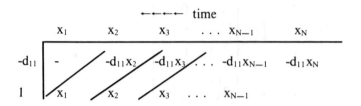

Figure 11.4 Convolution in the reverse direction of a time series with a two length prediction operator.

In this case the error series is

$$\tilde{\epsilon}_i = x_i - d_{11} x_{i+1} \qquad i = 1, \ldots N-1 \qquad 11.3\text{-}30$$

Following equation 2.1-2 or 6.2-5, the mean power is given by the autocorrelation with zero lag. The mean power in the forward and reverse direction is given by

$$P_1 = [2(N-1)]^{-1} \sum_{i=1}^{N-1} [\epsilon_i^2 + \tilde{\epsilon}_i^2]$$

or

$$P_1 = [2(N-1)]^{-1} \sum_{i=1}^{N-1} [(x_{i+1} - d_{11} x_i)^2 + (x_i - d_{11} x_{i+1})^2] \qquad 11.3\text{-}31$$

From maximum entropy, d_{11} is determined so that P_1 is a minimum.

$$\delta P_1 / \delta d_{11} = 0 \qquad 11.3\text{-}32$$

Maximum entropy spectral analysis

$$\delta P_1/\delta d_{11} = [2(N-1)]^{-1} \sum_{i=1}^{N-1} [(x_{i+1} - d_{11}x_i) 2(-x_i)$$

$$+ (x_i - d_{11} x_{i+1}) 2(-x_{i+1})]$$

$$= (N-1)^{-1} \sum_{i=1}^{N-1} [-(x_i x_{i+1} + x_{i+1} x_i) + d_{11}(x_{i+1}^2 + x_i^2)] = 0$$

This condition is satisfied if

$$d_{11} = \frac{2 \sum_{i=1}^{N-1} x_i x_{i+1}}{\sum_{i=1}^{N-1} (x_i^2 + x_{i+1}^2)} \qquad 11.3\text{-}33$$

Note that we are able to compute P_1 and d_{11} without using the autocovariance function directly and without making assumptions about the data beyond the end points. The matrix equation in 11.3-25 for m = 1 or a two point operator is

$$\begin{bmatrix} a_0 & a_1 \\ a_1 & a_0 \end{bmatrix} \cdot \begin{bmatrix} 1 \\ -d_{11} \end{bmatrix} = \begin{bmatrix} P_1 \\ 0 \end{bmatrix} \qquad 11.3\text{-}34$$

It is possible to solve this equation for a_1 using the estimates of d_{11} and a_0.

$$a_1 = a_0 d_{11} \qquad 11.3\text{-}35$$

In general the average power for a m + 1 long prediction error filter, run forwards and backwards, is

$$P_m = [2(N-m)]^{-1} \sum_{i=1}^{N-m} [(x_i - \sum_{k=1}^{m} d_{mk} x_{i+k})^2$$

$$+ (x_{i+m} - \sum_{k=1}^{m} d_{mk} x_{i+m-k})^2] \qquad 11.3\text{-}36$$

The higher order filter coefficients are obtained from the m lower order ones. Thus d_{22} is obtained by minimizing P_2 with respect to only d_{22}. The coefficient

d_{21} is found from a matrix equation due to Levinson (1947) which will be given as equation 11.3-46 later in this section. The Levinson recursion equation insures that $|d_{22}| < 1$ and that the filter, $[1, d_{21}, d_{22}]$, will be a *minimum phase* prediction error operator. The autocovariance matrix will then be Toeplitz and non-negative definite. In general the filter coefficients are

$$d_{mk} = d_{m-1\,k} - d_{mm} d_{m-1\,m-k} \qquad k = 1, \ldots, m-1$$

$$d_{m0} = -1 \qquad \qquad 11.3\text{-}37$$

$$d_{mk} = 0 \qquad k > m$$

Then equation 11.3-36 becomes

$$P_m = [2(N-m)]^{-1} \sum_{i=1}^{N-m} [(b_{mi} - d_{mm} b'_{mi})^2 + (b'_{mi} - d_{mm} b_{mi})^2] \qquad 11.3\text{-}38$$

where

$$b_{1i} = x_i \quad \text{and} \quad b'_{1i} = x_{1+i} \qquad *11.3\text{-}39$$

$$b_{mi} = \sum_{k=0}^{m} d_{m-1\,k} \, x_{i+k} = \sum_{k=0}^{m} d_{m-1\,m-k} \, x_{i+m-k} \qquad 11.3\text{-}40$$

$$b'_{mi} = \sum_{k=0}^{m} d_{m-1\,k} \, x_{i+m-k} = \sum_{k=0}^{m} d_{m-1\,m-k} \, x_{i+k} \qquad 11.3\text{-}41$$

Using equation 11.3-37, the recursion relations necessary to compute b and b' are

$$b_{mi} = b_{m-1\,i} - d_{m-1\,m-1} \cdot b'_{m-1\,i} \qquad *11.3\text{-}42$$

$$b'_{mi} = b'_{m-1\,i+1} - d_{m-1\,m-1} \cdot b_{m-1\,i+1} \qquad *11.3\text{-}43$$

It is necessary to find

$$\delta P_m / \delta d_{mm} = 0 \qquad 11.3\text{-}44$$

Maximum entropy spectral analysis

The minimum value of P_m is found to be given when

$$d_{mm} = \frac{2\sum_{i=1}^{N-m} b_{mi} b'_{mi}}{\sum_{i=1}^{N-m} (b_{mi}^2 + b'^2_{mi})} \qquad *11.3\text{-}45$$

Using a method which originated in the field of econometrics and introduced to Wiener filter theory by Levinson (1947), a recursion relation can be found for P_m by substituting 11.3-37 into equation 11.3-25. The Levinson recursion is in Chapter 14. Take the case for i = 3 as an example.

$$\begin{bmatrix} a_0 & a_1 & a_2 & a_3 \\ a_1 & a_0 & a_1 & a_2 \\ a_2 & a_1 & a_0 & a_1 \\ a_3 & a_2 & a_1 & a_0 \end{bmatrix} \cdot \begin{bmatrix} 1 \\ -d_{m1} \\ -d_{m2} \\ -d_{m3} \end{bmatrix} = \begin{bmatrix} P_m \\ 0 \\ 0 \\ 0 \end{bmatrix}$$

Inserting the recursion relations yields

$$\begin{bmatrix} a_0 & a_1 & a_2 & a_3 \\ a_1 & a_0 & a_1 & a_2 \\ a_2 & a_1 & a_0 & a_1 \\ a_3 & a_2 & a_1 & a_0 \end{bmatrix} \cdot \left\{ \begin{bmatrix} 1 \\ -d_{m-1\ 1} \\ -d_{m-1\ 2} \\ 0 \end{bmatrix} - d_{m3} \begin{bmatrix} 0 \\ d_{m-1\ 2} \\ d_{m-1\ 1} \\ 1 \end{bmatrix} \right\}$$

$$= \begin{bmatrix} P_{m-1} \\ 0 \\ 0 \\ d_{m3} P_{m-1}* \end{bmatrix} - d_{m3} \begin{bmatrix} d_{m3} P_{m-1} \\ 0 \\ 0 \\ P_{m-1} \end{bmatrix} = \begin{bmatrix} P_m \\ 0 \\ 0 \\ 0 \end{bmatrix} \qquad 11.3\text{-}46$$

The quantities marked with an * are obtained by symmetry. This balances the equation on the basis of the forward and backward running filter coefficients within the brackets, [ ], in the line above. Therefore the general recursion relation for i = m is

$$P_m = P_{m-1}(1 - d_{mm}^2) \qquad *11.3\text{-}47$$

Because $|d_{mm}| \leq 1$ from equation 11.3-37 it follows that P_m becomes successively smaller. That is, $0 \leq P_m \leq P_{m-1}$.

The following two subroutines may be used to compute a Burg maximum entropy spectral estimate.

```
      SUBROUTINE BURGAR(X,N,A,P,B1,B2,MM)
C
C     THIS SUBROUTINE BY D. C. GANLEY WILL CALCULATE THE BURG
C     ESTIMATES OF THE AUTOREGRESSIVE PARAMETERS FOR AR MODELS
C     OF ORDERS 1 TO N-1 WHICH ARE FITTED TO AN INPUT TIME SERIES
C     OF LENGTH N. THE ALGORITHM IS TAKEN FROM N. ANDERSON (1974).
C
C     INPUTS ARE:
C       X = INPUT TIME SERIES
C       N = LENGTH OF X
C       B1,B2 = WORK SPACES
C       MM = MAXIMUM ORDER CALCULATION FOR D.(DEFAULT IS N-1).
C
C     OUTPUTS ARE:
C       D = THE ESTIMATES OF THE AR PARAMETERS D(M,I) WHERE I GOES
C           FROM 1 TO M AND M IS THE ORDER OF THE PROCESS WHICH
C           RANGES FROM 1 TO MM. THESE PARAMETERS ARE STORED IN
C           THE ORDER D(1,1),D(2,1)D(2,2),D(3,1),...,D(N-1,N-1).
C           FOR AN AR PROCESS OF ORDER M WE HAVE
C           X(T) = E(T) + D(M,1)*X(T-1) + ... + D(M,M)*X(T-M)
C           WHERE E(T) IS WHITE NOISE.
C
C       P = OUTPUT POWER. P(I) IS THE ESTIMATE OF THE VARIANCE OF
C           THE WHITE NOISE OF THE AR PROCESS OF ORDER I-1.
C
C     DIMENSIONS ARE X(N),P(MM+1),D((MM+1)*MM/2),B1(N-1),B2(N-1)
C
      DIMENSION X(1),P(1),D(2),B1(1),B2(1)
      N1=N-1
      IF (MM.LE.0.OR.MM.GT.N1) MM=N1
      XN=0.0
      DO 3 I=1,N
    3 XN=XN+X(I)*X(I)
      P(1)=XN/N
      DO 5 I=1,N1
      B1(I)=X(I)
    5 B2(I)=X(I+1)
      XN=0.0
      XD=0.0
      DO 7 I=1,N1
      XN=XN+B1(I)*B2(I)
    7 XD=XD+B1(I)*B1(I)+B2(I)*B2(I)
```

Maximum entropy spectral analysis

```
      D(1)=2.0*XN/XD
      P(2)=P(1)*(1.0-D(1)*D(1))
      DO 19 I=2,MM
      JE=N-I
      LM=(I*I-I)/2
      K=LM+1
      LK=K-I
      KM=LM+I
      DO 13 J=1,JE
      B1(J)=B1(J)-D(LM)*B2(J)
   13 B2(J)=B2(J+1)-D(LM)*B1(J+1)
      XN=0.0
      XD=0.0
      DO 15 J=1,JE
      XN=XN+B1(J)*B2(J)
   15 XD=XD+B1(J)*B1(J)+B2(J)*B2(J)
      D(KM)=2.0*XN/XD
      P(I+1)=P(I)*(1.0-D(KM)*D(KM))
      JE=I-1
      DO 17 J=1,JE
   17 D(LM+J)=D(LK+J)-D(KM)*D(K-J)
   19 CONTINUE
      RETURN
      END
      SUBROUTINE SPECTR(A,M,PM,Y,N,F1,DF,DT)
C
C
C         THIS SUBROUTINE WILL COMPUTE THE MAXIMUM ENTROPY
C     SPECTRAL ESTIMATE OF AN AUTOREGRESSIVE PROCESS GIVEN THE
C     AR PARAMETERS. SPECIFICALLY IF D(1),D(2)...D(M) ARE THE
C     AR PARAMETERS THEN THE PROGRAM WILL CALCULATE THE SQUARE
C     OF THE MODULUS OF PM*DT / (1-D(1)*Z**2...-D(M)*Z**M)
C     EVALUATED ON THE UNIT CIRCLE.
C
C     INPUTS ARE:
C        D = AUTOREGRESSIVE PARAMETERS
C        M = ORDER OF THE AR PROCESS
C        PM = ESTIMATE OF THE VARIANCE OF THE WHITE NOISE
C        N = NUMBER OF FREQUENCIES TO CALCULATE POWER AT
C        F1 = FIRST FREQUENCY TO CALCULATE POWER AT
C        DF = FREQUENCY INCREMENT
C        DT = SAMPLE INTERVAL OF INPUT TIME SERIES IN MILLISECONDS
C
C     OUTPUTS ARE:
C        Y = POWER SPECTRUM
C
      DIMENSION D(1),Y(1)
      COMPLEX Z,SUM
      PFACT=PM*DT/1000.0
      DO 9 I=1,N
      F=F1+DF*(I-1)
      ARG=.0062831853*DT*F
      SUM=CMPLX(0.0,0.0)
```

```
      DO 5 J=1,M
      B=ARG*J
      C=COS(B)
      S=SIN(B)
      Z=CMPLX(C,-S)
    5 SUM=SUM+D(J)*Z
      C=1.0-REAL(SUM)
      S=-AIMAG(SUM)
      B=C*C+S*S
    9 Y(I)=PFACT/B
      RETURN
      END
```

Another approach to the determination of the filter coefficients has been evaluated by Ulrych and Clayton (1976). This procedure is a least squares estimation to minimize the prediction error power with respect to *all* the coefficients d_{mk} and not just the last one, d_{mm}, as in equation 11.3-37. This means that the filter $[1, d_{m1}, ... d_{mm}]$ will not be minimum phase and the autocovariance matrix is not necessarily Toeplitz. This technique appears to give a more stable spectral estimate for a case of signal consisting of a sinusoid with additive white noise. Other programs are given by Barrodale and Erickson (1980). Kromer (1969) has shown that for an autoregressive model of order m the variance of the estimate is given by

$$\text{Var}\,[P_E(f)] = N^{-1}\,2m\,\mathbf{P}_E^{\,2}(f) \qquad 11.3\text{-}48$$

where $\mathbf{P}_E$ is the true spectral value and P_E is the estimate from a finite set of data points.

11.4 Order of the AR process

A major problem in the use of the maximum entropy method is the determination of the order number, m = M. The most widely used criterion is the *final prediction error* (FPE) as given by Akaike (1969a,b,1970).

$$[\text{FPE}]_M = \frac{N+M+1}{N-M-1}\,P_M \qquad 11.4\text{-}1$$

The order of the autoregression is chosen so that the average error for a one-step prediction is minimized, considering both the error in the unpredictable part and the error due to inaccuracies for determining the filter coefficients. The prediction error power, P_m, is computed by equation 11.3-47 for successively higher values of m until a minimum is obtained for m = M.

Order of the AR process

A new *Akaike information criterion* called AIC has also been introduced by Akaike (1974,1976).

$$\text{AIC} = -2(\text{maximum log likelihood}) + 2(\text{number of independently adjusted parameters}) \quad 11.4\text{-}2$$

The choice of order number is made by an estimate called MAICE in which the smallest value of AIC is chosen from the models being fitted by the method of maximum likelihood. For a single time series the AIC criterion reduces to

$$\text{AIC}_m = N \log |P_m| + 2(1+M) \quad 11.4\text{-}3$$

Tests on FPE and AIC almost always result in the same order, M, being selected. They are asymptotically equivalent.

$$\text{AIC} \cong N \log (\text{FPE}) \quad 11.4\text{-}4$$

The one in 11.4-3 is often omitted because it is an additive constant that only takes into account the subtraction of the sample mean.

The minimizing by FPE or AIC does not give a consistent estimate of order when the process has a finite order structure. Ulrych and Bishop (1975) imposed a limit of $M = N/2$ to stabilize the spectral structure when sharp spectral lines are present. A modified version, BIC, of AIC was presented by Akaike (1976).

$$\text{BIC} = (N-m) \log_e[S(m)/(N-m)] + m \log_e[(S(0)-S(m)/m] \quad 11.4\text{-}5$$

where $S(m)$ is N times the maximum likelihood estimate of the variance of the white noise series, **n**, and m is the number of coefficients. Parzen (1976) has introduced another criterion called CAT.

$$\text{CAT}(m) = N^{-1} \sum_{i=1}^{m} \mathbf{P}_i^{-1} - \mathbf{P}_m^{-1} \quad 11.4\text{-}6$$

where

$$\mathbf{P}_i = P_i / (1-i/N) \quad 11.4\text{-}7$$

By minimizing CAT an *optimal* value of m is determined. A completely empirical criterion is suggested by Berryman (1978) on the basis of numerical modelling of seismic data. The length for the operator is chosen to be M where

$$M = 2N / \log_e 2N \quad 11.4\text{-}8$$

The proliferation of empirical and semi-theoretical criterion for determining an *optimal order* is perhaps less disturbing than the question of whether an autoregressive process is justified. This problem has already been discussed in the introduction to this chapter and is illustrated by examples presented by Treitel et al. (1977). If the desired spectra is from an ARMA process then it requires an infinitely long autoregressive process (m = ∞) to model it. Gersch and Sharpe (1973) have made model studies of the application of MEM to an ARMA process. They find that, indeed, the MEM spectrum fits the theoretical ARMA spectrum as the data length becomes great and a high order is used. They also find that Akaike's FPE estimate is useful provided there is a long data set.

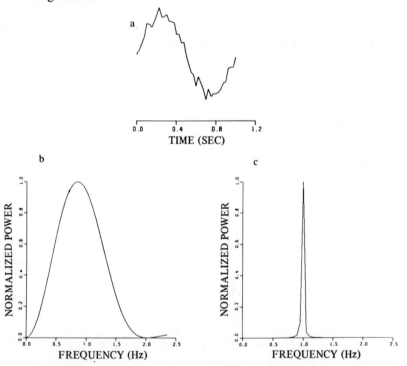

Figure 11.5 (a) A 1 Hz sinusoid with 10 percent white noise truncated with a 1 second window. (b) The power spectrum of the signal in (a) computed by using the square of the modulus of the Fourier transform. (c) The maximum entropy power spectrum of the signal in figure (a). (From Ulrych, 1972, courtesy of the J. Geophys. Res.)

There are many examples in the literature of the application of maximum entropy methods to determine the spectrum of truncated sinusoids in the presence of additive noise. The first of these was obtained by Ulrych (1972) and one of his results is reproduced in figure 11.5. Chen and Stegan

Order of the AR process

(1974) found that when data sets of truncated sinusoids with additive noise are very short there are frequency shifts in the MEM spectrum as a function of initial phase and length of sinusoid. Fourgere (1975) has pointed out that spectral peaks have a tendency to split when the noise is low and long filters are used. This instablity is not surprising since a sinusoid with additive noise does not conform to any autoregressive model on which the MEM is based. Techniques for dealing with this by minimizing the energy with respect to all filter coefficients are discussed by Ulrych and Clayton (1979). However one no longer has a Toeplitz autocovariance matrix. If pure sinusoids or exponentially decaying sinusoids are definitely known to be generated in the physical process then one should consider these methods or, perhaps more logically, a direct least squares fitting (Bloomfield, 1976).

11.5 The autocovariance function

Burg (1970) shows how to set up a matrix equation for computing the autocovariance function in terms of the prediction error filter coefficients and the error powers. Let the elements of the j point prediction filter, d_{ij}, be denoted by an N by N matrix, **D**.

$$\mathbf{D} = \begin{bmatrix} 1 & 0 & \ldots & 0 & 0 \\ d_{2N} & 1 & \ldots & 0 & 0 \\ d_{3N} & d_{2,N-1} & \ldots & 0 & 0 \\ \cdot & \cdot & & \cdot & \\ \cdot & \cdot & & \cdot & \\ \cdot & \cdot & & \cdot & \\ d_{N-1,N} & d_{N-1,N-1} & \ldots & 1 & 0 \\ d_{N,N} & d_{N-1,N-1} & \ldots & d_{22} & 1 \end{bmatrix} \qquad 11.5\text{-}1$$

The mean output power for a j point prediction filter is given by an N by N matrix in which the elements labelled by a star are not needed.

$$\mathbf{P'} = \begin{bmatrix} P_N & * & \ldots & * & * \\ 0 & P_{N-1} & \ldots & * & * \\ 0 & 0 & \ldots & * & * \\ \cdot & \cdot & & & \\ \cdot & \cdot & & & \\ \cdot & \cdot & & & \\ 0 & 0 & & P_2 & * \\ 0 & 0 & & 0 & P_1 \end{bmatrix} \qquad 11.5\text{-}2$$

The equation for the autocovariance function in 11.3-14 may be rewritten as

$$\mathbf{AD} = \mathbf{P}' \qquad 11.5\text{-}3$$

If we now define an N by N matrix **P** with non-zero elements only along the diagonal

$$\mathbf{P} = \begin{bmatrix} P_N & 0 & \ldots & 0 \\ 0 & P_{N-1} & \ldots & 0 \\ \vdots & \vdots & & \vdots \\ 0 & 0 & \ldots & P_1 \end{bmatrix} \qquad 11.5\text{-}4$$

then equation 11.5-3 can be recast in the following form by multiplying by $\mathbf{D}^T$:

$$\mathbf{D}^{*T}\mathbf{AD} = \mathbf{P} \qquad 11.5\text{-}5$$

Note that $\mathbf{A} = \mathbf{A}^T$ and that **P** is Hermitian. Taking successive inverses of the matrices in 11.3-5 we have

$$\mathbf{D}^{-1T}*\ \mathbf{D}^{*T}\ \mathbf{AD} = \mathbf{D}^{-1T}*\ \mathbf{P}$$
$$\mathbf{AD} = \mathbf{D}^{-1T}*\ \mathbf{P}$$
$$\mathbf{D} = \mathbf{A}^{-1}\ \mathbf{D}^{-1T}*\ \mathbf{P}$$
$$1 = \mathbf{D}^{-1T}\ \mathbf{A}^{-1}\ \mathbf{D}^{-1T}\ \mathbf{P}$$

or

$$\mathbf{P}^{-1} = \mathbf{D}^{-1}\ \mathbf{A}^{-1}\ \mathbf{D}^{-1T}* \qquad 11.5\text{-}6$$

Then continuing on to a solution for $\mathbf{A}^{-1}$ we have

$$\mathbf{DP}^{-1} = \mathbf{A}^{-1}\ \mathbf{D}^{-1T}*$$
$$\mathbf{DP}^{-1}\ \mathbf{D}^{T}* = \mathbf{A}^{-1} \equiv \mathbf{B} \qquad 11.5\text{-}7$$

Therefore the inverse of the autocovariance function, B, is given in terms of the coefficients of the error filter and the mean error powers. For a three-point prediction error filter with real coefficients

$$\begin{bmatrix} B_0 & B_1 & B_2 \\ B_1 & B_0 & B_1 \\ B_2 & B_1 & B_0 \end{bmatrix} = \begin{bmatrix} 1 & 0 & 0 \\ d_{23} & 1 & 0 \\ d_{33} & d_{22} & 1 \end{bmatrix} \begin{bmatrix} P_3^{-1} & 0 & 0 \\ 0 & P_2^{-1} & 0 \\ 0 & 0 & P_1^{-1} \end{bmatrix} \begin{bmatrix} 1 & d_{22} & d_{33} \\ 0 & 1 & d_{23} \\ 0 & 0 & 1 \end{bmatrix} \qquad 11.5\text{-}8$$

The autocovariance function

In terms of the filter coefficients the components of the inverse of the autocovariance function are

$$B_{ij} = \sum_{k=0}^{n} d_{m-k\ i-k}\ d_{m-k\ i-k} / P_{m-k} \qquad i,j=0,1,...m \qquad 11.5\text{-}9$$

in which n is the smaller of the two indices i and j and

$$B_{ij} = B_{ji} = B_{m-i\ m-j} = B_{m-j\ m-i} \qquad 11.5\text{-}10$$

The maximum entropy method allows one to compute the power spectrum which has the maximum entropy from the N values of the data, x_n, without any assumption about the continuation of the data. During the course of the calculation we find the optimum N point prediction filter whose output, ϵ_n, is the difference between the true value and the predicted value of the time series, x_n. Since the mean value $\epsilon_{n-1}\ \epsilon_n$ is zero for any n, the ϵ_n are all uncorrelated. The output from the prediction error filter has a white spectrum and its density level is P_m/f_N. From the density level and the transfer function, $1/\mathbf{D}^T\mathbf{E}^*$, of the prediction error filter one can obtain the true power spectra, P(f), of the input data. The resolution obtained by this method will be very much better than that obtained by any other proposed method. From a knowledge of P_m and the filter coefficients, it is then possible to obtain a superior estimate of the autocovariance, whose Toeplitz matrix will be non-negative definite. The autocovariance function may also be obtained by a Fourier transform.

$$a_n = \int_{-f_N}^{f_N} P_E(f)\ e^{-2\pi i f n \Delta t}\ df \qquad n = 0, 1, ...\ N\text{-}1 \qquad 11.5\text{-}11$$

11.6 Extension of maximum entropy method

The maximum entropy method of Burg does not yield a direct estimate of the power from the height of the spectral estimate, $P_E(f)$. Lacoss (1971) pointed out that the peak values appear to be proportional to the square of the power. The spectral power can be estimated by determining the area under each peak but this is seldom possible in practice due to the presence of closely spaced peaks. The power, ρ, of an isolated peak will be found to be proportional to the power, $P_E(f)$, and the bandwidth, B.

$$\rho = \pi P_E B \qquad 11.6\text{-}1$$

A method for obtaining a power estimate, ρ, without numerical integration has been presented by Johnson and Anderson (1978). The technique automatically ensures that the sum of the power estimate is equal to the total integrated power.

Burg's equation, 11.3-15a and 11.3-26, is written in terms of z transforms [$z = \exp(-2\pi i f)$].

$$P_E(f) = P_m \, \Delta t \, / \, [D(z) \cdot D(1/z)] \qquad 11.6\text{-}2$$

$$D(z) = [1 - \sum_{n=1}^{m} d_{mn} z^{-n}] \, z^m \qquad 11.6\text{-}3$$

One can let $d_{m0} = -1$ and the denominator of equation 11.6-2 has a multiplier of z^m and z^n.

$$D(z) = -\sum_{n=0}^{m} d_{mn} \, z^{m-n} \qquad 11.6\text{-}4$$

$$D(z) = -(z-z_1)(z-z_2) \ldots (z-z_m) \qquad *11.6\text{-}5$$

Equations marked with an asterisk are required for a computer program. The zeros of 11.6-5

$$z_k = r_k \cdot \exp(-2\pi i f_k) \qquad r_k > 1 \qquad 11.6\text{-}6$$

are outside the unit circle while the zeros of $D(1/z)$ are inside the unit circle at radius $1/r_k$. If the signal is a pure sinusoid with no noise present then the poles of P_E are on the unit circle. The addition of noise moves the singularities outside the unit circle. If the poles are near the unit circle then peaks will appear in the spectrum.

The total power, P_0, in the time series is found by integrating over all the frequencies within the Nyquist limits or else by summing the power, ρ_k, in all the spectral components.

Extension of maximum entropy method

$$P_0 = \int_{-0.5}^{0.5} P_E(f)df = \sum_{k=1}^{m} \rho(f_k) \qquad 11.6\text{-}7$$

In the following sections it will be shown that the power in each spectral component is given by the residue of the function $P_E(z)/z$ evaluated at $z = z_k$. Thus the total power is also given by the sum of residues.

$$P_0 = (2\pi)^{-1} \oint P(z)\, dz/z = \sum_{k=1}^{m} Res(z_k) \qquad 11.6\text{-}8$$

The integration is taken around the unit circle in the complex z plane. Power is a symmetric function with respect to zero frequency. Since $Res(z_k^*) = Res^*(z_k)$ the power of each spectral component is given by twice the real part, R, of the residues.

$$\begin{aligned}\rho(f_k) &= Res(z_k) + Res(z_k^*) \\ &= 2R[Res(z_k)]\end{aligned} \qquad 0 < f_k < 0.5 \qquad *11.6\text{-}9$$

For the two end points on the frequency scale the power is given by the residues.

$$\rho(f_k) = Res(z_k) \qquad f_k = 0,\ 0.5 \qquad *11.6\text{-}10$$

From equations 11.6-2 and 11.6-5 the spectral estimate at any frequency, f, may be written with the denominator given by a product, Π, from $j = 1$ to m of distances from the poles to the unit circle at $z = \exp(-2\pi i f)$.

$$P_E(f) = P_m \cdot \Delta t / [\Pi_{j=1 \text{ to } m}(z-z_j)(1/z - z_j^*)] \qquad 11.6\text{-}11$$

Consider the special case when a single well-defined peak occurs at a particular frequency, f_k. A pair of conjugate poles exist close to the unit circle at this frequency. The contribution from these poles to the spectrum is determined by $(z - z_j)(1/z - z_j^*)$ where $z = \exp(-2\pi i f)$ varies along the unit circle in the band of frequencies near $f = f_k$. Because we have said this is an isolated peak, all other poles are far away and from figure 11.6 we can replace z by z_k in all the terms where j is not equal to k. The approximate value of the Burg maximum entropy spectral estimate is

$$P_E(f) \cong P_m \cdot \Delta t / [(z-z_k)(1/z - z_k^*)\Pi_{j \neq k} |z_k - z_j|^2] \qquad 11.6\text{-}12$$

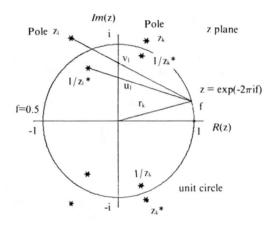

Figure 11.6 Poles inside and outside of the unit circle for the function $P_E(z)$. The contribution to the power at frequency, f, due to the pole at z_l is inversely proportional to the length v_l. To a good approximation v_l is equal to v_{lk} as is the length of u_l. The contribution to the power due to the pair of poles at j = 1 is inversely proportional to $v_l u_l$ or this can be approximated by the inverse square of v_{lk}, $v_{jk}^2 = |z_k - z_j|^2$, j = 1. Note that this replaces $1/z_j$ by z_j^*.

At frequencies in the vicinity of f_k the product of terms for j not equal to k is nearly independent of frequency and we can set them equal to D_k.

$$D_k = \Pi_{j \neq k} |z_k - z_j|^2 \quad j = 1, \ldots, m. \quad \text{11.6-13}$$

At frequencies near the poles at z_k and $1/z_k^*$ the distance from the unit circle to each pole is given by h in figure 11.7. From figure 11.7

$$h^2 = |\Delta z|^2 + (r_k - 1)^2 \approx |z - z_k| \, |1/z - z_k^*| \quad \text{11.6-14}$$

where

$$|\Delta z| = 2\pi |f - f_k| = 2\pi \cdot \Delta f \quad \text{11.6-15}$$

The approximate value of the maximum entropy spectral estimate is

$$P_E(f) \cong P_m \cdot \Delta t \, / \, [D_k \, (|\Delta z|^2 + (r_k - 1)^2)] \quad \text{11.6-16}$$

Extension of maximum entropy method

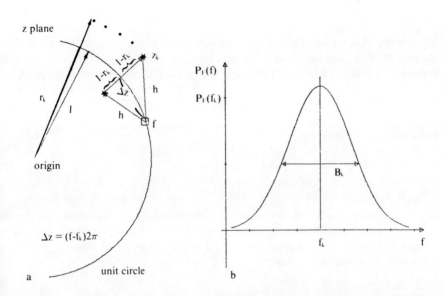

Figure 11.7 (a) An enlargement of the unit circle in figure 11.6 near the poles at z_k and $1/z_k^*$. The distance from the pole to z on the unit circle is denoted by h. (b) The shape of the power spectral estimate, P_E, near an isolated pole at z_k. The power in the harmonic at f_k is given by the area under the curve which is proportional to the product of the maximum value $P_E(f_k)$ and the bandwidth B_k. The bandwidth is related to the distance between the two conjugate poles by $B_k = |z_k - 1/z_k^*| / 2\pi$.

The distance between pair conjugate poles at z_k is

$$|z_k - 1/z_k^*| = (r_k - 1/r_k) = 2(r_k - 1) \qquad 11.6\text{-}17$$

The bandwidth of the peak, B_k (figure 11.7,b) may be shown to be related to the distance above.

$$B_k = (r_k - 1) / \pi \qquad *11.6\text{-}18$$

Substituting 11.6-15 for Δf and 11.6-18 for B_k into 11.6-16 gives a simple expression for P_E.

$$P_E(f) \cong P_m \cdot \Delta t / [D_k 4\pi^2 (\Delta f^2 + B_k^2/4)] \qquad 11.6\text{-}19$$

The peak value occurs when $\Delta f = 0$.

$$P_E(f_k) \cong P_m \cdot \Delta t / [D_k \pi^2 B_k^2] \qquad 11.6\text{-}20$$

Integration of equation 11.7-19 over all frequencies with D_k constant gives half the power of the isolated peak. The other half is added in from the poles in the lower half z plane at $f_k = -f_k$. The power of the spectral component is the $\rho(f_k)$.

$$\rho(f_k) = P_m \cdot \Delta t / [D_k \pi B_k] \qquad 11.6\text{-}21$$

Substituting 11.7-20 yields

$$\rho(f_k) = \pi P_E(f_k) B_k \qquad 11.6\text{-}22$$

which is just equation 11.6-1.

Finally it is necessary to relate the residue of the function $P_E(z)/z$ at $z = z_k$ to the power of the spectral component. For a simple pole at $z = a$ the residue of a function, f, is given by a relation found in any introductory text on functions of a complex variable.

$$Res(a) = [(z-a) f(z)]_{z=a} \qquad 11.6\text{-}23$$

Therefore the residue of $P_E(z)/z$ at $z = z_k$ is obtained directly from equation 11.6-11.

$$Res(z_k) = P_m \cdot \Delta t / [z_k \prod_{j \neq m} (z_k - z_j) \prod_{j=1 \text{ to } m} (1/z_k - z_j^*)] \qquad *11.6\text{-}24$$

Alternatively this may be written with the denominator having a product, Π of $j = 1, m, j \neq k$.

$$Res(z_k) = P_m \cdot \Delta t / [z_k(1/z_k - z_k^*) \prod_{j \neq k} [(z_k - z_j)(1/z_k - z_j^*)]] \qquad 11.6\text{-}25$$

Using the relation for D_k in 11.6-13 the residue simplifies to

$$Res(z_k) = P_m \cdot \Delta t / [z_k(1/z_k - z_k^*) D_k] \qquad 11.6\text{-}26$$

From the definition of the z transform and the relation for bandwidth in 11.6-17 and 11.6-18 we have that

$$z_k(1/z_k - z_k^*) = \exp(-2\pi i f_k) \cdot 2\pi B_k \cdot \exp(2\pi i f_k) = 2\pi B_k \qquad 11.6\text{-}27$$

Extension of maximum entropy method

The residue at frequency f_k is then

$$Res(z_k) = P_m \cdot \Delta t / [2\pi B_k D_k] \qquad 11.6\text{-}28$$

If we add the residue from the pole at $-f_k$ one obtains equation 11.6-21.

$$\rho(f_k) = 2Res(z_k) = P_m \cdot \Delta t / [\pi B_k D_k] \qquad 11.6\text{-}29$$

This establishes the identity of equation 11.6-8 and shows that the residue of a pole of $P_E(z)/z$ is equal to $\rho(f_k)$, the power of the harmonic component. In computing $\rho(f)$ it is necessary to obtain the Burg estimate of the prediction error filter $(1, -d_{m1}, ..., d_{mm})$ and determine the zeros, z_k, $k = 1$ to m of the polynomial in equation 11.6-5. Johnson and Anderson (1978) recommend the use of a double precision subroutine such as PA07PD in the Harwell subroutine library as a reliable method of computing the complex roots of a polynomial of order 100 or more. The power of the spectral components are obtained from equation 11.7-26 for the residue and equation 11.6-9 and 11.6-10 for $\rho(f_k)$. It is also useful to estimate the bandwidth as a function of frequency from equation 11.6-18.

Maximum entropy has been extended to cross-spectral analysis by Ulrych and Jensen (1974). Progress has also been made in extending the technique to two-dimensional problems by Ponsonby (1973) and to the multi-dimensional case by Morf et al. (1978). The topic will continue to be the subject of intense investigation. Its application to processes that are not autoregressive should be made with extreme caution. The references contain some of the more important articles on the subject, some of which have not been mentioned in this review of the subject.

References

Adams,J.A.(1976), Discussion on paper by R.N. McDonough, (1974). *Geophysics* **41**, 771-773.

Akaike,H.(1969a), Fitting autoregressive models for prediction. *Ann. Inst. Statist. Math.* **21**, 243-247.

Akaike,H.(1969b), Power spectrum estimation through autoregressive model fitting. *Ann. Inst. Statist. Math.* **21**, 407-419.

Akaike,H.(1970), Statistical predictor identification. *Ann. Inst. Statist. Math.* **22**, 203-217.

Akaike,H.(1974), A new look at the statistical model identification. *IEEE Transactions on Automatic Control.* **AC-19**, 716-720.

Akaike,H.(1976), Canonical correlation analysis of time series and the use of an information criterion. *System identification - advances and case studies.* Editors: R.K. Mehra and D.G. Laniotis. Academic Press, New York, 27-96.

Akaike,H.(1976), Time series analysis and control through parametric models. *Applied Time Series Analysis Symposium* University of Tulsa. Conveners: D.F. Findley and S.J. Laster, Tulsa, Oklahoma.

Anderson,N.(1974), On the calculation of filter coefficients for maximum entropy spectral analysis. *Geophysics* **39**, 69-72.

Anderson,N.(1980), A stepwise approach to computing the multi-dimensional fast Fourier transform of large arrays. *IEEE Transactions on Acoustics, Speech and Signal Processing* **ASSP 28**, 280-284.

Barrodale,I. and Erickson,R.E.(1980), Algorithms for least squares linear prediction and maximum entropy spectral analysis. *Geophysics* **45**, 420-446.

Bartlett,M.S.(1946), On the theoretical specifications and sampling properties of autocorrelated time series. *J. Royal Statistical Society* **B8**, 27-41.

Berryman,J.G.(1978), Choice of operator length for maximum entropy spectral analysis. *Geophysics* **43**, 1384-1391.

Bloomfield,P.(1976), Fourier analysis of time series. John Wiley, New York. 20-25.

Box,G.E.P. and Jenkins,G.M.(1976), Time series analysis: Forecasting and Control, Chapter 3. Holden-Day Inc., San Francisco, 46-84.

Brillouin,L.(1956), Science and information theory, Academic Press, New York, 159-161.

Burg,J.P.(1967), Maximum entropy spectral analysis. Paper presented at the 37th Annual Int. SEG Meeging, Oklahoma, October 31, 1967. Preprint - Texas Instruments, Dallas.

Burg,J.P.(1970), A new analysis technique for time series data. Presented at NATO Advanced Study Institute on Signal processing with Emphasis on Underwater Acoustics.

Burg,J.P.(1972), The relationships between maximum entropy spectra and maximum likelihood spectra. *Geophysics* **37**, 375-376.

Cadzow,J.A.(1980), High performance spectral estimator. A new ARMA method, *IEEE Transactions on Acoustics,Speech, and Signal Processing* **ASSP 28**, 524-529.

Capon,J.(1969), High-resolution frequency-wavenumber spectrum analysis. Proceeding of the IEEE **57**, 1408-1418.

Chen,W.Y. and Stegan,G.R.(1974), Experiments with maximum entropy power spectra of sinusoids. *J. Geophys. Res.* **79**, 3019-3022.

Fourgere,P.F.(1975), A solution to the problem of spontaneous line splitting in maximum entropy power spectrum analysis. *EOS Transactions, American Geophysical Union* **56**, 1054.

Fryer,G.J., Odegard,M.E. and Sutton,G.H.(1975), Deconvolution and spectral estimation using final prediction error. *Geophysics* **40**, 411-425.

Gersch,W.(1970), Estimation of the autoregressive parameters of a mixed autoregressive moving-average time series. *IEEE Transactions on Automatic Control* **AC-15**, 583-585.

Gersch,W. and Sharpe,D.R.(1973), Estimation of power spectra with finite order autoregressive models. *IEEE Transactions on Automatic Control* **AC-18**, 367-369.

Gersch,W. and Foutch,D.A.(1974), Least squares estimates of structural system parameters using covariance function data. *IEEE Transactions on Automatic Control* **AC-17**, 493-494.

Gutowski,P.R., Robinson,E.A. and Treitel,S.(1978), Spectral estimation: fact or fiction. *IEEE Transactions on Geoscience Electronics* **GE-16** 80-84.

Hubral,P.,Treitel,S. and Gutowski,P.R.(1980), A sum auto-regressive formula for the reflection response. *Geophysics* **45**, 1697-1905.

Jenkins,G.M and Watts,D.G.(1969), *Spectral Analysis and its Applications*. Holden Day, San Francisco, p 164.

References

Johnson,S.J. and Anderson,N.(1978), On power estimation in maximum entropy spectrum analysis. *Geophysics* **43**, 681-690.

Jones,R.H.(1974), Identification and autoregressive spectrum estimation. *IEEE Transactions on Automatic Control* **AC-19**, 894-898.

Jones,R.H.(1976), Autoregression order selection. *Geophysics* **41**, 771-773.

Kay,S.M.(1980), A new ARMA spectral estimator. *IEEE Transactions on Acoustics,Speech and Signal Processing* **ASSP 28**, 585-588.

Kay,S.M.(1980), Noise compensation for auto-regressive spectral estimates. *IEEE Transactions on Acoustics, Speech and Signal Processing* **ASSP 28**, 292-303.

Kromer,R.E.(1969), Asymptotic properties of the autoregressive spectral estimator. PhD Thesis. Department of Statistics, Stanford University, Stanford California. Technical Report 13.

Lacoss,R.T.(1971), Data adaptive spectral analysis methods. *Geophysics* **36**, 661-675.

Levinson,H.(1947), The Wiener RMS error criterion in filter design and prediction. *J. Math Physics* **25**, 261-278.

Morf,M., Vienna,A., Lee,D.T.L. and Kailath,T.(1978), Recursive multichannel maximum entropy spectral estimation. *IEEE Transactions on Geoscience Engineering* **GE-16**, 85-94.

McDonough,R.N.(1974), Maximum entropy spatial processing of array data. *Geophysics* **39**, 843-851.

Parzen,E.(1969), Multiple time series modeling in 'Multivariate Analysis II', P. R. Krishnaiah. Ed. New York, Academic Press, 389-409.

Peacock,K.L. and Treitel,S.(1969), Predictive deconvolution: Theory and Practice *Geophysics* **34**, 155-169.

Pisarenko,V.F.(1972), On the estimation of spectra by means of non-linear functions of the covariance matrix. *Geophysical Journal* **28**, 511-531.

Pisarenko,V.F.(1973), The retreival of harmonics from a covariance function. *Geophysical Journal* **33**, 347-366.

Ponsonby,J.E.(1973), An entropy measure for partially polarized radiation... Mon. Notices. *Roy. Astron. Soc.* **163**, 369-380.

Satorius,E.H. and Zeidler,J.R.(1978), Maximum entropy spectral analysis of multiple sinusoids in noise. *Geophysics* **43**, 1111-1128.

Shannon,C.E.(1948), A mathematical theory of communication. *Bell System Tech. J.* **27**, 379-423.

Smylie,D.E., Clarke,G.K.C. and Ulrych,T.J.(1973), Analysis of irregularities in the earth's rotation in 'Methods in Computational Physics' 13, Academic Press N.Y. 391-430.

Swinger,D.N.(1979), A comparison between Burg's maximum entropy method and a non-recursive technique for the spectral analysis of deterministic signals. *J. Geophys. Res.* **84**, 679-685.

Thomas,J.B.(1968), An Introduction to Statistical Communication Theory. J. Wiley & Sons, New York, 670.

Treitel,S., Gutowski,P.R. and Robinson,E.A.(1977), Empirical spectral analysis revisited. *Topics in Numerical Analysis* 3, Academic Press, New York, 429-446.

Ulrych,T.J.(1972), Maximum entropy power spectrum of truncated sinusoids. *J. Geophysical Research* 77 1396-1400. (&RE>

Ulrych,T.J. and Jensen,O.(1974), Cross spectral analysis using maximum entropy. *Geophysics* **39**, 353-354.

Ulrych,T.J. and Bishop,T.N.(1975), Maximum entropy spectral analysis and autoregressive decomposition. *Rev. Geophysics and Space Physics* **13**, 183-200.

Ulrych,T.J. and Clayton,R.W.(1976), Time series modelling and maximum entropy. *Physics of the Earth and Planetary Interiors* **12**, 188-200.

Van den Bos,A.(1971), Alternative interpretation of maximum entropy spectral analysis. *IEEE Transactions on Information Theory* **AC-17**, 493-494.

Walker,G.(1931), On periodicity in series of related terms. *Proc. Royal Society* **A131**, 518-532.

Yule,G.U.(1927), On a method of investigating periodicities in disturbed series, with special reference to Wolfer's sunspot numbers. *Phil. Trans. Roy. Soc. London, Ser. A* **226**, 267-298.

12 The maximum likelihood method of spectral estimation

12.1 Introduction
The maximum likelihood method of obtaining the power spectral density estimation was formulated by Capon (1969) for use in the frequency-wavenumber domain. It was applied very successfully to teleseismic data from the large aperture seismic array (LASA) in Montana. The technique has been extended to the frequency domain on single channel data by Lacoss (1971). Instead of using a fixed time domain window on the autocovariance function or a fixed frequency domain window on the periodogram, the window shape is changed as a function of wavenumber or frequency. The window is designed to reject all frequency component in an optimal way except for the one frequency component which is desired.

Let the input to the filter with impulse response, W, be y and let the output be x.

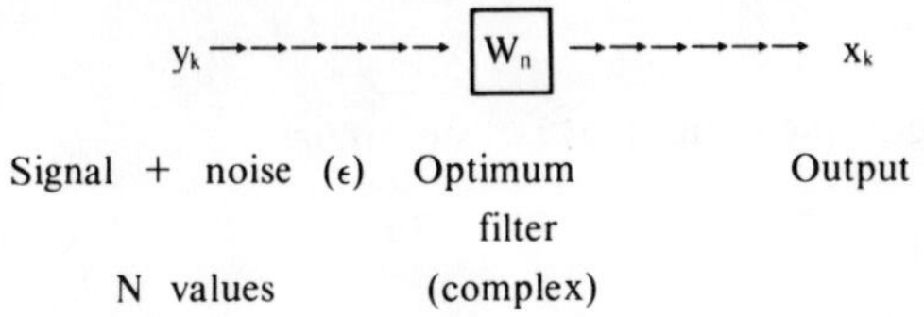

$y_k \longrightarrow \boxed{W_n} \longrightarrow x_k$

Signal + noise (ϵ)   Optimum filter (complex)   Output
N values

Figure 12.1 Symbols used in deriving a maximum likelihood operator.

The input consists of a sinusoidal component with amplitude A and a zero mean noise component produced by a random process. Assume the sampling interval is Δt and k is a time index.

$$y_k = A\, e^{+2i\pi f \Delta t k} + \epsilon_k \qquad 12.1\text{-}1$$

The output from the filter is given by a convolution equation.

$$x_k = \sum_{n=1}^{N} W_n\, y_{k+1-n} \qquad 12.1\text{-}2$$

Let us demand that the filter coefficients satisfy the condition that a pure signal without noise be passed without alteration.

The maximum likelihood method of spectral estimation

$$A\,e^{i2\pi fk\Delta t} = \sum_{n=1}^{N} W_n\, A\, e^{i2\pi f\Delta t(k+1-n)} \qquad 12.1\text{-}3$$

Dividing both sides by the input signal, $A\exp(i2\pi fk\Delta t)$ yields

$$1 = \sum_{n=1}^{N} W_n\, e^{i2\pi f\Delta t(1-n)} \qquad 12.1\text{-}4$$

In matrix notation with **W** and **E** being the column vectors

$$\mathbf{W} = \begin{bmatrix} W_1 \\ W_2 \\ \vdots \\ W_N \end{bmatrix} \qquad \mathbf{E} = \begin{bmatrix} 1 \\ e^{i2\pi f\Delta t} \\ \vdots \\ e^{2i\pi f(N-1)\Delta t} \end{bmatrix}$$

the condition in 12.1-4 may be written as

$$\boxed{1 = \mathbf{E}^{*T}\mathbf{W}} \qquad 12.1\text{-}5$$

The filter should also pass the complex conjugate of the signal

$$x_k = A\, e^{-i2\pi fk\Delta t}$$

without change. This condition may be written as

$$\boxed{1 = \mathbf{E}^{T}\mathbf{W}} \qquad 12.1\text{-}6$$

The variance of a pure noise input will need to be minimized if the signal power is to be determined with reliance. Since the mean of the noise is zero, the variance of the output is given very simply by the expectation, E, of the squared output.

$$\sigma^2 = E[x_k]^2 \qquad 12.1\text{-}7$$

For $N = 3$ we obtain

$$\sigma^2 = E\left[\sum_{n=1}^{3} W_n\, \epsilon_{k+1-n}\right]^2$$

Introduction

$$= E[(W_1^* \epsilon_k^* + W_2^* \epsilon_{k-1}^* + W_3^* \epsilon_{k-2}^*)(W_1 \epsilon_k + W_2 \epsilon_{k-1} + W_3 \epsilon_{k-2})]$$

$$= (k-1)^{-1} \sum_{k=1}^{K} [W_1^* W_1 \epsilon_k^* \epsilon_k + W_1^* W_2 \epsilon_k^* \epsilon_{k-1} + W_1^* W_3 \epsilon_k^* \epsilon_{k-2}$$

$$+ W_2^* W_1 \epsilon_{k-1}^* \epsilon_k + \ldots \ldots \quad W_3^* W_3 \epsilon_{k-2}^* \epsilon_{k-2}]$$

$$= [W_1^* W_1 a_0 + W_1^* W_2 a_1 + W_1^* W_3 a_2 + W_2^* W_1 a_1 + \ldots \quad W_3 W_3 a_0]$$

In matrix notation the equation is given compactly as

$$\boxed{\sigma^2 = \mathbf{W}^{*T} \mathbf{A} \mathbf{W}} \qquad 12.1\text{-}8$$

where **A** is the N by N Toeplitz correlation matrix formed from the autocovariance coefficients.

The problem as posed by Capon is to minimize the variance subject to the condition given in 12.1-5. The technique used is to relate the two equations with an undetermined multiplier, λ, and to minimize with respect to the unknown filter coefficients, W_n.

$$(\partial/\partial \mathbf{W})[\, \mathbf{W}^{*T} \mathbf{A} \mathbf{W} - \lambda(\mathbf{E}^{*T} \mathbf{W} - 1)\,] = 0 \qquad 12.1\text{-}9$$

The solution is obtained very easily if all the terms are real. For the case when N = 3, it is easily seen that the operation with $\partial/\partial W_1$, $\partial/\partial W_2$ and $\partial/\partial W_3$ yields

$$\begin{array}{llllll}
2W_1 a_0 & + & 2W_2 a_1 & + & 2W_3 a_2 & - \lambda = 0 \\
2W_1 a_1 & + & 2W_2 a_0 & + & 2W_3 a_1 & - \lambda e^{-2\pi i f \Delta t} = 0 \\
2W_1 a_2 & + & 2W_2 a_1 & + & 2W_3 a_0 & - \lambda e^{-4\pi i f \Delta t} = 0
\end{array}$$

or

$$\begin{bmatrix} a_0 & a_1 & a_2 \\ a_1 & a_0 & a_1 \\ a_2 & a_1 & a_0 \end{bmatrix} \cdot \begin{bmatrix} W_1 \\ W_2 \\ W_3 \end{bmatrix} = \lambda/2 \cdot \begin{bmatrix} 1 \\ e^{-2\pi i f \Delta t} \\ e^{-4\pi i f \Delta t} \end{bmatrix} \qquad 12.4\text{-}10$$

In matrix notation

$$\mathbf{A}\mathbf{W} = \lambda \mathbf{E}^*/2 \qquad 12.1\text{-}11$$

Multiplying by the inverse of **A** yields a solution

$$\mathbf{W} = \lambda \mathbf{A}^{-1} \mathbf{E}^*/2 \qquad 12.1\text{-}12$$

This equation is also valid when the filter coefficients are complex. To determine the unknown multiplier, λ, substitute **W** into the auxiliary condition on the signal (equation 12.1-6).

$$1 = E^T A^{-1} E^* \lambda / 2 \qquad 12.1\text{-}13$$

or

$$\lambda/2 = 1/[E^T A^{-1} E^*] \qquad 12.1\text{-}14$$

The filter designed to pass a single frequency component and reject noise in an optimum way is then

$$W = A^{-1} E^* / [E^T A^{-1} E^*] \qquad 12.1\text{-}15$$

The variance of the output is given by substituting **W** into equation 12.1-8.

$$\sigma^2 = (A^{-1} E^*)^{*T} A A^{-1} E^* / [E^T A^{-1} E^*]^2$$

or

$$\sigma^2 = 1/[E^T A^{-1} E^*] \qquad 12.1\text{-}16$$

This value is a power estimate and is just the maximum likelihood spectral estimator as given by Lacoss (1971).

$$P_L = 1/[E^T A^{-1} E^*] \qquad 12.1\text{-}17$$

12.2 Comparison of MEM and MLM methods

A theoretical example of two sine waves with frequencies 0.15 and 0.2 Hz respectively together with unity power white noise has been given by Lacoss (1971). The autocovariance function for the signal and noise is

$$a_n = 1 + 5.33 \cos(0.3 \pi n) + 10.66 \cos(0.4 \pi n) \quad n = 0, 1, \ldots N\text{-}1 \qquad 12.2\text{-}1$$

The illustration in figure 12.2 shows the results for N equal to only 11. The maximum entropy method of Burg resolves the two spectral components completely. However the peak values of this estimate are 6 db apart instead of the 3 db in the input signal. The peak values do not appear to be proportional to the square of the signal amplitude but the area under each peak is proportional to power. The maximum likelihood method due to Capon just detects the two peaks and gives a good estimate of the relative power. The power spectral estimate using a Bartlett window fails completely in detecting the presence of two sinusoidal components.

Comparison of MEM and MLM methods

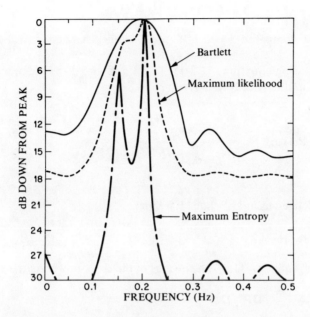

Figure 12.2 Relative spectral estimates of a signal consisting of white noise plus two sine waves with frequencies 0.15 and 0.2 Hz. (From Lacoss, 1971, courtesy of the Society of Exploration Geophysicists).

The theoretical relationship between the maximum likelihood method is given by Burg (1971) for the wave-number case. Let **E**(k) be an N high *beam-steer* column vector.

$$\mathbf{E} = \begin{bmatrix} 1 \\ e^{+2\pi k \Delta xi} \\ \vdots \\ e^{+2\pi kn \Delta xi} \\ \vdots \\ e^{2\pi k(N-1) \Delta xi} \end{bmatrix} \quad\quad 12.2\text{-}2$$

If **A** is the cross-power matrix at frequency f, the wavenumber is k and Δx is the spacing between N detectors, then in the Nyquist band

$$|k| \leq 1/2\Delta x$$

The maximum likelihood method yields a power spectral estimate given by

$$P_L(k) = N\Delta x/[\mathbf{E}^T(k)\,\mathbf{A}^{-1}(f)\,\mathbf{E}^*(k)] \qquad 12.2\text{-}3$$

This may be compared to 12.1-15 for a single channel frequency domain estimate.

Following equations 11.3-15, 11.5-1 and 11.5-3 the maximum entropy spectrum for a filter with N terms is given by

$$P_E(k,n) = \frac{P_n \Delta x}{\left|\sum_{j=1}^{n} d_{jn}\,e^{i2\pi(j-1)k\Delta x}\right|^2} \qquad 12.2\text{-}4$$

where P_n is defined as in 11.5-2 or 11.5-4 and

$$d_{1n} \equiv 1$$

From equation 11.5-7 the inverse of the cross-power matrix is given by

$$\mathbf{A}^{-1} = \mathbf{D}\mathbf{P}^{-1}\mathbf{D}^{*T} \qquad 12.2\text{-}5$$

This may be substituted into the denominator of 12.2-3.

$$\mathbf{E}^T \mathbf{A}^{-1} \mathbf{E}^* = \mathbf{E}^T \mathbf{D} \mathbf{P}^{-1} \mathbf{D}^{*T} \mathbf{E}^*$$

$$= (\mathbf{D}^T \mathbf{E})^T \mathbf{P}^{-1} (\mathbf{D}^* \mathbf{E}^*) \qquad 12.2\text{-}6$$

$$= \sum_{n=1}^{N} P_n^{-1} \left|\sum_{j=1}^{n} d_{jn}\,e^{-i2\pi(j-1)k\Delta x}\right|^2$$

$$= \sum_{n=1}^{N} \Delta x / P_E(k,n) \qquad 12.2\text{-}7$$

Therefore the relation between the maximum likelihood method and the maximum entropy method is

$$1/P_L(k) = N^{-1} \sum_{n=1}^{N} 1/P_E(k,n) \qquad 12.2\text{-}8$$

It is seen that the maximum likelihood estimate is equal to the mean of the reciprocals of the maximum entropy estimates found from the one point to N point prediction error filter. This explains why the maximum entropy method gives a higher resolution since the MLM estimate is equivalent to a parallel resistor system of averaging from the lowest to the highest possible resolutions obtainable by the MEM. The maximum likelihood spectral estimate is not consistent with the covariance matrix. Because of the relationship in equation 12.2-8 it is usual to compute the maximum likelihood estimate from maximum entropy spectral estimate computer programs. The MLM estimate has not found much favor since the investigators have generally sought spectra with sharply defined peaks even though these may not necessarily be generated by the physical process. If pure sinusoids are known to be present in the physical model and the background is white noise it would seem that the maximum likelihood method is the preferred spectral estimator.

References

Burg,J.P.(1971), The relationship between maximum entropy spectra and maximum likelihood spectra. Preprint - Texas Instruments, Dallas.

Capon,J.(1969), High-resolution frequency-wave-number spectrum analysis. *Proceeding of the IEEE* **57**, 1408-1418.

Lacoss,R.T.(1971), Data adaptive spectral analysis methods. *Geophysics* **36**, 661-675.

Pendrel,J.V. and Smylie,D.E.(1979), The relationship between maximum entropy and maximum likelihood spectra. *Geophysics* **44**, 1738-1739.

13 Minimum delay and properties of filters

13.1 Introduction

For a discrete time sequence we have seen that the convolution equation is

$$y_L = \sum_{t=0}^{m} W_t \, x_{L-t} \qquad \text{2.1-8}$$

where $\mathbf{W} = (W_0, W_1, \ldots W_m)$ is the row vector or wavelet representing the impulse response function. The input signal is $\mathbf{x}$ and the output is $\mathbf{y}$. The autocovariance of the impulse response function could be defined as follows.

$$a_j = \sum_{i=0}^{\infty} W_{i+j} \, W_i^* \qquad \text{13.1-1}$$

If the row vector W_i has $m + 1$ terms running from 0 to m, then j goes from -m to +m.

The *reverse of the wavelet* is defined as the reverse complex conjugate. The reverse of $\mathbf{W}$ is

$$\mathbf{W_R} = (W_m^*, W_{m-1}^* \ldots W_0^*) \qquad \text{13.1-2}$$

The autocovariance may be viewed as the convolution of a wavelet with its reverse (Robinson, 1964).

$$\mathbf{a} = \mathbf{W} * \mathbf{W_R} \qquad \text{13.1-3}$$

If we take the Fourier transform of the impulse response function we obtain the *transfer function* $Y(\omega)$ which describes the filter or circuit in the frequency domain.

$$Y(\omega) = \sum_{t=0}^{\infty} W_t \, e^{-i\omega t} \qquad \text{13.1-4}$$

Introduction

where

$$W_t = \frac{1}{2\pi} \int_{-\pi}^{\pi} Y(\omega) \, e^{i\omega t} \, d\omega \qquad 13.1\text{-}5$$

The frequency interval $-\pi$ to π encompasses all frequencies which the digitizing period allows. That is, the maximum frequency is $\pm 1/2$ cycles per unit time while the maximum angular frequency, ω, is $\pm\pi$ radians/unit time. The power for the transfer function is

$$P(\omega) = Y(\omega) \, Y(-\omega) = Y(\omega) \, Y^*(\omega) \qquad 13.1\text{-}6$$

In polar co-ordinates the transfer function is

$$Y(\omega) = |Y(\omega)| \, e^{-i\Theta(\omega)} = |Y(\omega)| \, e^{i\phi(\omega)} \qquad 13.1\text{-}7$$

where

$$|Y(\omega)| = \text{GAIN}$$
$$\Theta(\omega) = \text{PHASE LAG} = -\phi(\omega) \qquad 3.1\text{-}10$$

13.2 Z transform

The *z transformation* is obtained by letting

$$z = e^{-i\omega} \qquad 13.2\text{-}1$$

As was seen in 2.5-4 this represents a unit of delay time of the type $\exp(-i\omega(\Delta t))$ where Δt is the digitizing interval and is set equal to one unit of time. Each point on the unit circle $|z| = 1$ represents an infinitely long sinusoidal oscillation with frequency $f = \omega/(2\pi)$. The unit circle is shown in figure 13.1 with the real and imaginary parts obtained from the following relation:

$$|z| = 1 \quad \text{where} \quad z = \cos\omega - i\sin\omega \quad \Delta t = 1$$

The z transform was known as far back as Laplace (1779, 1812) when he introduced it with the coefficients for a Taylor series which he called the *generating function* and which we now call the impulse response, W_t.

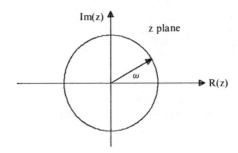

Figure 13.1 The unit circle in the z plane. Pure harmonics with angular frequency ω, that is those without attenuation, are represented by points on the unit circle.

$$Y(z) = \sum_{t=0}^{\infty} W_t \, z^t \qquad 13.2\text{-}2$$

The z transform was reintroduced by Hurewicz (1947) using z^{-1} instead of z as defined in 13.2-1. This sign convention is common in engineering texts and such authors as Jury (1964), Ford and Hearne (1966), Bath (1974), Kulhanek (1976) and Claerbout (1976), etc. The classical definition is used in publications by Robinson (1963), Robinson and Treitel (1965), Van Nostrand (1971) and in this book. This ambiguity in sign should not cause confusion as long as a consistent definition is used throughout.

Instead of the Fourier transform of equation 13.1-4 it is possible to take the Laplace transform (Sherwood and Trorey, 1965). The conversion to a z transform is made, from 4.4-3 by

$$z = e^{-s} \qquad 13.2\text{-}3$$

The Laplace transform (or complex Fourier transform) may be plotted on the s plane (s = σ + iω). The process of sampling the original analog signal introduces a periodicity into the transform (figure 13.2).

The complex function in region A is folded into regions B and C and is duplicated in regions D and E. In taking the z transform all the complex frequency data is folded on top of each other and the data on the right of the ordinate is mapped inside the unit circle while the data on the left is mapped outside the unit circle in the z plane.

The z transform of the transfer function was shown in 13.2-2. This is part of a Laurent series and by the use of the Cauchy-Hadamard theorem it is possible to show that the condition for stability of the filter is

(1) $Y(z)$ is analytic for $|z| < 1$.

Z transform

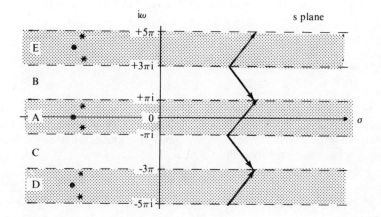

Figure 13.2 Periodicity of the transform of digitized or sampled data. Singularities or paths along the arrows in the principal domain, A($-\pi$ to $+\pi$), are indistinguishable from those in aliased domains.

(2) All singularities are outside the unit circle. If a singularity is on the unit circle, the operator may be stable or unstable. The condition for stability has been given by Paley and Wiener for analog filters (Wiener, 1942, 1949) while Kolmogorov (1941) and Hurewicz (1947) have determined it for a digital filter. Some of these aspects on stable operators were covered in Chapter 4.

The Kolmogorov condition (also known as the Paley-Wiener criterion, Paley and Wiener, 1934) states that the necessary and sufficient condition that a realizable linear operator, W_t, is stable is if its power function $P(\omega)$ is real,

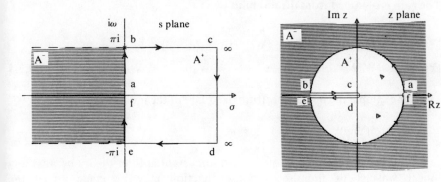

Figure 13.3 Mapping of the s plane into the z plane.

non-negative, integrable, and non-zero almost everywhere in the interval $-\pi$ to π and if

$$\int_{-\pi}^{\pi} \log P(\omega) \, d\omega > -\infty \qquad 13.2\text{-}4$$

This condition implies that the cumulative spectral distribution function (equation 7.2-12) does not have a plateau. If this occurred, then the power spectrum, $P(\omega)$, would be zero over a band of frequencies since the log of $P(\omega)$ can only be minus infinity if $P(\omega)$ is zero.

13.3 Dipole presentation of filtering

Dipole. A wavelet or row vector with only two terms is called a dipole (Robinson, 1964). An example is the wavelet $\mathbf{W} = (2,-1) = (W_0, W_1)$.

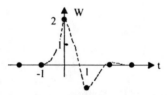

Figure 13.4 An example of a wavelet which is a dipole when digitized.

The transfer function of the wavelet above is given by 13.1-4.

$$Y(\omega) = 2 - e^{-i\omega}$$

The gain is obtained from the modulus

$$|Y(\omega)| = [(W_0 + W_1 e^{-i\omega})(W_0^* + W_1^* e^{i\omega})]^{1/2}$$
$$|Y(\omega)| = (W_0 W_0^* + W_0 W_1^* e^{i\omega} + W_1 W_0^* e^{-i\omega} + W_1 W_1^*)^{1/2} \qquad 13.3\text{-}1$$

For our example the power spectrum or square of the modulus is

$$|Y(\omega)|^2 = 5 - 4 \cos \omega$$

The concept of a dipole is most important in filter theory because any longer wavelet or impulse response function may be made up of the

Dipole presentation of filtering

convolution of dipoles by virtue of the convolution theorem. Let (α_0, α_1), (β_0, β_1), ... (ω_0, ω_1) be n dipoles. Then if the signal is transmitted through n filters, this is equivalent to the operation of a single filter with transfer function, **W**.

$$\mathbf{W} = (\alpha_0, \alpha_1) * (\beta_0, \beta_1) * (\gamma_0, \gamma_1) * \ldots * (\omega_0, \omega_1) \qquad 13.3\text{-}2$$

Taking the Fourier transforms and the z transform we can evaluate this convolution algebraically.

$$W(z) = (\alpha_0 + \alpha_1 z)(\beta_0 + \beta_1 z)(\gamma_0 + \gamma_1 z) \ldots (\omega_0 + \omega_1 z) \qquad 13.3\text{-}3$$

or

$$W(z) = W_0 + W_1 z + W_2 z^2 + \ldots W_n z^n \qquad 13.3\text{-}4$$

A black box representation of this n stage filter operation is shown in figure 13.5.

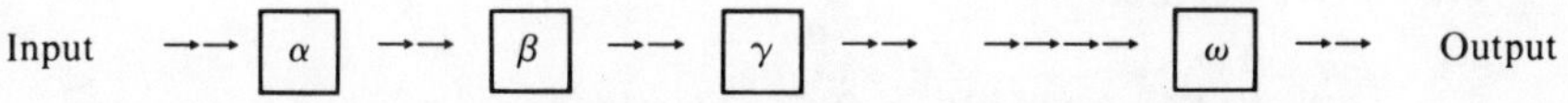

Input $\to$ [α] $\to$ [β] $\to$ [γ] $\to \cdots \to$ [ω] $\to$ Output

Figure 13.5 An n-stage filter in which each impulse response is a dipole.

This is equivalent to a single filter operation.

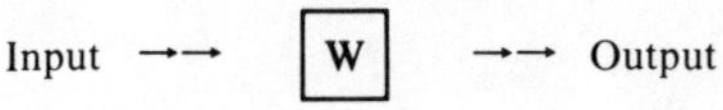

Input $\to$ [W] $\to$ Output

Figure 13.6 The equivalent single-stage filter with n + 1 terms in the impulse response.

The coefficients $W_0, W_1, \ldots W_n$ are real in all cases of physical interest. Therefore if there are any imaginary or complex zeros in W(z), these must occur in complex conjugate pairs. For example, the impulse response (5, -2, 1) has the following z transform:

$$W_A = 5 - 2z + z^2 = (-1 + 2i + z)(-1 - 2i + z) \qquad 13.3\text{-}5$$

Note that the three length wavelet is composed of a pair of complex dipoles, which are conjugates of one another. As another example, the impulse response

(864 , -144 , 186 , -55 , -79 , 4 , 4)

has the z transform

$$W_B = 864 - 144z + 186z^2 - 55z^3 - 79z^4 + 4z^5 + 4z^6$$
$$= (4+z)(4-z)(-3+2iz)(-3-2zi)(2-z)(3+z)$$
13.3-6

A third impulse response, formed from the product of W_A and W_B would be

$$(4320, -2448, 2082, -791, -99, 123, -67, -4, 4)$$

Its z transform would be

$$W_C = 4320 - 2448z + 2082z^2 - 791z^3 - 99z^4 + 123z^5 - 67z^6 - 4z^7 + 4z^8$$
$$= (4+z)(4-z)(-3+2iz)(-3-2iz)(2-z)(3+z)(-1+2i+z)(-1-2i+z)$$
13.3-7

The zeros of W_C occur at

$$z = \pm 4, \pm(3/2)i, 2, -3, 1\pm 2i$$

A fourth example is

$$W_D = (2+z)(4+z)(3-z)(1+2i-z)(1-2i-z)$$

or

$$W_D = 120 + 2z - 11z^2 + 11z^3 - 5z^4 - z^5$$
13.3-8

The wavelets and the singularities are illustrated in figure 13.7.

13.4 Normalized wavelet

Any two-length wavelet may be written so its first coefficient is equal to one. Thus

$$W = (W_0, W_1)$$
13.4-1

may be written

$$W = (1, W_1/W_0) = (1, W_n)$$
13.4-2

Normalized wavelet

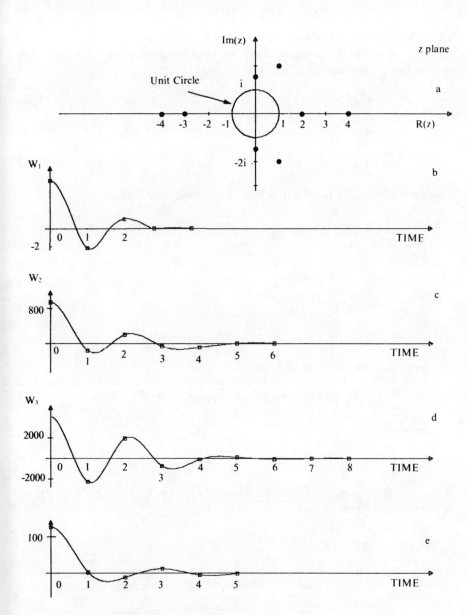

Figure 13.7 (a) Zeros of Impulse response W_C.
(b) Impulse response of W_A (equation 11.3-5).
(c) Impulse response of W_B (equation 13.3-6).
(d) Impulse response of W_C (equation 13.3-7).
(e) Impulse response of W_D (equation 13.3-8).

13.5 Minimum delay wavelet

A dipole in which the first coefficient has the largest magnitude is a minimum delay wavelet. See chapter 21 for more on minimum delay.

$$|W_0| > |W_1| \qquad 13.5\text{-}1$$

The z transform of a dipole was seen to be

$$W(z) = W_0 + W_1 z \qquad 13.5\text{-}2$$

The transfer function for *real* dipoles was obtained from 13.1-7.

$$Y(\omega) = |Y(\omega)|\, e^{i\phi(\omega)} \qquad 13.1\text{-}7$$

$$= W_0 + W_1\, e^{-i\omega}$$
$$= (W_0 + W_1 \cos \omega) - i\, W_1 \sin \omega \qquad 13.5\text{-}3$$

The gain of the dipole from 13.3-1 is

$$|Y(\omega)| = (W_0^2 + 2 W_0 W_1 \cos \omega + W_1^2)^{1/2} \qquad 13.5\text{-}4$$

and the phase spectrum is

$$\phi(\omega) = -\tan^{-1}(W_1 \sin \omega / [W_0 + W_1 \cos \omega]) \qquad 13.5\text{-}5$$

The normalized minimum delay wavelet from section 13.3 is $W = (1, -1/2)$ and it has gain and phase spectrum

$$|Y(\omega)| = (5/4 - \cos \omega)^{1/2} \qquad 13.5\text{-}6$$

$$\phi(\omega) = +\tan^{-1}(\sin \omega /[2 - \cos \omega]) \qquad 13.5\text{-}7$$

These functions are plotted in figure 13.8.

The dipole of a complex wavelet with real and imaginary parts may be written as

$$W = (W_{0R} + W_{0I} i,\ W_{1R} + i W_{1I}) \qquad 13.5\text{-}8$$

Minimum delay wavelet

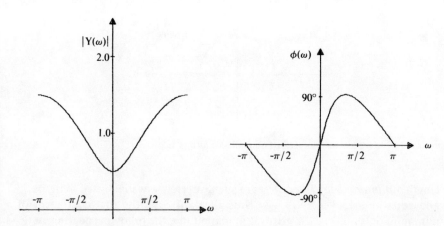

Figure 13.8 Amplitude spectrum or gain, $|Y(\omega)|$, and phase spectrum, $\phi(\omega)$, of a minimum delay dipole $(1, -1/2)$.

The gain of the dipole for this more general case is

$$|Y(\omega)| = [W_{0R}^2 + W_{0I}^2 + W_{1R}^2 + W_{1I}^2 + 2\cos\omega(W_{0R}W_{1R} + W_{0I}W_{1I}) + 2\sin\omega(W_{0R}W_{1I} - W_{0I}W_{1R})]^{1/2} \qquad 13.5\text{-}9$$

The phase spectrum is

$$\phi(\omega) = \tan^{-1}[(W_{0I} + W_{1I}\cos\omega - W_{1R}\sin\omega) / (W_{0R} + W_{1R}\cos\omega + W_{1I}\sin\omega)] \quad 13.5\text{-}10$$

For the normalized wavelet $(1, W_n)$ where $|W_n| < 1$, the z transform is

$$W(z) = 1 + W_n z \qquad 13.5\text{-}11$$

The zero of the z transform is found by setting the equation equal to zero.

$$W(z) = 0 \qquad 13.5\text{-}12$$

The zero occurs at

$$z = -1 / W_n \qquad 13.5\text{-}13$$

Since $|W_n| < 1$ the zero is outside the unit circle in the z plane. The case of a real zero is shown in figure 13.9.

If all the n dipoles making up a longer wavelet in equation 13.3-3 are minimum delay, then the convolution of n such dipoles produces an n + 1

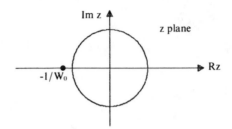

Figure 13.9 Zero of a real normalized minimum delay dipole.

length *minimum delay filter*. A set of such wavelets was illustrated in figure 13.7 and equations 13.3-5 to 13.3-8. Note that although the first coefficient of a minimum delay wavelet is greatest in magnitude, the following coefficient need not decrease uniformly. As a simple example the wavelet (16, 0, -1) is minimum delay because its z transform is $(4+z)(4-z)$.

13.6 Maximum delay wavelet

A dipole in which the last coefficient is largest in magnitude is a maximum delay wavelet. A convolution of n maximum delay wavelets produces a maximum delay filter. The reverse of a minimum delay wavelet $(1, W_n)$ is a maximum delay dipole $(W_n^*, 1)$. The z transform is

$$W(z) = W_n^* + z \qquad 13.6\text{-}1$$

The zero is at

$$z = -W_n^* \qquad |W_n^*| < 1 \qquad 13.6\text{-}2$$

and lies inside the unit circle as in figure 13.10.

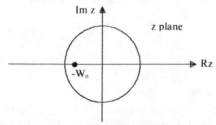

Figure 13.10 Zero for a normalized real maximum delay dipole.

Maximum delay wavelet

The maximum delay wavelet which is related to the minimum delay dipole (1, -1/2) is (-1/2, 1). The transfer function is

$$Y(\omega) = -1/2 + e^{-i\omega} \qquad 13.6\text{-}3$$

The gain is the same as that given by equation 13.5-6

$$|Y(\omega)| = (5/4 - \cos \omega)^{1/2} \qquad 13.6\text{-}4$$

but the phase spectrum is different

$$\phi(\omega) = -\tan^{-1}(\sin \omega / [\cos \omega - 0.5]) \qquad 13.6\text{-}5$$

These functions are plotted in figure 13.11. Notice that the phase spectrum is everywhere larger than that obtained for the related minimum delay wavelet.

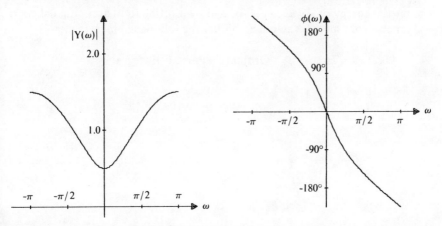

Figure 13.11 Amplitude spectrum gain, $|Y(\omega)|$, and phase spectrum, $\phi(\omega)$, of a maximum delay dipole (-1/2, 1).

13.7 Mixed delay filter

A convolution of n maximum and minimum delay dipoles produces an n + 1 length filter called a mixed delay filter. If we choose either the dipole or its reverse to obtain *n* two length wavelets then it is possible to produce 2^n filters with n + 1 terms. Each filter is part of a set and will have the same gain and same autocovariance function. The phase spectrum of each will be different.

Example: The following 3 minimum delay wavelets may be used to produce a set of 2^3 or 8 filters.

(10,4) , (5,3) , (2,1)

(1) (10,4)*(5,-3)*(2,1) = (100,30,-34,-12) - minimum delay wavelet.
(2) (4,10)*(-3,5)*(1,2) = (-12,-34,30,100) - maximum delay wavelet.
(3) (10,4)*(5,-3)*(1,2) = (50,90,-32,24) ⎤
(4) (4,10)*(-3,5)*(2,1) = (24,-32,90,50) ⎥
(5) (10,4)*(-3,5)*(2,1) = (-60,48,78,20) ⎥
(6) (4,10)*(5,-3)*(1,2) = (20,78,48,-60) ⎥ mixed delay wavelets
(7) (10,4)*(-3,5)*(1,2) = (-30,-22,96,40) ⎥
(8) (4,10)*(5,-3)*(2,1) = (40,96,-22,-30) ⎦

13.8 The partial energy

The partial energy of a wavelet is formed by plotting total accumulated energy as a function of time. For an n length filter the following plot is made:

Abscissa (Time)	Ordinate (Partial Energy)
0	W_0^2
1	$W_0^2 + W_1^2$
.	.
.	.
.	.
m	$\sum_{i=0}^{m} W_i^2$

Example: Obtain the partial energy for a convolution of 2 dipoles or their reverses as listed in the table below. The results are shown in figure 13.12.

Convolution	z Transform		Wavelet	Type
(2,1)*(3,-1)	(2+z) (3-z)	= $6+z-z^2$	(6,1,-1)	minimum delay
(1,2)*(-1,3)	(1+2z) (-1+3z)	= $-1+z+6z^2$	(-1,1,6)	maximum delay
(2,1)*(-1,3)	(2+z) (-1+3z)	= $-2+5z+3z^2$	(-2,5,3)	mixed delay
(1,2)*(3,-1)	(1+2z) (3-z)	= $3+5z-2z^2$	(3,5,-2)	mixed delay

Maximum, minimum and mixed delay wavelets may be distinguished by this partial energy diagram. Reference should be made to the chapter on Hilbert transformation for further properties of physically realizable filter operators.

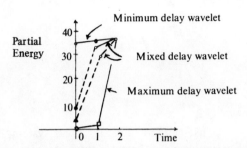

Figure 13.12 Partial energy of minimum, maximum and mixed delay wavelets.

References

Bath,M.(1974), Spectral analysis in Geophysics. Elsevier Scientific Publishing Co., New York, 176.

Claerbout,J.F.(1976), Fundamentals of geophysical data processing. McGraw-Hill Inc., New York, 1-274.

Ford,W.T. and Hearne,J.W.(1966), Least-squares inverse filtering. *Geophysics* **31**, 917-926.

Hurewicz,W.(1947), Filters and servo systems, Ch. 5 in *Theory of Servomechanisms*, E. H.M. James, N.B. Nichols and R.S. Phillips. McGraw-Hill, New York 231-261.

Jury,E.I.(1964), Theory and application of the z transform method. John Wiley & Sons, New York.

Kulhanek,O.(1976), Introduction to digital filtering in geophysics. Elsevier Scientific Publishing Co., New York, 1-168.

Kuo,B.C.(1964), Automatic control systems. Prentice-Hall, Inc., Englewood Cliffs, New Jersey.

Laplace,P.S.(1779), Ouevres Completes. See also his book Analytic Theory of Probability - The Calculus of Generating Functions. Published 1882 by Gautier-Villars,Paris.

Kolmogorov,A.N.(1941), Stationary sequences in Hilbert space. *Bull. Moscow State Univ. Math.* **2**, 40.

Robinson,E.A.(1963), Structural properties of stationary stochastic processes with applications. Time series analysis. Ed. M. Rosenblat, John Wiley & Sons, New York.

Robinson,E.A. and Treitel,S.(1964), Principles of digital filtering. *Geophysics* **29**, 395-404.

Robinson,E.A.(1964), Wavelet composition of time series. Ch. 2. Recursive decomposition of stochastic processes. Ch. 3. Econometric model building. Ed. H.O. Wold, North Holland Publishing Co., Amsterdam.

Robinson,E.A. and Treitel,S.(1965), Dispersive digital filters. *Reviews of Geophysics* **3**, 433-461.

Robinson,E.A. and Silvia,M.T.(1978), Digital signal processing and time series analysis. Holden-Day Inc., San Francisco, 1-405.

Sherwood,J.W.C. and Trorey,A.W.(1965), Minimum phase and related properties of a horizontally stratified absorptive earth to plane acoustic waves. *Geophysics* **30**, 191-197.

Van Nostrand,R.(1966), Seismic Filtering. Translated by N Rothenburg from Le Filtrage en Sismique . An edition by the *French Institute of Petroleum. Edition by Society of Exploration Geophysicists,* Tulsa, Okla., 1971.

Wiener,N.(1949), Extrapolation, interpolation and smoothing of stationary time series. John Wiley & Sons, New York. See also - Time series; The M.I.T. Press,Cambridge,Mass., 1966.

14 Deconvolution

14.1 Inverse filtering

If **W** is the impulse response function for any linear system, then the output may be obtained by a convolution of the input signal with the impulse response.

$$y = W * x$$

$$x \longrightarrow \boxed{W} \longrightarrow y \qquad 2.1\text{-}7$$

Figure 14.1   Black box representation of convolution or filtering.

If we are given **W** and **y**, it is possible to solve the inverse problem and find **x**. This process is inverse filtering or deconvolution. The system necessary to perform this operation is called the inverse filter and is denoted W^{-1}.

$$y \longrightarrow \boxed{W^{-1}} \longrightarrow x$$

Figure 14.2   Black box representation of deconvolution or inverse filtering.

An example is the recovery of ground displacement or velocity after the data has been recorded on a seismograph with impulse response **W**. The deconvolution equation is

$$x = W^{-1} * y \qquad 14.1\text{-}1$$

Substituting 2.1-7 into 14.1-1, one obtains the original input signal.

$$x = W^{-1} * W * x \qquad 14.1\text{-}2$$

It is possible to obtain the signal **x** by a convolution of itself with the unit impulse wavelet (see 2.1, Appendix 2).

$$x = \delta * x \qquad 14.1\text{-}3$$

Inverse filtering

Comparing 14.1-2 and 14.1-3 it is seen that

$$\delta = \mathbf{W}^{-1} * \mathbf{W} = \sum_{t=0}^{\infty} w_t w^{-1}_{L-t} \qquad 14.1\text{-}4$$

To solve equation 14.1-4 for the inverse filter, a z transformation is made. The impulse response, **W**, is assumed to be normalized. The z transform of δ is one.

$$1 = (w_0^{-1} + w_1^{-1}z + \ldots)(1 + w_1z + w_2z^2 + \ldots w_nz^n) \qquad 14.1\text{-}5$$

or

$$(w_0^{-1} + w_1^{-1}z + \ldots) = 1 / (1 + w_1z + w_2z^2 + \ldots w_nz^n) \qquad 14.1\text{-}6$$

For an impulse response consisting of a normalized dipole, $(1, w_n)$, the inverse filter is

$$\mathbf{W}^{-1} = 1/(1 + w_nz) = 1 - w_nz + w_n^2z^2 - w_n^3z^3 + \ldots \qquad 14.1\text{-}7$$

or

$$\mathbf{W}^{-1} = (1, -w_n, w_n^2, -w_n^3, \ldots) \ .$$

It is necessary to have an infinitely long wavelet to carry out this process exactly. If **W** is a normalized minimum delay dipole ($|w_n|<1$), then $\mathbf{W}^{-1}$ is stable. If **W** is a maximum delay dipole, then $\mathbf{W}^{-1}$ is unstable since the series in 14.1-6 does not converge. If a stable inverse or deconvolution filter is found, the original data in figure 14.2 is given by

$$x_L = \sum_{t=0}^{\infty} w^{-1}_t y_{L-t} \qquad 14.1\text{-}8$$

Note that w_t^{-1} is a deconvolution coefficient and the -1 does *not* denote $1/w_t$.

14.2 Truncated deconvolution
If the inverse or deconvolution filter is stable the series in 14.1-8 converges and it is possible to obtain a reasonable approximation by truncating at an arbitrary point.

$$x_L = \sum_{t=0}^{m} w_t^{-1} y_{L-t} \qquad 14.2\text{-}1$$

This is not the best inverse that can be constructed because other least square approximate inverses of the same length give smaller errors between the true input, **x**, to the filter and the approximate input, x_a, as obtained from a deconvolution operation.

The error squared is obtained from a difference equation.

$$(\text{Error})^2 = [\text{Desired Output - Real Output}]^2$$
$$E = [\delta - \mathbf{W} * \mathbf{W}^{-1}]^2 \qquad 14.2\text{-}2$$

E is called the squared norm. This equation follows from 14.1-4 or 14.2-1. It is assumed that a delta Dirac function was fed into the filter, **W**, originally. If the inverse filter functions properly its output will be a delta Dirac function; however, since the filter is truncated, the output will be somewhat different. The exact inverse for an impulse response consisting of a dipole, $\mathbf{W} = (1, w_n)$, was found to be $(1, -w_n, w_n^2, -w_n^3, ...)$ by 14.1-7.

The squared error if **W** is a dipole is given below for various length inverse filters.

Table 14.1 Squared norm for truncated inverse filters.

Length	Truncated Inverse Filter	E
1	(1)	w_n^2
2	$(1, -w_n)$	w_n^4
3	$(1, -w_n, w_n^2)$	w_n^6
4	$(1, -w_n, w_n^2, -w_n^3)$	w_n^8

14.3 Least squares deconvolution dipole

Deconvolution filters can be constructed for which the squared norm is less than for the equivalent length truncated deconvolution filter. These are generated by approximating the infinite length deconvolution operator in a least squares or optimum sense. To illustrate the techniques for computing such an operator a two-length inverse filter will be derived first.

The impulse response will be assumed to be a normalized dipole.

$$\mathbf{W} = (1, w_n) \qquad |w_n| < 1 \qquad 14.3\text{-}1$$

The autocovariance function of the impulse response is obtained from the coefficients of the product of the z transforms of W and W_R, the reverse dipole

Least squares deconvolution dipole

$$(1 + w_n z)(w_n + z) = w_n + (1 + w_n^2)z + w_n z^2$$
$$\mathbf{a} = (a_{-1}, a_0, a_1) = (w_n, 1+w_n^2, w_n) \qquad 14.3\text{-}2$$

If the input to the filter is a unit impulse or Dirac delta function, the output is the impulse response.

$$1, 0, 0 \longrightarrow \boxed{W} \longrightarrow 1, w_n, 0 \ldots$$

Figure 14.3 Convolution with a unit impulse.

The desired output from the inverse filter is the Dirac delta function

$$\mathbf{x}_d = \delta = (1, 0, 0, \ldots) \qquad 14.3\text{-}3$$

A deconvolution with the approximate inverse filter yields an approximation to this signal (see figure 14.4)

$$\mathbf{x}_a = \mathbf{W}^{-1} * \mathbf{W} = (w_0^{-1}, w_0^{-1}w_n + w_1^{-1}, w_1^{-1}w_n) \qquad 14.3\text{-}4$$

whereas the desired set is $\delta = (1,0,0)$. The error is

$$\mathbf{e} = (1,0,0) - (w_0^{-1}, w_0^{-1}w_n + w_1^{-1}, w_1^{-1}w_n)$$
$$= (1 - w_0^{-1}, -w_0^{-1}w_n - W_1^{-1}, -W_1^{-1}W_n) \qquad 14.3\text{-}5$$

The squared norm is

$$E = (1 - w_0^{-1})^2 + (-w_0^{-1}w_n - w_1^{-1})^2 + (-w_1^{-1}w_n)^2$$
$$= 1 - 2w_0^{-1} + (w_0^{-1})^2 + (w_0^{-1})^2 w_n^2 + 2w_0^{-1}w_n w_1^{-1} + (w_1^{-1})^2 + (w_1^{-1}w_n)^2$$
$$= 1 - 2w_0^{-1} + (w_0^{-1})^2(1+w_n^2) + 2w_0^{-1}w_1^{-1}w_n + (w_1^{-1})^2(1+w_n^2) \qquad 14.3\text{-}6$$

Substituting the autocovariance terms for the appropriate lags one has

$$E = 1 - 2w_0^{-1} + (w_0^{-1})^2 a_0 + 2w_0^{-1}w_1^{-1}a_1 + (w_1^{-1})^2 a_0 \qquad 14.3\text{-}7$$

To minimize E the equation above is differentiated with respect to the unknown coefficient and then equated to zero.

$$\partial E / \partial w_0^{-1} = -2 + 2w_0^{-1}a_0 + 2w_1^{-1}a_1 = 0 \qquad 14.3\text{-}8$$

$$\partial E / \partial w_1^{-1} = 2w_0^{-1}a_1 + 2w_1^{-1}a_0 = 0 \qquad 14.3\text{-}9$$

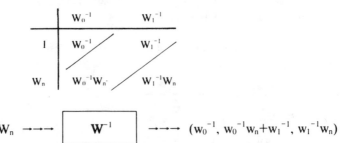

Figure 14.4 Deconvolution with a dipole. The computation can be made as a folding operation as shown in the upper diagram (see section 2.3). Note that the -1 in w_i^{-1} indicates an inverse or deconvolution coefficient and not $1/w_i$.

The following two simultaneous equations must be solved.

$$a_0 w_0^{-1} + a_1 w_1^{-1} = 1 \qquad \text{14.3-10}$$

$$a_1 w_0^{-1} + a_0 w_1^{-1} = 0 \qquad \text{14.3-11}$$

The solutions are most easily written in terms of determinants

$$w_0^{-1} = \frac{\begin{vmatrix} 1 & a_1 \\ 0 & a_0 \end{vmatrix}}{\begin{vmatrix} a_0 & a_1 \\ a_1 & a_0 \end{vmatrix}} = \frac{a_0}{a_0^2 - a_1^2} \qquad \text{14.3-12}$$

$$w_1^{-1} = \frac{\begin{vmatrix} a_0 & 1 \\ a_1 & 0 \end{vmatrix}}{\begin{vmatrix} a_0 & a_1 \\ a_1 & a_0 \end{vmatrix}} = \frac{-a_1}{a_0^2 - a_1^2} \qquad \text{14.3-13}$$

Least squares deconvolution dipole

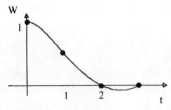

A dipole forming the impulse response.

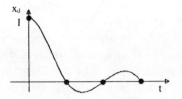

The desired output is a unit impulse at time 0. $\mathbf{x}_d = (1,0,0)$

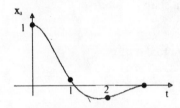

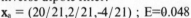

Output from least squares inverse dipole filter.
$\mathbf{x}_a = (20/21, 2/21, -4/21)$; E=0.048

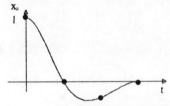

Output from a two-length truncated inverse filter.
$\mathbf{x}_a = (1, 0, -1/4)$; E=0.062

Figure 14.5 Actual output signals from inverse filter. The desired output is a unit spike at time 0.

or in terms of the coefficients of the dipole.

$$\mathbf{W}^{-1} = (1+w_n^2) / (1+w_n^2+w_n^4) \, , \, -w_n / (1+w_n^2+w_n^4) \qquad 14.3\text{-}14$$

As an example let the impulse response $\mathbf{W}$ be the dipole $(1, 1/2)$ - see figure 14.5. Then

$$\mathbf{W}^{-1} = (20/21, -8/21)$$

$$\mathbf{x}_a = (20,21, +2/21, -4/21)$$

$$E = 1/21 \cong 0.048$$

The error squared is 1/21 compared to 1/16 for the equivalent truncated inverse filter. The least squares inverse filter is better constructed so that the squared error between the desired output and actual output of the deconvolution filter is minimized. Truncating the exact inverse is not a desirable procedure.

14.4 General equations for deconvolution

The impulse response of the linear operator will no longer be restricted to a dipole but will be assumed to have $n + 1$ terms. The input to the approximate inverse filter, $\mathbf{W}^{-1} = (w_0^{-1}, w_1^{-1}, \ldots w_m^{-1})$, will be $\mathbf{y}$ and the actual output will be $\mathbf{x}_a$. The desired output will be the wavelet $\mathbf{d} = (d_0, d_1, \ldots d_{m+n})$. The desired output may be a spike at zero time ($\delta = 1, 0, 0, \ldots$) or it may have any other shape. In this case the inverse filter will act as a shaping filter.

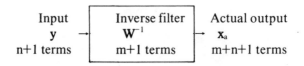

Figure 14.6 Input and output from a deconvolution filter.

$$\mathbf{x}_a = \mathbf{W}^{-1} * \mathbf{y} \qquad 14.4\text{-}1$$

The z transform of the impulse response, $\mathbf{W}$, may be factored into p factors, each representing a dipole.

$$\begin{aligned} W(z) &= w_0 + w_1 z + w_2 z^2 + \ldots w_p z^p \\ &= (\alpha_0 + \alpha_1 z)(\beta_0 + \beta_1 z)(\gamma_0 + \gamma_1 z) \ldots (\omega_0 + \omega_1 z) \end{aligned} \qquad 14.4\text{-}2$$

That is, $\mathbf{W}$ is a convolution of p dipoles.

$$\mathbf{W} = (\alpha_0, \alpha_1) * (\beta_0, \beta_1) * (\gamma_0, \gamma_1) * \ldots * (\omega_0, \omega_1) \qquad 14.4\text{-}3$$

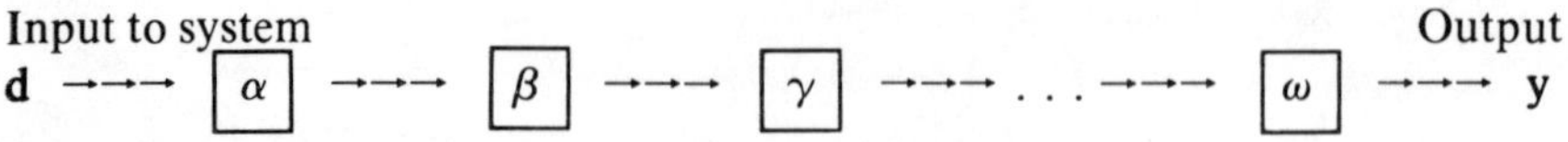

Input to system
$\mathbf{d} \longrightarrow \boxed{\alpha} \longrightarrow \boxed{\beta} \longrightarrow \boxed{\gamma} \longrightarrow \ldots \longrightarrow \boxed{\omega} \longrightarrow \mathbf{y}$
Output

Figure 14.7 Convolution by a series of p dipoles.

The exact inverse of $\mathbf{W}$ is obtained from the z transform.

General equations for deconvolution

$$\mathbf{W}^{-1}(z) = 1 / [(\alpha_0+\alpha_1 z)(\beta_0+\beta_1 z)(\gamma_0+\gamma_1 z)\ldots(\omega_0+\omega_1 z)] \quad 14.4\text{-}4$$

If each component dipole is a minimum delay wavelet, then the entire wavelet is minimum delay and the inverse is stable. If the wavelet, **W**, is maximum delay, then the stable inverse is an antiwavelet; however one needs future values of the signal to use this form.

$$\mathbf{W}^{-1}(z) = \frac{w_{-1}^{-1}}{z} + \frac{w_{-2}^{-1}}{z^2} + \frac{w_{-3}^{-1}}{z^3} + \ldots \quad 14.4\text{-}5$$

If the dipoles making up **W** are a mixture of minimum and maximum delay wavelets, the inverse is a mixed delay wavelet involving the entire Laurent series. Both past and future values of the signal must be known to use a mixed delay deconvolution filter.

$$\mathbf{W}^{-1}(z) = \ldots + \frac{w_{-2}^{-1}}{z^2} + \frac{w_{-1}^{-1}}{z} + w_0^{-1} + w_1^{-1}z + w_2^{-1}z^2 + \ldots \quad 14.4\text{-}6$$

For this section it is supposed that **W** is a minimum delay wavelet.

The squared norm is the square of the difference between the desired output, **d**, and the actual output $\mathbf{x}_a$.

$$E = \sum_{j=0}^{m+n} (d_j - x_{aj})^2 \quad 14.4\text{-}7$$

The convolution equation $\mathbf{x}_a = \mathbf{W}^{-1} * \mathbf{y}$ is

$$x_{aj} = \sum_{i=0}^{m} w_i^{-1} y_{j-i} \quad 14.4\text{-}8$$

so

$$E = \sum_{j=0}^{m+n} (d_j - \sum_{i=0}^{m} w_i^{-1} y_{j-i})^2 \quad 14.4\text{-}9$$

To minimize the squared norm, E, we set the partial derivatives of E with respect to the unknown filter coefficients, w_i^{-1}, equal to zero.

$$\frac{\partial E}{\partial w_k^{-1}} = \sum_{j=0}^{m+n} 2(d_j - \sum_{i=0}^{m} w_i^{-1} y_{j-i})(-y_{j-k}) = 0 \qquad 14.4\text{-}10$$

or

$$\sum_{j=0}^{m+n} (\sum_{i=0}^{m} w_i^{-1} y_{j-i}) y_{j-k} = \sum_{j=0}^{m+n} d_j y_{j-k} \qquad 14.4\text{-}11$$

The order of summation may be changed.

$$\sum_{i=0}^{m} w_i^{-1} \sum_{j=0}^{m+n} y_{j-i} y_{j-k} = \sum_{j=0}^{m+n} d_j y_{j-k} \qquad 14.4\text{-}12$$

The autocovariance of the output, y, is

$$a_k = \sum_{J=0}^{m+n} y_{J+k} y_J \qquad 14.4\text{-}13$$

$$a_{k-i} = \sum_{J=0}^{m+n} y_{J+k-i} y_J \qquad 14.4\text{-}14$$

Let $j = J+k$ and let k and i take on all values from 0 to m.

$$a_{k-i} = \sum_{j=0}^{m+n} y_{j-i} y_{j-k} \qquad 14.4\text{-}15$$

Also define the cross covariance of the wavelet **d** with the wavelet **y** (desired output and input) as

General equations for deconvolution

$$c_k = \sum_{j=0}^{m+n} d_j \, y_{j-k} \qquad k = 0, \ldots m \qquad 14.4\text{-}16$$

Then the equation defining the inverse filter is

$$\boxed{\sum_{i=0}^{m} w_i^{-1} \, a_{k-i} = c_k} \qquad k = 0, 1, \ldots m \qquad 14.4\text{-}17$$

From these $m + 1$ equations the approximate least squares filter or a shaping filter may be obtained in terms of the autocovariance of the input data, y, and the cross covariance between the input data and the desired output, d.

If the desired output is a Dirac delta function for zero time then

$$\begin{aligned} y &= W \text{ the impulse response function} \\ d &= \delta = (1, 0, 0, \ldots) \end{aligned} \qquad 14.4\text{-}18$$

In this case the inverse filter is defined by equation 14.4-16 where $j = 0$ is the only non-zero value

$$\sum_{i=0}^{m} w_i^{-1} \, a_{k-i} = W_{-k} \qquad 14.4\text{-}19$$

or since W is zero for negative time, and also $a_i = a_{-i}$, the solution is defined by the following set of linear equations:

$$\begin{aligned} w_0^{-1} a_0 + w_1^{-1} a_1 + w_2^{-1} a_2 + \ldots w_m^{-1} a_m &= w_0^* \\ w_0^{-1} a_1 + w_1^{-1} a_0 + w_2^{-1} a_1 + \ldots w_m^{-1} a_{m-1} &= 0 \\ \vdots \qquad \vdots \qquad \vdots & \\ w_0^{-1} a_m + w_1^{-1} a_{m-1} + w_2^{-1} a_{m-2} + \ldots w_m^{-1} a_0 &= 0 \end{aligned} \qquad 14.4\text{-}20$$

The star indicates a complex conjugate and reminds us that the w_0 on the right hand side arises from a negative indexed value of W in equation 14.4-19. For real filters w_0 and w_0^* are identical. In matrix notation, equation 14.4-19 may be written

$$\boxed{AW^{-1} = W_0} \qquad 14.4\text{-}21$$

where W^{-1} and W_0 are column vectors and A is the coefficient matrix in 14.4-22.

Notice the high degree of symmetry in the matrix. All elements in any super-diagonal or sub-diagonal are equal. Since there are only m+1 autocovariance functions and they are shifted one term from one equation to the next, this may be used to save computation time. This matrix is known as the Toeplitz type and calculations involving it are given in section 14.6.

$$\begin{bmatrix} a_0 & a_1 & a_2 & \ldots, & a_m \\ a_1 & a_0 & a_1 & \ldots & a_{m-1} \\ a_2 & a_1 & a_0 & \ldots & a_{m-2} \\ \cdot & \cdot & \cdot & & \cdot \\ \cdot & \cdot & \cdot & & \cdot \\ \cdot & \cdot & \cdot & & \cdot \\ a_{m-1} & a_{m-2} & a_{m-3} & \ldots & a_1 \\ a_m & a_{m-1} & a_{m-2} & \ldots & a_0 \end{bmatrix} \begin{bmatrix} w_0^{-1} \\ w_1^{-1} \\ w_2^{-1} \\ \cdot \\ \cdot \\ \cdot \\ w_{m-1}^{-1} \\ w_m^{-1} \end{bmatrix} = \begin{bmatrix} w_0^* \\ 0 \\ 0 \\ \cdot \\ \cdot \\ \cdot \\ 0 \\ 0 \end{bmatrix} \qquad 14.4\text{-}22$$

14.5 Z transform of inverse filter

If one has a sampled portion, W, of a minimum phase function then its z transform may be written as follows:

$$W(z) = w_0 + w_1 z + w_2 z^2 + \ldots w_m z^m \qquad 14.5\text{-}1$$

The complex conjugate is the function $W^*(1/z)$.

$$W^*(1/z) = w_0^* + w_1^*/z + w_2^*/z^2 + \ldots w_m^*/z^m \qquad 14.5\text{-}2$$

The product of these two polynomials is the z transform of the autocovariance function

Z transform of inverse filter

$$A(z) = W(z) W^*(1/z)$$
$$= a_{-n}z^{-n} + \ldots a_{-1}z^{-1} + a_0 + a_1z^1 + \ldots a_m z^m \qquad 14.5\text{-}3$$

For instance, the coefficients of z^0 give the autocovariance for zero lag.

$$a_0 = w_0 w_0^* + w_1 w_1^* + \ldots + w_m w_m^* \qquad 14.5\text{-}4$$

Similarly the coefficients of z^{-1} and z^1 give autocovariances for unit lag.

$$a_{-1} = w_0 w_1^* + w_1 w_2^* + \ldots w_{m-1} w_m^*$$
$$a_1 = w^*_0 w_1 + w_1^* w_2 + \ldots w^*_{m-1} w_m \qquad 14.5\text{-}5$$

If the filter, **W**, is real then the autocovariance coefficients for positive and negative lags are identical.

$$a_k = a_{-k}^* \qquad 14.5\text{-}6$$

The power spectrum of the filter is obtained from 14.5-3 if $\exp(-i\omega \cdot \Delta t)$ is substituted in place of z. The inverse minimum phase function, $\mathbf{W}^{-1}$, will have the same autocovariance and power spectrum as **W**.

$$A(z) = W^{-1}(z) W^{-1*}(1/z) \qquad 14.5\text{-}7$$

Since W(z) is minimum phase, the inverse filter may be computed.

$$W(z) = 1/W^{-1}(z) \qquad 14.5\text{-}8$$

Substituting this into 14.5-3 gives an alternate form for the autocovariance.

$$A(z) = W^*(1/z) / W^{-1}(z) \qquad 14.5\text{-}9$$

An equation for the solution of the inverse filter is obtained by multiplying both sides by $W^{-1}(z)$.

$$A(z) W^{-1}(z) = W^*(1/z) \qquad 14.5\text{-}10$$

Let us write out the z transform in full.

$$(a_{-m}/z^m + \ldots + a_{-1}/z + a_0 + a_1 z + a_1 z^2 + \ldots + a_m z^m)(w_0^{-1} + w_z^{-1} z + \ldots + w_m^{-1} z^m)$$
$$= w_0^* + w_1^*/z + w_2^*/z^2 + \ldots + w_m^*/z^m \qquad 14.5\text{-}11$$

Deconvolution

The coefficients of powers of z such as $z^0, z^1, \ldots z^m$ yield the following set of m equations.

$$a_0 w_0^{-1} + a_{-1} w_1^{-1} + \ldots a_{-m} w_m^{-1} = w_0^*$$
$$a_1 w_0^{-1} + a_0 w_1^{-1} + \ldots a_{1-m} w_m^{-1} = 0$$
$$\vdots$$
$$a_m w_0^{-1} + a_{m-1} w_1^{-1} + \ldots a_0 w_m^{-1} = 0$$

The set of simultaneous equations for the solutions of $\mathbf{W}^{-1}$ may be written in matrix form.

$$\begin{bmatrix} a_0 & a_{-1} & a_{-2} & \ldots & a_{-m} \\ a_1 & a_0 & a_{-1} & \ldots & a_{-m-1} \\ \vdots & & & & \\ a_m & a_{m-1} & a_{m-2} & \ldots & a_0 \end{bmatrix} \begin{bmatrix} w_0^{-1} \\ w_1^{-1} \\ \vdots \\ w_m^{-1} \end{bmatrix} = \begin{bmatrix} w_0^* \\ 0 \\ \vdots \\ 0 \end{bmatrix} \qquad 14.5\text{-}12$$

For real coefficients, $a_1 = a_{-1}$, $a_m = a_{-m}$, etc., and $w_0^* = w_0$, this equation is identical to 14.4-22 derived in the previous section.

14.6 Levinson recursion for inverse filter

A practical method of solving for the least squares inverse filter was presented by Levinson (1947). Wiener's solution involving continuous or analog signals, as derived in Appendix 3, was reformulated in discrete form as shown in sections 14.4 or 14.5. The first filter coefficient is normalized to unity by division by w_0^{-1}.

$$W_i = w_i^{-1} / w_0^{-1} \qquad i = 1, \ldots j \qquad 14.6\text{-}1$$

The capitalized italic symbol W will be used to denote the inverse filter in this section to avoid writing the superscript in $\mathbf{W}^{-1}$.

$$\begin{bmatrix} a_0 & a_1 & a_2 & \ldots & a_m \\ a_1 & a_0 & a_1 & \ldots & a_{m-1} \\ \vdots & & & & \\ a_m & a_{m-1} & a_{m-2} & \ldots & a_0 \end{bmatrix} \begin{bmatrix} 1 \\ W_1 \\ \vdots \\ W_m \end{bmatrix} = \begin{bmatrix} w_0^*/w_0^{-1} \\ 0 \\ \vdots \\ 0 \end{bmatrix} \qquad 14.6\text{-}2$$

Levinson recursion for inverse filter

Most Levinson algorithms are made more universal by substituting the autocorrelation function for the autocovariance. This is obtained by dividing both sides of the equation by the autocovariance with zero lag.

$$A_{j+1} = a_j / a_0 \qquad j = 0, \ldots m \qquad \qquad 14.6\text{-}3$$

Computer programs with a Fortran language do not use the subscript zero in DO loops so it is convenient to reformulate equation 14.6-2 with subscripts running from 1 to $n = m + 1$. The inverse filter is determined by successive iterations denoted by the index j. The filter length is increased by one with each iteration until it has $m + 1$ terms so it is convenient to set out the theory with a double index, ij, to indicate the coefficient and the order number.

$$\begin{bmatrix} A_1 & A_2 & A_3 & \ldots & A_j \\ A_2 & A_1 & A_3 & \ldots & A_{j-1} \\ \cdot & & & & \\ \cdot & & & & \\ \cdot & & & & \\ A_j & A_{j-1} & A_{j-2} & \ldots & A_1 \end{bmatrix} \begin{bmatrix} W_{ij} \\ W_{2j} \\ \cdot \\ \cdot \\ \cdot \\ W_{jj} \end{bmatrix} = \begin{bmatrix} V_j \\ 0 \\ \cdot \\ \cdot \\ \cdot \\ 0 \end{bmatrix} \qquad j = 1, \ldots n \qquad 14.6\text{-}4$$

The cross correlations coefficient on the right-hand side is denoted by V_j. Since $w_0 = 1/w_0^{-1}$ the final value is

$$V_n = w_0 w_0^* / a_0 \qquad \qquad 14.6\text{-}5$$

For the first iteration, the filter has only one term.

$$[A_1] \, [W_{11}] = [V_1] \qquad j = 1 \qquad \qquad 14.6\text{-}6$$

Due to the normalizations the solution is very easy. Equation numbers with stars are required in the computer program.

$$A_1 = 1 \; ; \quad V_1 = 1 \quad W_{11} = 1 \quad j = 1 \qquad \qquad *14.6\text{-}7$$

For the second iteration, $j = 2$, the matrix equation to be solved is as follows:

$$\begin{bmatrix} A_1 & A_2 \\ A_2 & A_1 \end{bmatrix} \begin{bmatrix} W_{12} \\ W_{22} \end{bmatrix} = \begin{bmatrix} V_2 \\ 0 \end{bmatrix} \qquad \qquad 14.6\text{-}8$$

An iterative solution is initiated by adding a new row to 14.6-6 and defining a new constant, E_j. The two simultaneous equations and their matrix are written as follows:

$$A_1W_{11} + A_20 = V_1 \qquad \begin{bmatrix} A_1 & A_2 \\ A_2 & A_1 \end{bmatrix}\begin{bmatrix} W_{i1} \\ 0 \end{bmatrix} = \begin{bmatrix} V_1 \\ E_2 \end{bmatrix} \qquad 14.6\text{-}9$$
$$A_2W_{11} + A_10 = E_2$$

Notice that the first equation is exactly the same as 14.6-6. The pair of equations may be reversed and the matrix for them written down.

$$A_10 + A_2W_{11} = E_2 \qquad \begin{bmatrix} A_1 & A_2 \\ A_2 & A_1 \end{bmatrix}\begin{bmatrix} 0 \\ W_{11} \end{bmatrix} = \begin{bmatrix} E_2 \\ V_1 \end{bmatrix} \qquad 14.6\text{-}10$$
$$A_20 + A_1W_{11} = V_1$$

The key step in obtaining an iterative solution is to form a new equation by subtracting a ratio, R_j, of reversed matrix equation, 14.6-10, from the original construction, 14.6-9.

$$\begin{bmatrix} A_1 & A_2 \\ A_2 & A_1 \end{bmatrix}\left\{\begin{bmatrix} W_{11} \\ 0 \end{bmatrix} - R_2\begin{bmatrix} 0 \\ W_{11} \end{bmatrix}\right\} = \begin{bmatrix} V_1 \\ E_2 \end{bmatrix} - R_2\begin{bmatrix} E_2 \\ V_1 \end{bmatrix} \qquad 14.6\text{-}11$$

The parameter E_2 is obtained from the top row of 14.6-10.

$$E_2 = A_2W_{11} \qquad 14.6\text{-}12$$

Equations 14.6-8 and 14.6-11 may be made equivalent. For this to be true the right-hand side of the bottom row of 14.6-11 must be equal to the zero on the bottom right-hand side of 14.6-8.

$$E_2 - R_2V_1 = 0 \qquad 14.6\text{-}13$$

Solving for the ratio, R_2, one has

$$R_2 = E_2 / V_1 \qquad 14.6\text{-}14$$

A solution may now be obtained for V_2 from the top right-hand side of equation 14.6-8 and 14.6-11.

$$V_2 = V_1 - R_2E_2 \qquad 14.6\text{-}15$$

The new coefficients for a two-length inverse filter are obtained by equating the equivalent parts of the left-hand side of equations 14.6-8 and 14.6-11.

$$W_{12} = W_{11} = 1 \qquad 14.6\text{-}16$$

Levinson recursion for inverse filter

$$W_{22} = -R_2 W_{11} \qquad 14.6\text{-}17$$

For the third iteration, j = 3, the equations to be solved can be expressed in terms of a 3 by 3 autocorrelation matrix.

$$\begin{bmatrix} A_1 & A_2 & A_3 \\ A_2 & A_1 & A_2 \\ A_3 & A_2 & A_1 \end{bmatrix} \begin{bmatrix} W_{13} \\ W_{23} \\ W_{33} \end{bmatrix} = \begin{bmatrix} V_3 \\ 0 \\ 0 \end{bmatrix} \qquad 14.6\text{-}18$$

The equations in 14.6-8, from the previous iteration, are augmented by one row together with a new parameter, E_3.

$$\begin{bmatrix} A_1 & A_2 & A_3 \\ A_2 & A_1 & A_2 \\ A_3 & A_2 & A_1 \end{bmatrix} \begin{bmatrix} W_{12} \\ W_{22} \\ 0 \end{bmatrix} = \begin{bmatrix} V_2 \\ 0 \\ E_3 \end{bmatrix} \qquad 14.6\text{-}19$$

The reverse set of equations may also be written down.

$$\begin{bmatrix} A_1 & A_2 & A_3 \\ A_2 & A_1 & A_2 \\ A_3 & A_2 & A_1 \end{bmatrix} \begin{bmatrix} 0 \\ W_{22} \\ W_{12} \end{bmatrix} = \begin{bmatrix} E_3 \\ 0 \\ V_2 \end{bmatrix} \qquad 14.6\text{-}20$$

A ratio, R_3, of the reversed matrix equation, 14.6-20, is subtracted from the original set in 14.6-19.

$$\begin{bmatrix} A_1 & A_2 & A_3 \\ A_2 & A_1 & A_2 \\ A_3 & A_2 & A_1 \end{bmatrix} \left\{ \begin{bmatrix} W_{12} \\ W_{22} \\ 0 \end{bmatrix} - R_3 \begin{bmatrix} 0 \\ W_{22} \\ W_{12} \end{bmatrix} \right\} = \begin{bmatrix} V_2 \\ 0 \\ E_3 \end{bmatrix} - R_3 \begin{bmatrix} E_3 \\ 0 \\ V_2 \end{bmatrix} \qquad 14.6\text{-}21$$

The parameter E_3 is obtained from the top row of 14.6-20.

$$E_3 = A_2 W_{22} + A_3 W_{12} \qquad 14.6\text{-}22$$

Equations 14.6-18 and 14.6-21 are equivalent. From the right-hand side bottom row of each equation it is possible to solve for R_3.

$$E_3 - R_3 V_2 = 0 \qquad 14.6\text{-}23$$

$$R_3 = E_3 / V_2 \qquad 14.6\text{-}24$$

From the top right-hand side of the equivalent set a new value is obtained for V_3.

$$V_3 = V_2 - R_3 E_3 \qquad 14.6\text{-}25$$

The new coefficients for a three-length inverse filter are obtained by equating the equivalent parts of the left-hand side of equations 14.6-18 and 14.6-21.

$$W_{13} = W_{12} - R_3 W_{32} = 1 \qquad W_{32} = 0 \qquad 14.6\text{-}26$$

$$W_{23} = W_{22} - R_3 W_{22} \qquad 14.6\text{-}27$$

$$W_{33} = W_{32} - R_3 W_{12} \qquad W_{32} = 0 \qquad 14.6\text{-}28$$

The recursion equations can be written down for the general case.

$$E_j = \sum_{i=2}^{j} A_i\, W_{j-i+1\ j-1} \qquad j = 2,\ldots,n \qquad *14.6\text{-}29$$

$$R_j = E_j / V_{j-1} \qquad *14.6\text{-}30$$

$$V_j = V_{j-1} - R_j E_j \qquad *14.6\text{-}31$$

$$W_{ij} = W_{i\ j-1} - R_j W_{j-i+1\ j-1} \qquad i = 1,\ldots,j \qquad *14.6\text{-}32$$

where

$$W_{j\ j-1} = 0 \qquad *14.6\text{-}33$$

In terms of the z transform the equations for the inverse filter coefficients can be written as follows. The original indexing notation, starting from zero, is used.

Levinson recursion for inverse filter

$$W_{(j)}^{-1}(z) = W_{(j-1)}^{-1}(z) - R_j z^{j-1} W_{(j-1)}^{-1}(1/z) \qquad 14.6\text{-}34$$

As an example, for $j = 3$, the z transform for powers of z equal to 0, 1 and 2 are given below:

$$\begin{aligned} w_{03}^{-1} z^0 &= w_{02}^{-1} z^0 - R_3 z^2 (1/z)^2 \, w_{22}^{-1} & i &= 0 \\ w_{13}^{-1} z^1 &= w_{12}^{-1} z^1 - R_3 z^2 (1/z)^1 \, w_{12}^{-1} & i &= 1 \\ w_{23} z^2 &= w_{22}^{-1} z^2 - R_3 z^2 (1/z)^0 \, w_{02}^{-1} & i &= 2 \end{aligned}$$

In the general case the coefficient of z^i is as follows:

$$w_{ij}^{-1} z^i = w_{i\ j-1}^{-1} z^i - R_j z^{j-1} (1/z)^{j-i-1} w_{j-i-1\ j-1} \qquad 14.6\text{-}35$$

Note that from equation 14.6-30 and 14.6-31

$$V_j = V_{j-1} - E_j^2 / V_{j-1}$$

or

$$V_j = V_{j-1} [1 - (E_j / V_{j-1})^2] \qquad 14.6\text{-}36$$

Since $V_1 = 1$ and $V_n = 1/a_0 a_0^*$ is positive it follows that $R_j = E_j/V_{j-1}$ is always bounded by -1 and $+1$.

$$-1 < E_j / V_{j-1} < 1 \qquad 14.6\text{-}37$$

By examining the z transform of the inverse filter in 14.6-34 it is seen that, since $w^{-1}(j-1)$ is always minimum phase, the new filter, $w^{-1}(j)$, will also be minimum phase. The Levinson recursion method always yields a minimum phase inverse filter if the original wavelet or filter, from which the autocovariance function was computed, is minimum phase. It is also possible to monitor the error

$$e = \delta - WW^{-1} \qquad 14.6\text{-}38$$

or the squared norm.

$$EE^t = ee^t \qquad 14.6\text{-}39$$

The recursion algorithm can be modified to truncate the computation if the squared norm falls below some specified value.

A subroutine for solving equation 14.6-2 with the Levinson method is presented below. The input is the autocorrelation function.

```
      SUBROUTINE LEVREC(A,W,V,N)
C     A IS THE AUTOCORRELATION FUNCTION WITH
C     ZERO LAG TERM UNITY.
C     W IS THE INVERSE OR SPIKING FILTER
C     N IS THE NUMBER OF AUTOCORRELATION TERMS
      REAL A(50), W(50), R(50), V(50), D(50), E(50)
      W(1) = 1
      V(1) = 1
      D(1) = A(1)
      DO 2 K = 2, N
    2 D(K) = 0.0
      DO 5 J = 2, N
      EE = 0.0
      DO 3 I = 2, J
    3 EE = EE + A(I) * W(J-I+1)
      E(J) = EE
      R(J) = E(J)/V(J-1)
      V(J) = V(J-1) - E(J) * R(J)
      DO 4 I = 1, J
    4 W(I) = D(I) - R(J) * D(J+1-I)
      DO 5 I = 1, J
    5 D(I) = W(I)
      RETURN
      END
```

14.7 An inverse shaping filter

An inverse filter may be designed in which the desired output has some shape other than a unit impulse at time zero. The output may be a wavelet with any desired form or it may be a unit impulse at a delayed time. The physical operation is illustrated in figure 14.8.

The shaping filter is an m+1 row vector, S, or column matrix **S**.

$$S = (S_0, \ S_1, \ S_2, \ \ldots \ S_m) \equiv \mathbf{S}^t \qquad 14.7\text{-}1$$

The t indicates a transform of the row vector into a column vector. The actual output from this filter is an $m + n + 1$ row vector $\mathbf{x}_a$ (or a column matrix, **X**).

$$\mathbf{x}_a = (x_0, \ x_1, \ x_2, \ \ldots \ x_{m+n}) \equiv \mathbf{X}^t \qquad 14.7\text{-}2$$

The desired output is also an $m + n + 1$ row vector, d, or column matrix, **D**.

An inverse shaping filter

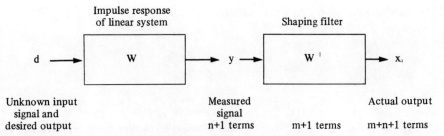

Figure 14.8 Flow diagram for a signal through a filter and its inverse.

$$d = (d_0, d_1, \ldots, d_{m+n}) \equiv D^t \qquad 14.7\text{-}3$$

The deconvolution giving the actual output is

$$x_a = y * S \qquad 14.7\text{-}4$$

which may be written as the following matrix equation

$$X = Y^t S \qquad 14.7\text{-}5$$

where **Y** is a set of numbers called the *regressor matrix* having $(m + 1)$ rows by $(m + n + 1)$ columns.

$$Y = \begin{bmatrix} y_0 & y_1 & y_2 & \ldots & y_n & 0 & 0 & \ldots & 0 & 0 \\ 0 & y_0 & y_1 & \ldots & y_{n-1} & y_n & 0 & \ldots & 0 & 0 \\ . & . & . & & . & . & & . & . \\ . & . & . & & . & . & & . & . \\ 0 & 0 & 0 & \ldots & & . & & \ldots & y_{n-1} & y_n \end{bmatrix} \qquad 14.7\text{-}6$$

For example, if $n = 4, m = 3$, the regressor matrix has 4 rows and 8 columns and the deconvolution is $X = Y^t S$.

$$\begin{bmatrix} x_0 \\ x_1 \\ x_2 \\ x_3 \\ x_4 \\ x_5 \\ x_6 \\ x_7 \end{bmatrix} = \begin{bmatrix} y_0 & 0 & 0 & 0 \\ y_1 & y_0 & 0 & 0 \\ y_2 & y_1 & y_0 & 0 \\ y_3 & y_2 & y_1 & y_0 \\ y_4 & y_3 & y_2 & y_1 \\ 0 & y_4 & y_3 & y_2 \\ 0 & 0 & y_4 & y_3 \\ 0 & 0 & 0 & y_4 \end{bmatrix} \begin{bmatrix} S_0 \\ S_1 \\ S_2 \\ S_3 \end{bmatrix} \qquad 14.7\text{-}7$$

Matrix multiplication may be summarized as

$$a_{ij} = \sum_{k=1}^{n} b_{ik} c_{kj} \qquad 14.7\text{-}8$$

where **B** is a m x n matrix, **C** is a n x p matrix and **A** is a m x p matrix. The element in the ith row and j^{th} column is labelled a_{ij} in this more general case in which the array of numbers is **A**.

$$\mathbf{A} = \begin{bmatrix} a_{11} & \cdots & a_{1p} \\ a_{21} & \cdots & \cdot \\ \cdot & \cdots & a_{ij} & \cdot \\ a_{m1} & \cdots & a_{mp} \end{bmatrix} \qquad 14.7\text{-}9$$

The error in the output from the filter is

$$e = d - x_a$$

or

$$\mathbf{E} = \mathbf{D} - \mathbf{X} \qquad 14.7\text{-}10$$

The squared norm of the error in matrix notation is the product of **E** with its transpose.

$$\mathbf{EE}^t = ee^t \qquad 14.7\text{-}11$$

where the t indicates the matrix transpose. The transpose of a matrix is obtained by interchanging rows and columns. Therefore

$$\mathbf{EE}^T = \begin{bmatrix} d_0 - x_0 \\ d_1 - x_1 \\ \cdot \\ \cdot \\ \cdot \\ d_{m+n} - x_{m+n} \end{bmatrix} \begin{bmatrix} d_0 - x_0 & d_1 - x_1 & \cdots & d_{m+n} - x_{m+n} \end{bmatrix} \qquad 14.7\text{-}12$$

The squared norm is a minimum in a least squares sense if the regressor, **Y**, is normal to the error, **e**.

An inverse shaping filter

$$\mathbf{YE} = \mathbf{Y}(\mathbf{D} - \mathbf{X}) = 0 \qquad 14.7\text{-}13$$

or

$$\mathbf{YX} = \mathbf{YD} \qquad 14.7\text{-}14$$

Substituting the deconvolution, for the actual output one has

$$\mathbf{YY^T S} = \mathbf{YD} \qquad 14.7\text{-}15$$

This is equivalent to equation 14.4-17 as the following analysis will show:

$$\sum_{i=0}^{m} \mathbf{W}_i^{-1} a_{k-i} = c_k$$

The product of the 4 x 8 regressor matrix and its transpose is

$$\mathbf{YY^t} = \begin{bmatrix} y_0 & y_1 & y_2 & y_3 & y_4 & 0 & 0 & 0 \\ 0 & y_0 & y_1 & y_2 & y_3 & y_4 & 0 & 0 \\ 0 & 0 & y_0 & y_1 & y_2 & y_3 & y_4 & 0 \\ 0 & 0 & 0 & y_0 & y_1 & y_2 & y_3 & y_4 \end{bmatrix} \begin{bmatrix} y_0 & 0 & 0 & 0 \\ y_1 & y_0 & 0 & 0 \\ y_2 & y_1 & y_0 & 0 \\ y_3 & y_2 & y_1 & y_0 \\ y_4 & y_3 & y_2 & y_1 \\ 0 & y_4 & y_3 & y_2 \\ 0 & 0 & y_4 & y_3 \\ 0 & 0 & 0 & y_4 \end{bmatrix} \qquad 14.7\text{-}16$$

which is just the autocovariance matrix.

$$\mathbf{YY^t} = \begin{bmatrix} a_0 & a_1 & a_2 & a_3 \\ a_{-1} & a_0 & a_1 & a_2 \\ a_{-2} & a_{-1} & a_0 & a_1 \\ a_{-3} & a_{-2} & a_{-1} & a_0 \end{bmatrix} = \mathbf{A} \qquad 14.7\text{-}17$$

Matrix multiplication of **Y** and **D** gives the cross covariance, **C**, of the desired output and the input signal to the shaping (inverse) filter.

$$\mathbf{C} = \begin{bmatrix} c_0 \\ c_1 \\ c_2 \\ c_3 \end{bmatrix} = \begin{bmatrix} y_0 & y_1 & y_2 & y_3 & y_4 & 0 & 0 & 0 \\ 0 & y_0 & y_1 & y_2 & y_3 & y_4 & 0 & 0 \\ 0 & 0 & y_0 & y_1 & y_2 & y_3 & y_4 & 0 \\ 0 & 0 & 0 & y_0 & y_1 & y_2 & y_3 & y_4 \end{bmatrix} \begin{bmatrix} d_0 \\ d_1 \\ d_2 \\ d_3 \\ \cdot \\ \cdot \\ \cdot \\ d_7 \end{bmatrix} = \mathbf{YD} \qquad 14.7\text{-}18$$

The equation for the inverse or shaping filter is

$$\mathbf{AS} = \mathbf{C} \qquad 14.7\text{-}19$$

That is, the least squares inverse filter times the autocovariance of the input is equal to the cross covariance of the input and desired output. The inverse or shaping filter is found from

$$\mathbf{S} = \mathbf{A}^{-1}\mathbf{C} \qquad 14.7\text{-}20$$

The shaping filter may be solved by a modification of the Levinson recursion method, making use of equation 14.7-19 and the solution for the inverse filter in section 14.6. The autocovariance is converted to an autocorrelation matrix by division with the zero lag term. The indexing is initiated at unity for ease in setting up the computer algorithms.

$$\begin{aligned} A_{i+1} &= a_i / a_0 \\ C_{i+1} &= c_i / a_0 \end{aligned} \qquad i = 0, \ldots n \qquad 14.7\text{-}21$$

As in the previous section a double subscript, ij; $i = 1,\ldots j$; $j = 1,\ldots n+1$, is used to indicate the filter coefficients and the order number. For the first iterations, $j = 1$, the equation is very simple.

$$[A_1][S_{11}] = [C_1] \qquad 14.7\text{-}22$$

Since $A_1 = 1$ the solution is readily obtained.

$$S_{11} = C_1 \qquad *14.7\text{-}23$$

For the second iteration, $j = 2$, the equations must be solved for a pair of shaping coefficients, S_{12} and S_{22}.

$$\begin{bmatrix} A_1 & A_2 \\ A_2 & A_1 \end{bmatrix} \begin{bmatrix} S_{12} \\ S_{22} \end{bmatrix} = \begin{bmatrix} C_1 \\ C_2 \end{bmatrix} \qquad 14.7\text{-}24$$

Rewrite equations 14.6-8 for the inverse spiking filter in reverse order.

$$\begin{bmatrix} A_1 & A_2 \\ A_2 & A_1 \end{bmatrix} \begin{bmatrix} W_{22} \\ W_{12} \end{bmatrix} = \begin{bmatrix} 0 \\ V_2 \end{bmatrix} \qquad 14.7\text{-}25$$

A new set of equations is generated from 14.7-22 by adding a new row and

An inverse shaping filter

defining a new constant, E_2.

$$\begin{bmatrix} A_1 & A_2 \\ A_2 & A_1 \end{bmatrix} \begin{bmatrix} S_{11} \\ 0 \end{bmatrix} = \begin{bmatrix} C_1 \\ E_2 \end{bmatrix} \qquad 14.7\text{-}26$$

Note that the first equation in 14.7-26 is identical to 14.7-22. A new relation may be defined by subtracting a ratio, R_2, of equation 14.7-25 from 14.7-26.

$$\begin{bmatrix} A_1 & A_2 \\ A_2 & A_1 \end{bmatrix} \left\{ \begin{bmatrix} S_{11} \\ 0 \end{bmatrix} - R_2 \begin{bmatrix} W_{22} \\ W_{12} \end{bmatrix} \right\} = \begin{bmatrix} C_1 \\ E_2 \end{bmatrix} - R_2 \begin{bmatrix} 0 \\ V_2 \end{bmatrix} \qquad 14.7\text{-}27$$

The parameter E_2 is obtained from the bottom row of 14.7-26.

$$E_2 = A_2 S_{11} \qquad 14.7\text{-}28$$

Equations 14.7-24 and 14.7-27 may be made equal. For this to be true the right hand side of the bottom row must be made equal.

$$C_2 = E_2 - R_2 V_2 \qquad 14.7\text{-}29$$

This may be solved for the ratio R_2.

$$R_2 = (E_2 - C_2) / V_2 \qquad 14.7\text{-}30$$

The filter coefficient are obtained by equating the equivalent parts of the left hand side of equations 14.7-24 and 14.7-27.

$$S_{12} = S_{11} - R_2 W_{22} \qquad 14.7\text{-}31$$

$$S_{22} = 0 - R_2 W_{12} \qquad 14.7\text{-}32$$

To obtain a general set of equations the next iteration, j = 3, will be written down in detail. The equation to be solved for the 3-length shaping coefficients, S_{13}, S_{23} and $S_{33}1$, is as follows.

$$\begin{bmatrix} A_1 & A_2 & A_3 \\ A_2 & A_1 & A_2 \\ A_3 & A_2 & A_1 \end{bmatrix} \begin{bmatrix} S_{13} \\ S_{23} \\ S_{33} \end{bmatrix} = \begin{bmatrix} C_1 \\ C_2 \\ C_3 \end{bmatrix} \qquad 14.7\text{-}33$$

Rewrite equations 14.6-18 from the inverse spiking filter in reverse order.

$$\begin{bmatrix} A_1 & A_2 & A_3 \\ A_2 & A_1 & A_2 \\ A_3 & A_2 & A_1 \end{bmatrix} \begin{bmatrix} W_{33} \\ W_{23} \\ W_{13} \end{bmatrix} = \begin{bmatrix} 0 \\ 0 \\ V_3 \end{bmatrix} \qquad 14.7\text{-}34$$

Another set of equations is formed by augmenting the previous iteration, 14.7-24, for the shaping filter and defining a new constant, E_3.

$$\begin{bmatrix} A_1 & A_2 & A_3 \\ A_2 & A_1 & A_2 \\ A_3 & A_2 & A_1 \end{bmatrix} \begin{bmatrix} S_{12} \\ S_{22} \\ 0 \end{bmatrix} = \begin{bmatrix} C_1 \\ C_2 \\ E_3 \end{bmatrix} \qquad 14.7\text{-}35$$

A ratio, R_3, of the reversed set in 14.7-34 is subtracted from 14.7-35.

$$\begin{bmatrix} A_1 & A_2 & A_3 \\ A_2 & A_1 & A_2 \\ A_3 & A_2 & A_1 \end{bmatrix} \left\{ \begin{bmatrix} S_{12} \\ S_{22} \\ 0 \end{bmatrix} - R_3 \begin{bmatrix} W_{33} \\ W_{23} \\ W_{13} \end{bmatrix} \right\} = \begin{bmatrix} C_1 \\ C_2 \\ E_3 \end{bmatrix} - R_3 \begin{bmatrix} 0 \\ 0 \\ V_3 \end{bmatrix} \qquad 14.7\text{-}36$$

The parameter E_3 is solved from the bottom row of 14.7-35.

$$E_3 = A_2 S_{22} + A_3 S_{12} \qquad 14.7\text{-}37$$

Equations 14.7-33 and 14.7-36 will be equivalent if the right-hand side of the bottom row is equal.

$$C_3 = E_3 - R_3 V_3 \qquad 14.7\text{-}38$$

This yields a solution for the ratio R_3.

$$R_3 = (E_3 - C_3) / V_3 \qquad 14.7\text{-}39$$

The filter coefficients are obtained by equating the equivalent parts of the left hand side of equations 14.7-33 and 14.7-36.

$$S_{13} = S_{12} - R_3 W_{33} \qquad 14.7\text{-}40$$

$$S_{23} = S_{22} - R_3 W_{23} \qquad 14.7\text{-}41$$

$$S_{33} = S_{32} - R_3 W_{13} \qquad (S_{32} \equiv 0) \qquad 14.7\text{-}42$$

An inverse shaping filter

It is now possible to write the general recursion equations to solve for the coefficients of a shaping filter of any desired length.

$$E_j = \sum_{i=2}^{j} A_i \, S_{j-i+1 \; j-1} \qquad j = 2, \ldots n \qquad \text{*14.7-43}$$

$$R_j = (E_j - C_j) / V_j \qquad \text{*14.7-44}$$

$$S_{ij} = S_{ij-1} - R_j \, W_{j-i+1 \; j} \qquad i = 1, 2 \ldots j; \qquad j = 2, \ldots n \qquad \text{*14.7-45}$$

where
$$S_{j \; j-1} = 0 \qquad \text{*14.7-46}$$

Note that the z transform for the shaping filter may be written in a similar fashion to equation 14.6-34. The notation for the inverse filter coefficients is used here because the original indexing notation, starting from zero, is employed.

$$s_{(j)}(z) = s_{(j-1)}^{-1}(z) - R_j z^j \, w_j^{-1}(1/z) \qquad 14.7\text{-}47$$

? Student Problem: It is left as an exercise for the student to write an algorithm incorporating the Levinson equations for the inverse spiking filter and those of the shaping filter using the equations marked with an asterisk.

14.8 Rice's inverse convolution filter

Rice (1962) developed a shaping filter which increased the resolution of seismic reflection records but was not restricted to minimum phase wavelets. The seismogram is assumed to be made up of Ricker wavelets y(t). A least squares inverse filter is then designed which will transform each Ricker wavelet into an approximation of the unit impulse or Dirac delta function. This process of *dividing* out the reflection wavelet making up the seismogram should leave a record composed of the reflection coefficients. This could be interpreted more directly in terms of velocity and density discontinuities. Rice considers the case when the input signal is a symmetric Ricker wavelet (figure 14.9). The inverse filter will normally have the same number of points. Note the dominance of high frequencies in the inverse's amplitude spectrum. The input pulse or Ricker wavelet for 75 Hz signal is shown in figure 14.9 while the input pulse for a 37.5

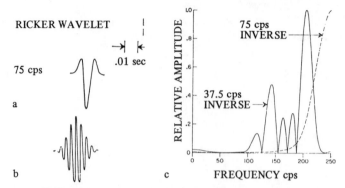

Figure 14.9 (a) 75 Hz Ricker wavelet. (b) The inverse Ricker wavelet for 75 cps and a sampling interval of 2 ms. (c) Amplitude spectrum of a 37.5 and a 75 Hz inverse Ricker wavelet. The phase spectrum is linear with slope depending on the choice of zero time.(From Rice, 1960, courtesy of the Society of Exploration Geophysicists.)

Hz signal is shown in figure 14.10. This also has the output from a convolution of the input pulse and the inverse filter. The calculated least squares inverse and the convolution $Y*W^{-1}$ is shown in figure 14.10.

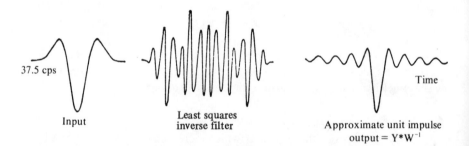

Figure 14.10 Input and output from a deconvolution filter with response W^{-1}. (From Rice, 1960, courtesy of the Society of Exploration Geophysicists.)

Notice that the approximate unit impulse has above half the width of the original Ricker wavelet. Figure 14.11 shows the improvement obtained if the inverse filter is applied to a synthetic seismogram. It is seen that the resolution is increased if the noise on the traces is small.

The success of a deconvolution operation depends greatly on the stationarity of the autocovariance function. It also depends on the noise present and its frequency content as compared to that of the desired wavelet. Figure 14.12 illustrates the effect of non-stationarity on the operation and figure 14.13 is an impressive example of the success of deconvolution in reducing reverberations on marine seismic data.

The Wiener-Hopf optimum filter

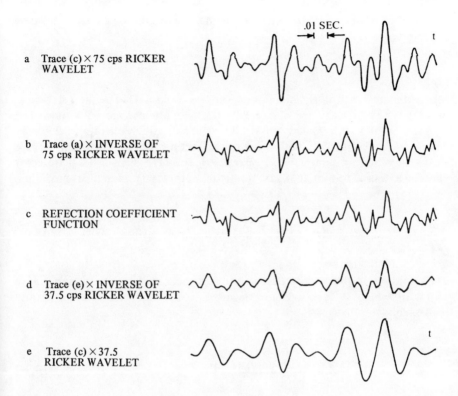

Figure 14.11 Results of applying a 37.5 Hz deconvolution filter (figure 14.13) to a synthetic seismogram computed from a continuous velocity log on a Nebraska well.(From Rice, 1960, courtesy of the Society of Exploration Geophysicists.)

14.9 The Wiener-Hopf optimum filter

Wiener (1942,1949) obtained an integral equation which could be solved to yield the impulse response for an optimum filter to be used to extract signals from a noisy signal, an operation termed *smoothing*. It could also be used as a prediction operator to predict the signal in the future on the basis of the past statistical behavior. Kolmogorov (1941) has obtained a similar equation which could be used for the discrete case of prediction. Wiener's original work was published in 1942 as a classified, yellow-bound defence report. The mathematics is somewhat difficult and it was known as *the yellow peril* among engineers trying to follow it. The work became available in the open literature in 1949. An article by Bode and Shannon (1950) and the theoretical formulation given by Lee (1964) did much to bring the work to wider recognition.

Deconvolution

Let x(t) be an input signal consisting of a desired signal, s(t), and some random noise, n(t).

$$x(t) = s(t) + n(t) \qquad 14.9\text{-}1$$

It is assumed that both signal and noise are stationary. That is, their statistical properties (mean, variance, autocovariance) do not change with time. The problem is to find a *linear*, physically realizable, filter operator, which we denote W^{-1} in conformity with the previous sections where this was an inverse or deconvolution operator. This operator is to make the mean square error between a desired output, d(t), and the actual output, y(t), as small as possible:

$$E^2 = \lim_{T \to \infty} (2T)^{-1} \int_{-T}^{T} [y(t) - d(t)]^2 \, dt \qquad 14.9\text{-}2$$

The signal y(t) is obtained by a convolution (or deconvolution) equation (figure 14.14) in which W^{-1} is zero for negative time to be causal or physically realizable.

$$y(t) = \int_{-\infty}^{\infty} W^{-1}(L) \, x(t-L) \, dL \qquad 14.9\text{-}3$$

The derivation of the integral equation known as the Wiener-Hopf equation is given in Appendix 3. It follows the techniques used in deriving the Euler-Lagrange equation for the minimization of a functional. The solution is found to be given in terms of the cross covariance, c_{xd}, between the input signal and the desired output and a_x, the autocovariance of the input signal.

$$c_{xd}(L) = \int_{0}^{\infty} W^{-1}(t) \, a_x(L-t) \, dt \qquad L \geq 0 \qquad 14.9\text{-}4$$

Following Lee (1964) the solution of this equation for the Fourier transform, $Y_{opt}(\omega)$, of the impulse response, $W^{-1}()t$, is given by

$$Y_{opt}(\omega) = [2\pi P^+(\omega)^{-1}] \int_{0}^{\infty} e^{-i\omega t} \left[\int_{-\infty}^{\infty} c_{xd}(\omega_1) \, e^{i\omega_1 t} (P^-(\omega_1))^{-1} \, d\omega_1 \right] dt \qquad 14.9\text{-}5$$

The Wiener-Hopf optimum filter

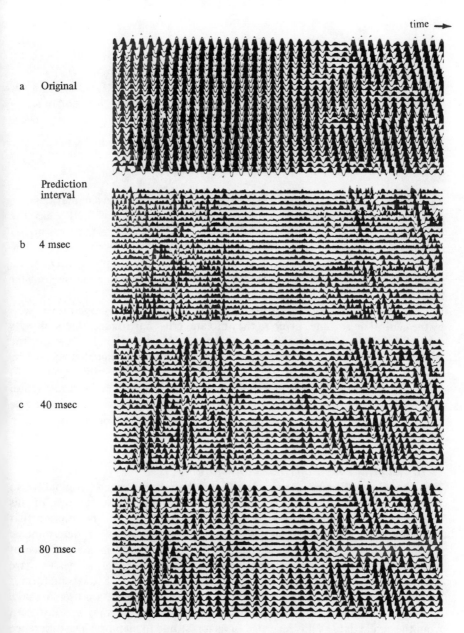

Figure 14.12 Examples of predictive deconvolution with different prediction intervals. A: Original data. B: Deconvolved data with a prediction interval of 4 milliseconds. C: Deconvolved data with a prediction interval of 40 milliseconds. D: Deconvolved data with a prediction interval of 80 milliseconds. Note that the shortest prediction interval gives reflection events with maximum resolution but some loss of individual character. (Courtesy of Prakla-Seismos, GMBH, Hanover.)

where

$$P_x(\omega) = P^+(\omega)\ P^-(\omega) \qquad 14.9\text{-}6$$

is the autopower spectral density of the input data. The function $P_x(\omega)$ is factored by finding the zeros and poles and forming a rational function of ω.

$$P_x(\omega) = K^2 \frac{(\omega-z_1)(\omega-z_1^*)(\omega-z_2)(\omega-z_2^*)\ \cdots}{(\omega-p_1)(\omega-p_1^*)(\omega-p_2)(\omega-p_2^*)\ \cdots} \qquad 14.9\text{-}7$$

The factoring is then carried out so that P^+ has all the singularities (z_1, $z_2\ldots$, $p_1,p_2\ldots$) in the upper complex frequency domain while P^- has the singularities ($z_1^*,z_2^*\ldots$, $p_1^*, p_2^*\ldots$) in the lower domain. $C_{xd}(\omega)$ is the cross power spectral density obtained from the input data and the desired output. The integration of the inner integral is generally done as a contour integral over the upper plane. The equation in 14.9-5 will yield a minimum phase transfer function.

The statistical nature of the prediction or smoothing operator is seen by the central importance given to the auto and cross covariance functions in its derivation. Since all computations will be done on the digital computer, the Wiener-Hopf equation is seldom derived for continuous signal. Instead, the discrete formulation given by Rice, Robinson or Levinson (section 14.5 to 14.8) are used instead. The restriction that this be a linear filter may be a restriction that limits the usefulness of the operator. In fact, the physical process may be known so that more restrictions can be placed on the design of the operator. In this case a non-linear operator may be more suitable.

14.10 Alternate methods of deconvolution

This chapter has discussed deconvolution as based on Wiener's original work. Two different approaches have been used in applying a discrete form of this theory. In the first an inverse operator is obtained under the assumption that we know the source wavelet, **W**. This is sometimes called deterministic deconvolution. This form of Wiener filtering is successful in the presence of additive noise since the operator is obtained by minimizing the mean square error between the input signal and the actual output signal. The second form is predictive deconvolution as presented in Chapter 12. Past data is used to make a prediction of the signal in the future. This predicted value may be subtracted from the actual data as it is recorded or as it is stored in memory. The difference is a prediction error function which may be closely related to the earth's reflectivity in the case of a seismic reflection signal (see Chapter 16). The reflection coefficients from a sedimentary section often resemble a white noise series. Predictive deconvolution will give reliable results only if the recorded

Alternate methods of deconvolution

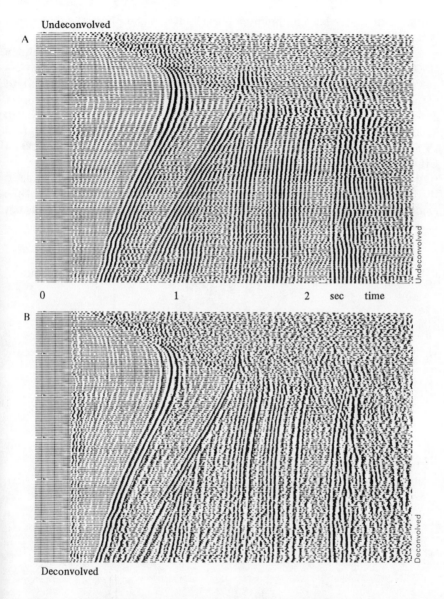

Figure 14.13 A - Display of marine field seismic reflection data before deconvolution. B - Display of the same data after deconvolution. (Courtesy of Prakla-Seismos, GMBH, Hanover.)

data is relatively noise free. The filter for a prediction distance of one unit is just the optimal, zero lag, inverse filter derived in this chapter. It assumes that the seismic source wavelet is a minimum phase function and physically this should be a good approximation for an impulsive source.

A state-space model of deconvolution was introduced by Kalman (1960) and also by Stratonovich (1960) and applied to seismology by Bayless and Brigham (1970). The method is useful when we know the physical details of the system generating a signal and can describe them by a set of first order differential equations. The output is generally given by a linear dynamic system which is driven by white noise. A derivation of a useful set of equations is given by Sorenson (1966) and the technique is described by Ott and Meder (1972) and Crump (1974). Kalman filtering is ideally suited to data sets which are not stationary. For example, the source wavelet may change substantially due to the effects of attenuation and dispersion in the earth. Since the details of the change and the physical components are seldom known, the technique has not been used much up to the present time.

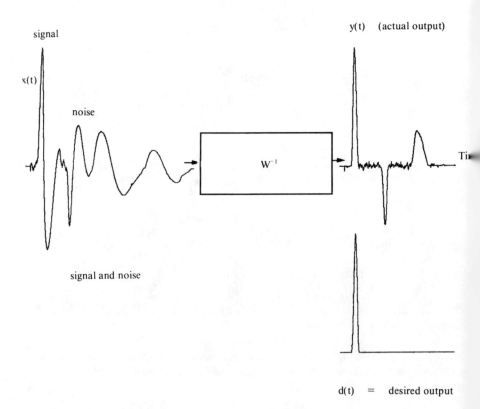

Figure 14.14 Deconvolution with an optimum filter.

Alternate methods of deconvolution

A time domain deconvolution operator which distributes the shaping error away from the region of interest rather than in a least squares sense has been presented by Mereu (1976,1978). Let the input signal be **W**, an impulse response that we want to be shaped into **D**, a desired signal. Usually **D** will be a spike, δ, but it can be any other function. The deconvolution operator will be **F**. A convolution of **W** with **F** in the time domain is to yield **D**.

$$\mathbf{D} = \mathbf{W} * \mathbf{F} \qquad \text{14.10-1}$$

The filter **F** is composed of convolutions of **W**, its reverse, $\mathbf{W_R}$, and a multiplication by a rectangular reversing function, $\mathbf{R}(n\Delta t)$, which is centered on the autocovariance of **W** to produce alternate sign reversals (figure 14.15). That is, odd indexed values of the autocovariance function are reversed in sign. An N stage filter **F** is given by

$$\mathbf{F} = \mathbf{W_R} * \mathbf{F_1} * \mathbf{F_2} * \ldots \mathbf{F_N} * \mathbf{D} \qquad \text{14.10-2}$$

where

$$\mathbf{F_1} = \mathbf{R} \cdot [\mathbf{W} * \mathbf{W_R}] \qquad \text{14.10-3}$$

$$\mathbf{F_2} = \mathbf{R} \cdot [\mathbf{W} * \mathbf{W_R} * \mathbf{F_1}] = \mathbf{R}[\mathbf{W} * \mathbf{W_R} * \mathbf{R}(\mathbf{W} * \mathbf{W_R})] \qquad \text{14.10-4}$$

$$\mathbf{F_N} = \mathbf{R} \cdot [\mathbf{W} * \mathbf{W_R} * \mathbf{F_1} * \ldots \mathbf{F_{N-1}}] \qquad \text{14.10-5}$$

Consider the effect of a single stage filter in which the desired output is a spike in the centre, δ. The actual output is then

$$\mathbf{y} = \mathbf{W} * \mathbf{F} \qquad \text{14.10-6}$$

$$\mathbf{y} = \mathbf{W} * \mathbf{W_R} * \mathbf{R} \cdot [\mathbf{W} * \mathbf{W_R}] \qquad \text{14.10-7}$$

Each convolution of **W** with $\mathbf{W_R}$ is an autocovariance function, **A**.

$$\mathbf{y} = \mathbf{A} * \mathbf{R} \cdot \mathbf{A} \qquad \text{14.10-8}$$

As an example, let us normalize the autocovariance so that the zero-lag term, a_0, is unity. In this autocorrelation function ($\mathbf{A} = a_1, 1, a_1$) has only three terms the convolution in 14.10-8 has an output with 5 terms, 2 of which are zero.

$$\mathbf{y} = (a_1, 1, a_1) * (-a_1, 1, -a_1)$$
$$\mathbf{y} = (-a_1^2, 0, 1-2a_1^2, 0, -a_1^2) \qquad 14.10\text{-}9$$

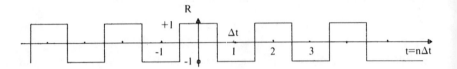

Figure 14.15 A rectangular reversing function. It is also identical to a Walsh function, WAL(n-2).

If the filter is extended to N stages the number of null values or zeros on each side of the spike becomes n.

$$n = 2^N - 1 \qquad 14.10\text{-}11$$

Figure 14.16 compares an inverse least squares filter with 253 points to a similar length Mereu shaping filter where $N = 7$. With this number of filter coefficients the two filter operators resemble each other but they are not identical. The inverse filter has minimized the error energy in a least squares sense but the desired output is not a perfect spike. The Mereu filter has a perfect spike in the centre but it has distributed the errors beyond the region shown in the plot.

A non-linear method, called homomorphic deconvolution will be discussed in a later chapter. Another method introduced by Wiggins (1977) is called minimum entropy deconvolution. This technique tries to find the least number of large spikes consistent with the data and makes no assumptions about the phase of the source function. It attempts to minimize the entropy or introduce the most order or simplicity into a set of seismograms. Basically it is a statistical method, making use of the kurtosis of the probability distribution of the reflectivity to simplify and emphasize strong reflections in a seismic recording. Further evaluation of the method has been carried out by Ulrych et al. (1978). Other research and reviews on the general subject of deconvolution may be found in the references and the one by Kailath (1974) which lists 376 references is the most comprehensive to date. Barrodale and Erickson (1980) developed an algorithm for an inverse filter using a least squares solution without a Toeplitz structure on the autocovariance matrix. The inverse or deconvolution filter may have poles inside the unit circle in the z domain, that is, it will not be minimum phase, but this can be remedied.

References

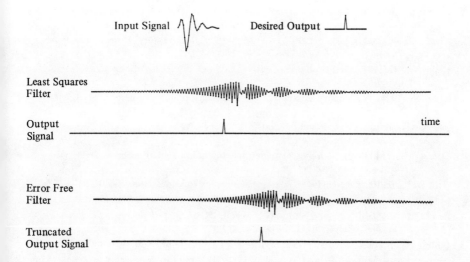

Figure 14.16 A comparison between the shape of a 253 point inverse least squares filter and a truncated 253 point shaping filter by Mereu (1976) for the case where N = 7. (Courtesy of the Society of Exploration Geophysicists.)

References

Arya,V.K.(1978), Deconvolution of seismic data - an overview. *IEEE Transactions on Geoscience Electronics,* **GE-16,** 95-98.

Barrodale,I. and Erickson,R.E.(1980), Algorithms for least squares linear prediction and maximum entropy spectral analysis, Part I and II. *Geophysics* **45,** 420-446.

Bayless,J.W. and Brigham,E.D.(1970), Application of Kalman filtering to continuous signal restoration. *Geophysics* **35,** 2-23.

Berkhout,A.J. and Zaanen,P.R.(1976), A comparison between Wiener filtering Kalman filtering and deterministic least squares estimation. *Geophysical Prospecting* **24,** 141-197.

Berkhout,A.J.(1977), Least squares inverse filters and wavelet deconvolution. *Geophysics* **42,** 1369-1383.

Bode,H.W. and Shannon,C.E.(1950), A simplified derivation of linear least square smoothing and prediction theory. *Proc. I.R.E.* **38,** 417-425.

Brigham,E.O., Smith,H.W., Bostick Jr.,F.X. and Duesterhoeft Jr.,W.C.(1968), An iterative technique for determining inverse filters. *IEEE Transactions of Geoscience Electronics* **GE6,** 86-96.

Claerbout,J.F. and Robinson,E.A.(1964), The error in least squares inverse filtering. *Geophysics* **29,** 118-120.

Claerbout,J.F.(1976), Fundamentals of geophysical data processing. McGraw-Hill, New York. 1-274.
Clayton,R.W. and Ulrych,T.J.(1977), A restoration method for impulsive functions. *IEEE Transactions on Information Theory, 262-264*.
Conaway,J.G.(1981), Deconvolution of gamma-ray logs in the case of dipping radioactive zones. *Geophysics* **46**, 198-202.
Crump,N.(1974), A Kalman filter approach to the deconvolution of seismic signals. *Geophysics* **39**, 1-13.
Deregowski,S.M.(1971), Optimum digital filtering and inverse filtering in the frequency domain. *Geophysical Prospecting* **19**, 729-768.
Deregowski,S.M.(1978), Self-matching deconvolution in the frequency domain. *Geophysical Prospecting* **26**, 252-290.
Ford,W.T. and Hearne,J.H.(1966), Least squares inverse filtering. *Geophysics* **31**, 917-926.
Ford,W.T.(1978), Optimum mixed delay spiking filters. *Geophysics* **43**, 197-215.
Fryer,G.J., Odegard,M.E. and Sutton,G.H.(1975), Deconvolution and spectral estimation using final prediction error. *Geophysics* **40**, 411-425.
Gailbraith,J.N.(1971), Prediction error as a criterion for operator length. *Geophysics* **35**, 251-265.
Griffiths,L.G., Smolka,F.R. and Trembly,L.D.(1977), Adaptive deconvolution A new technique for processing time varying seismic data. *Geophysics* **42**, 742-759.
Kailath,T.(1974), A view of three decades of linear filtering theory. *IEEE Transactions of Information Theory* **IT-20**, 146-180.
Kalman,R.E.(1960), A new approach to linear filtering and prediction problems. *Transactions of ASME (Journal of Basic Engineering)* **82D**, 34-45.
Kolmogorov,A.N.(1939), Sur l'interpolation et extrapolation des suites stationnaires. *Compte les Rendues Academy Science* **208**, 2043.
Kolmogorov,A.N.(1941), Stationary sequences in Hilbert space. Bulletin of Mathematics, University of Moscow, **2**, .
Kolmogorov,A.N.(1941), Interpolation and extrapolation, Bulletin de l'academie des Sciences de U.S.S.R., Ser. 5, 3-14.
Krein,M.G.(1945), On a generalization of some investigations of G. Szego, W.M. Smirnov and A.N. Kolmogorov. Doklady, Akademy Nawk, SSR, **46**, 91-94, 306-309.
Kulhanek,O.(1976), Introduction to digital filtering in geophysics. Elsevier Scientific Publishing Co., Amsterdam. 1-168.
Kunetz,G. and Fourmann,J.M.(1968), Efficient deconvolution of marine seismic records. *Geophysics* **33**, 412-423.
Lee,Y.W.(1964), Statistical theory of communication. John Wiley & Sons Inc.
Levinson,N.(1947), The Wiener RMS error criterion in filter design and prediction. *Journal of Mathematics and Physics* **25**, 261-278.
Lines,L.R. and Ulrych,T.J.(1977), The old and the new in seismic deconvolution and wavelet estimation. *Geophysical Prospecting* **25**, 512-540.
Mendel,J.M. and Kormylo,J.(1978), Single-channel white noise estimators for deconvolution. *Geophysics* **43**, 102-124.
Mereu,R.F.(1976), Exact wave-shaping with a time-domain digital filter of finite length. *Geophysics* **41**, 659-672.
Mereu,R.F.(1978), Computer program to obtain the weights of a time-domain wave shaping filter which is optimum in an error distribution sense. *Geophysics* **43**, 197-215.

References

Negi,J.G. and Dimri,V.P.(1979), On Wiener filter and maximum entropy method for multichannel complex systems. *Geophysical Prospecting* **27**, 156-167.

Ooe,M. and Ulrych,T.J.(1979), Minimum entropy deconvolution with an exponential transformation. *Geophysical Prospecting* **27**, 512-540.

Oppenheim,A.V.(1978), Applications of digital signal processing. Prentice-Hall Inc., Englewood Cliffs, New Jersey. 1-499.

Ott,N. and Meder,H.G.(1972), The Kalman filter as a prediction error filter. *Geophysical Prospecting* **20**, 549-560.

Peacock,K.L. and Treitel,S.(1969), Predictive deconvolution, theory and practice. *Geophysics* **34**, 155-169.

Rice,R.B.(1962), Inverse convolution filters. *Geophysics* **27**, 4-18.

Ristow,D. and Kosbahn,B.(1979), Time varying prediction filtering by means of updating. *Geophysical Prospecting* **27**, 40-61.

Robinson,E.A.(1957), Prediction deconvolution of seismic traces. *Geophysics* **22**, 539-586.

Robinson,E.A.(1963), Predictive decomposition of discrete filters for the detection of nuclear explosions. *Journal of Geophysical Research* **68**, 5559-5567.

Robinson,E.A.(1963), Mathematical development of discrete filters for the detection of nuclear explosions. *J. Geophysical Research* **68**, 5559-5567.

Robinson,E.A.(1967), Predictive decomposition of time series with application to seismic exploration. *Geophysics* **32**, 418-484.

Robinson,E.A.(1964), Wavelet composition of time series. In(F 2> Econometric model building. Edited by Herman Wold, Amsterdam, North Holland Publishing Co., 37-106.

Robinson,E.A.(1966), Collection of Fortran II Programs for filtering and spectral analysis of single channel time series. *Geophy. Prosp.* **14**, Supplement No. 1.

Silvia,M.T. and Robinson,E.A.(1979), Deconvolution of geophysical time series in the exploration for oil and natural gas. Elsevier Scientific Publishing Co., 1-251.

Sims,G.S. and D'Mello,M.R.(1978), Adaptive deconvolution of seismic signals. *IEEE Transactions on Geoscience Electronics* **GE16**, 99-103.

Sinton,J.B.Ward,R.W. and Watkins,J.S.(1978), Suppression of long-delayed multiple reflections by predictive deconvolution. *Geophysics* **43**, 1352-1367.

Sorenson,H.W.(1966), Kalman filtering techniques. *Advances in Control Systems, Theory and Application* **3**, Edited by C.T. Leondes. Academic Press, New York.

Stratonovich,R.L.(1960), Application of the theory of Markov processes for optimum filtration of signals. *Radio Engineering and Electronical Physics (USSR)* **1**, 1-19.

Stratonovich,R.L.(1970), Detection and estimation of signals in noise when one or both are non-Gaussian. *Proceeding IEEE* **58**, 670-679.

Taylor,H.L., Banks,S.C. and McCoy,J.F.(1979), Deconvolution with the l_1 norm. *Geophysics* **44**, 39-52.

Treitel,S. and Robinson,E.A.(1964), The stability of digital filters. *IEEE Transactions on Geoscience Electronics* **GE-2**, 6-18.

Treitel,S. and Robinson,E.A.(1966), The design of high resolution digital filters. *IEEE Transactions on Geoscience Electronics* **GE4**, 25-38.

Treitel,S.(1974), The complex Wiener filter. *Geophysics* **39**, 169-173.

Ulrych,T.J., Walker,C.J. and Jurkevics,A.J.(1978), A filter length criterion for maximum entropy deconvolution. *Journal Canadian Society of Exploration Geophysicists* **14**, 21-28.

Wadsworth,G.P. and Robinson,E.A.(1953), Detection of reflections on seismic records by linear operators. *Geophysics* **16**, 539-586.

Wang,R.J. and Treitel,S.(1971), Adaptive signal processing through stochastic approximation. *Geophysical Prospecting* **19**, 718-728.

Wang,R.J.(1977), Adaptive predictive deconvolution of seismic data. *Geophysical Prospecting* **25**, 342-381.

White,R.E. and Mereu,R.F.(1972), Deconvolution of refraction seismograms from large underwater explosions. *Geophysics* **37**, 431-444.

Wiener,N. and Hopf.E.(1931), On a class of singular integral equations. *Proceedings of Prussian Academy, Mathematics - Physics Series*. 696.

Wiener,N.(1942,1949), *Extrapolation, interpolation and smoothing of stationary time series*. The M.I.T. Press, Cambridge, Mass., 163.

Wiggins,R.A. and Robinson,E.A.(1965), Recursive solution to the multichannel filtering problem. *J. Geophysical Research* **70**, 1885-1891.

Wiggins,R.A.(1977), Minimum entropy deconvolution. *Proceedings International Symposium on Computer Aided Seismic Analysis and Discrimination, Falmouth Mass. IEEE Computer Society. 7-14*.

Wood,L.C.,Heiser,R.C.,Treitel,S. and Riley,P.L.(1978), The debubbling of marine source signatures. *Geophysics* **43**, 715-729.

15 Band pass filters

15.1 Ideal filters and their truncated approximation

An ideal low pass filter is one with a response of unity in the desired band pass region and zero elsewhere. The modulus of the transfer function, $|Y(\omega)|$, for such an ideal filter is illustrated in figure 15.1. The transfer function, $Y(\omega)$,

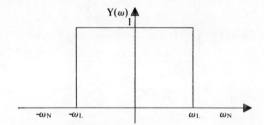

Figure 15.1 Gain of an ideal low pass filter.

has been normalized requiring the gain to be unity within the band pass region which lies between 0 and ω_L radians per second.

Approximations to such ideal filters as the one illustrated above have been designed in the digital domain with a finite number of coefficients in the impulse response. An example is given by Tooley (1955) and Fortran programs have been published by Robinson (1966). The impulse response is given by the Fourier transform of the desired transfer function. Since the impulse response function must be real, the real part of $Y(\omega)$ must have even symmetry while the imaginary part must have odd symmetry. It is convenient to also demand that the phase shift be zero in which case the imaginary part must be zero by equation 3.1-10. Under these conditions it will be sufficient to use a Fourier cosine transform as given by equation 1.8 (Appendix 1). Let ω_N be the Nyquist frequency (figure 15.1). The Nyquist frequency is given by the sampling interval, Δt, from the expression below (see 8.1-2).

$$\omega_N/2\pi = f_N = 1/2\Delta t \qquad 15.1\text{-}1$$

The linear operator, W_n, representing the impulse response is then

$$W_n = 2(\omega_N)^{-1} \int_0^{\omega_N} Y(\omega) \cos n\pi\omega/\omega_N \, d\omega \qquad n = \pm 1, \pm 2, \pm 3, \ldots \qquad 15.1\text{-}2$$

$$W_0 = (\omega_N)^{-1} \int_0^{\omega_N} Y(\omega) \, d\omega \qquad 15.1\text{-}3$$

For the low pass filter in figure 15.1 the impulse response is derived below.

$$W_0 = (\omega_N)^{-1} \int_0^{\omega_L} 1 \, d\omega = \omega_L/\omega_N \qquad 15.1\text{-}4$$

$$W_n = 2(\omega_N)^{-1} \int_0^{\omega_L} 1 \, \cos n\pi\omega/\omega_N \, d\omega \qquad 15.1\text{-}5$$

$$W_n = \frac{2\omega_L}{\omega_N} \cdot \frac{\sin n\pi\omega_L/\omega_N}{n\pi\omega_L/\omega_N} \qquad 15.1\text{-}6$$

The function above has the form $(\sin x)/x$.

The transfer function for a *high pass filter* is illustrated in figure 15.2. The dotted parts are obtained by consideration of symmetry so that

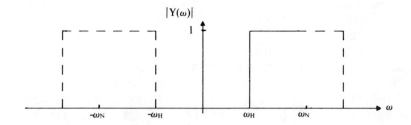

Figure 15.2   Gain of an ideal high pass filter.

$|Y(\omega)|$ is an even function. The coefficients in the linear operator are given by substituting the appropriate units in 15.1-2 and 15.1-3.

Ideal filters and their truncated approximation

$$W_0 = (\omega_N)^{-1} \int_{\omega_N}^{\omega_N} d\omega = (\omega_N - \omega_H)/\omega_N \qquad 15.1\text{-}7$$

$$W_n = 2(\omega_N)^{-1} \int_{\omega_H}^{\omega_N} \cos n\pi\omega/\omega_N \, d\omega$$
$$= 2(n\pi)^{-1} \cdot [\sin n\pi\omega_N/\omega_N - \sin n\pi\omega_H/\omega_N]$$

$$W_n = -2\omega_H(\omega_N)^{-1} \cdot [\sin n\pi\omega_H/\omega_N] \, / \, [n\pi\omega_H/(\omega_N)] \quad n=\pm1, \pm2,.. \qquad 15.1\text{-}8$$

The transfer function for an ideal band pass filter is given in figure 15.3. It can be considered a combination of a low and a high pass filter.

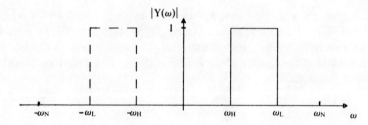

Figure 15.3 An ideal band pass filter.

The impulse response will consist of the following coefficients.

$$W_0 = \frac{\omega_H - \omega_L}{\omega_N} \qquad 15.1\text{-}9$$

$$W_n = 2(n\pi)^{-1}[\sin n\pi\omega_H/\omega_N - \sin n\pi\omega_L/\omega_N] \quad n=\pm1, \pm2, \pm3, ... \qquad 15.1\text{-}10$$

In practice, it is necessary to truncate the impulse response after a certain number of terms. An example of the impulse response for a high pass filter with a digitizing interval of 2.5 milliseconds is given in figure 15.4. The transfer functions of two truncated versions of the filter are shown in figure

15.5. The oscillation and overshooting which is a very serious defect in these types of filters is called Gibbs' effect. Errors incurred by truncating the series can be reduced by multiplying the impulse response by a weighting function. Kaiser (1966) recommends the use of a modified Bessel type window function

$$W(t) = \begin{bmatrix} \dfrac{I_0(\omega_1(\tau^2-t^2)^{1/2})}{I_0(\omega_1\tau)} & |t| < \tau \\ 0 & |t| > \tau \end{bmatrix} \qquad 15.1\text{-}11$$

where I_0 is a modified Bessel function and $\omega_1\tau$ is a parameter that controls the energy in the central lobe. The Fourier transform of 15.1-11 is

$$Y(\omega) = \frac{2}{I_0(\omega_1\tau)} \frac{\sin(\tau(\omega^2-\omega_1^2)^{1/2})}{(\omega^2-\omega_1^2)^{1/2}} \qquad 15.1\text{-}12$$

Another approach is to eliminate discontinuities in the first derivative of the transfer function of an ideal filter (Martin, 1959; Ormsby, 1961). Neither of these approaches yields completely satisfactory operators and it is recommended that the Butterworth function be used as described in detail in the following sections.

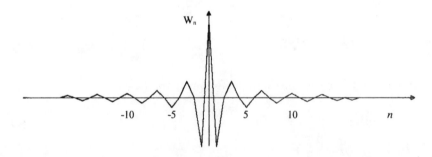

Figure 15.4   Impulse response for high pass filter with a cut-off frequency of 100 Hz.

15.2 Recursion filters

Digital filters may be applied in a number of different ways by the use of the convolution integral. Direct application in the time domain is generally inefficient on a computer and it is usual to seek some transformation such as the fast Fourier transform. This approach has great merit for it allows one to

Recursion filters

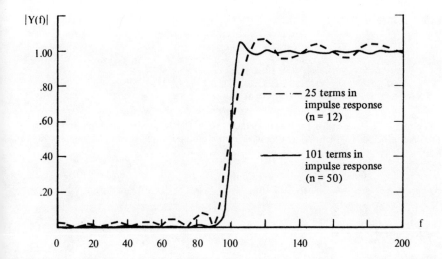

Figure 15.5 Gain of a truncated high pass filter in which the number of terms in the linear operator is 25 or 101.

design highly specialized transfer functions based on the known peculiarities of the signal and noise. Convolution with a z transform is also very efficient on a digital computer as was seen in Chapter 2. The z transform method can be made even more rapid and precise if the transfer function can be expressed as the ratio of two polynomials for it can then be evaluated by a recursive process. Recursion filtering (Golden and Kaiser, 1964; Shanks, 1967) involves a feedback loop using the polynomial in the denominator of the transform.

Let the desired z transform of the impulse response be a function of the type

$$W(z) = [N_0 + N_1 z + N_2 z^2] / (1 + D_1 z + D_2 z^2) \qquad 15.2\text{-}1$$

In the z plane convolution is a multiplication of two polynomials representing the impulse response and the input data, $X(z)$ (figure 15.6). If $Y(z)$ is the z transform of the output then it is given by

$$Y(z) = \frac{N_0 + N_1 z + N_2 z^2}{1 + D_1 z + D_2 z^2} \cdot X(z) \qquad 15.2\text{-}2$$

The right- and left-hand sides of equation 15.2-2 may be multiplied by the denominator of $W(z)$.

$$Y(z) + zY(z) \cdot (D_1 + D_2 z) = (N_0 + N_1 z + N_2 z^2) \cdot X(z)$$

Band pass filters

```
Input Data           Linear Operator            Output data

                  ┌─────────────────┐
x₀,x₁,x₂    →  →  │ N₀ + N₁z + N₂z² │    →   y₀,y₁,y₂...
                  │ ─────────────── │
                  │  1 + D₁z + D₂z² │
                  └─────────────────┘
x₀+x₁z+x₂z²+...      convolution            y₀+y₁z+y₂z²+...
```

Figure 15.6 The output from a linear operator, W(z), which specifies the impulse response of a filter, is a series of coefficients, y_n, obtained from a multiplication of two polynomials, X(z) and W(z).

or

$$Y(z) = (N_0 + N_1 z + N_2 z^2) \cdot X(z) - zY(z) \cdot (D_1 + D_2 z) \qquad 15.2\text{-}3$$

This equation may be programmed on a digital computer so that it simulates the operation of an analog feedback amplifier. Whenever z occurs the input or output is delayed by one unit and similarly a z^2 causes a delay of two units. Notice that the denominator of W(z) operates on the output from the filter (see figure 15.7).

In a digital computer the operations represented by equation 15.2-3 are given explicitly by the following equation:

$$y_n = N_0 x_n + N_1 x_{n-1} + N_2 x_{n-2} - [D_1 y_{n-1} + D_2 y_{n-2}] \qquad 15.2\text{-}4$$

where it is assumed that the output is calculated in the sequence y_0, y_1, y_2.... It is also assumed that the filter does not respond until a non-zero signal has arrived, therefore x_{-1}, x_{-2}, y_{-1} and y_{-2} are all zero.

A somewhat more general transfer function in the z domain is given by equation 15.2-5.

$$W(z) = \frac{N(z)}{D(z)} = \frac{N_0 + N_1 z + N_2 z^2 + \ldots N_n z^n}{1 + D_1 z + D_2 z + \ldots D_m z^m} \qquad 15.2\text{-}5$$

The recursive relation for the output is

$$y_n = \sum_{i=0}^{n} N_i x_{n-i} - \sum_{j=1}^{m} D_j y_{n-j} \qquad 15.2\text{-}6$$

Recursion filters

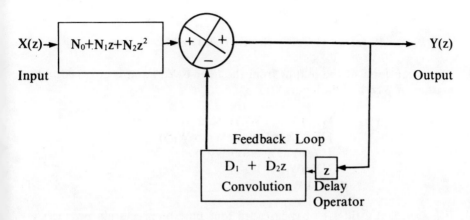

Figure 15.7 A flow diagram indicating the operation of a simple recursion filter.

The impulse response may be obtained easily for these filters by letting the input be the unit impulse wavelet.

$$\mathbf{x} = (1, 0, 0, 0 \ldots)$$

As an example of the application of recursive techniques we shall examine the band-pass Butterworth filter whose transfer function will be derived in section 15.10. The z transform of the impulse response has the form

$$W(z) = \frac{(1 - z^2)^4}{B_1(z)B_2(z)B_3(z)B_4(z)} \qquad 15.2\text{-}7$$

where

$$B_j = 1 - D_{2j-1}z + D_{2j}z^2 \qquad j = 1,2,3,4 \qquad 15.2\text{-}8$$

and the coefficients D_{2j-1} and D_{2j} are determined by the low and high pass cut-off frequencies. Equation 15.2-7 may be written as a cascaded product of four filters

$$W(z) = W_1(z) \, W_2(z) \, W_3(z) \, W_4(z) \qquad 15.2\text{-}9$$

where terms like W_1 have the form

$$W_1(z) = (1 - z^2) / (1 - D_1(z) + D_2 z^2) \qquad 15.2\text{-}10$$

The z transform of the output from the filter is $Y(z)$ and is obtained from a product of $W(z)$ and the input, $X(z)$.

$$\begin{aligned} Y(z) &= W_1(z)\, W_2(z)\, W_3(z)\, W_4(z)\, X(z) \\ &= [W_1(z) X(z)]\, W_2(z)\, W_3(z)\, W_4(z) \\ &= [C(z) W_2(z)]\, W_3(z)\, W_4(z) \\ &= [D(z) W_3(z)]\, W_4(z) \\ &= E(z)\, W_4(z) \end{aligned} \qquad 15.2\text{-}11$$

It is seen that a filter that is cascaded four times in succession produces a z transform output which we call $C(z)$, $D(z)$, $E(z)$ and finally $Y(z)$. The flow diagram for the cascaded system is shown in figure 15.8. The recursive equations for programming are

$$c_n = x_n - x_{n-2} + D_1\, c_{n-1} - D_2\, c_{n-2} \qquad 15.2\text{-}12$$

$$d_n = c_n - c_{n-2} + D_3\, d_{n-1} - D_4\, d_{n-2} \qquad 15.2\text{-}13$$

$$e_n = d_n - d_{n-2} + D_5\, e_{n-1} - D_6\, e_{n-2} \qquad 15.2\text{-}14$$

$$y_n = e_n - e_{n-2} + D_7\, y_{n-1} - D_8\, y_{n-2} \qquad 15.2\text{-}15$$
$$n = 0, 1, 2, \dots N$$

In multichannel filtering, the z transform can be written in the form of a sum of z transforms. Then a parallel form may be programmed (figure 15.9).

$$W(z) = W_0(z) \sum_{j=1}^{n} W_{Nj}(z) / [1 + W_{Dj}(z)] \qquad 15.2\text{-}16$$

The advantage of either the cascade or parallel form of digital filter is that the parameters need not be determined as precisely. The cascade form is often

Recursion filters

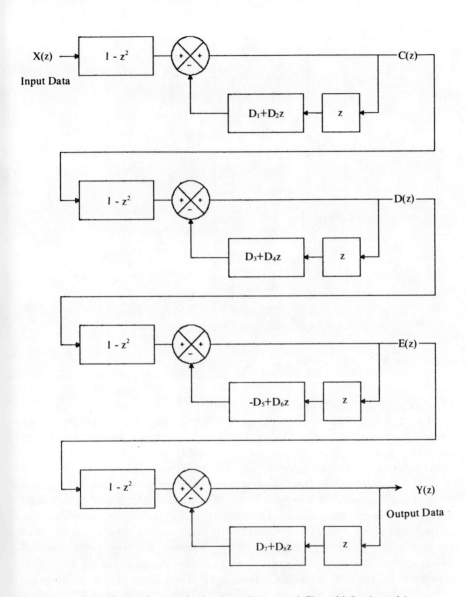

Figure 15.8 Flow diagram for recursive band pass Butterworth filter with 8 poles and 4 zeros.

convenient when using the bilinear z transform (section 15.10). The parallel form may be convenient when using the standard z transform method since it is often amenable to an expansion of the transform function in partial fractions.

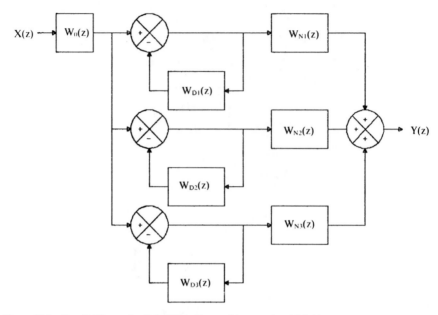

Figure 15.9 Parallel form of a digital filter for n = 3 in equation 15.2-16.

15.3 Zero phase shift filters - cascade form

A zero phase shift filter is easily obtained when the data is in digital form. Two methods are possible, each having a different amplitude response and zero phase shift at all frequencies. The first is a cascade form and the second involves a parallel superposition of recursive filters (see Chapter 17).

The data may be filtered using a convolution or recursive technique in the normal manner. The output from this process is reversed and passed through the same filter again. The output vector is reversed to obtain the desired zero phase shift data. It should be noticed that the input is filtered twice so that the amplitude response is $|Y|^2$. This alteration, involving two sections of filtering, is usually desirable since it increases the sharpness of the cut-off. It may even be feasible to design a filter with fewer poles (i.e. a lower order polynomial in the denominator) in the original transfer function. The cascade operation is illustrated in figure 15.10. A sufficient number of *zeros* must be *added* to the end of the *input data* to allow the output from the initial stage of convolution to become significantly small. The typical symmetric shape of a zero phase impulse is illustrated later in figure 15.29.

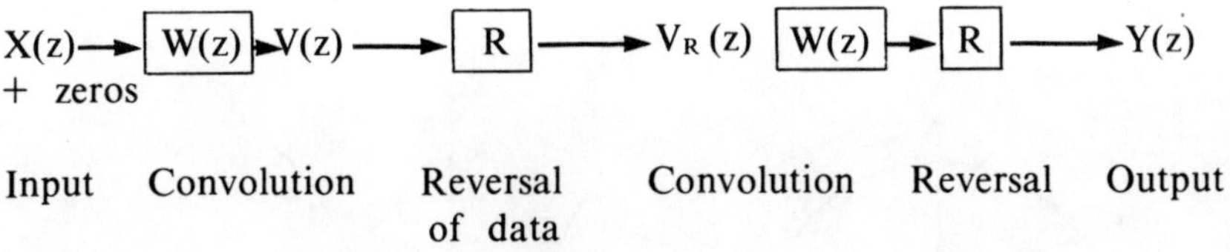

Rejection and narrow pass filters

```
X(z)→┤W(z)├→V(z)────→┤ R ├────→V_R(z) ┤W(z)├→┤ R ├────→Y(z)
 + zeros
```

Input   Convolution   Reversal   Convolution   Reversal   Output
                        of data

Figure 15.10 Cascade operation to produce filtered data with zero phase shift.

15.4 Rejection and narrow pass filters

Rejection or notch filters (Truxal, 1955) may be designed directly with the aid of a z diagram on which frequencies are plotted around the unit circle. The simplest filter has a zero on the unit circle at the frequency which is to be rejected. For example, the D.C. level, or zero frequency, will be removed by

$$W(z) = 1 - z \qquad 15.4\text{-}1$$

As illustrated in figure 15.11 the amplitude at any other frequency, f_1, will be proportional to the length of the vector from the zero to the frequency on the unit circle. The response is obviously not very good because it distorts the spectrum at all other frequencies between 0 and the Nyquist frequency, f_N, where there is a gain of two.

The filter can be improved by adding a pole very close to the zero but outside the unit circle. If a pole is placed at $z = 1/D$ where a suitable choice of z is 1.01 and $D = 0.9901$ then the z transform of the impulse response becomes

$$W(z) = \frac{G(1-z)}{1-Dz} \qquad 15.4\text{-}2$$

The gain factor, G, is obtained by normalizing the amplification to unity at some point, say the Nyquist frequency $\omega_N = \pi$, $z = \exp(-i\omega) = -1$. Setting $W(z)W^*(z)$ equal to one at $\omega = \pi$, the gain factor becomes

$$G = (1 + D)/2 \qquad 15.4\text{-}3$$

Therefore the impulse response, for $D = 0.9901$, is

$$W(z) = \frac{0.99505(1-z)}{1 - 0.9901\,z} \qquad 15.4\text{-}4$$

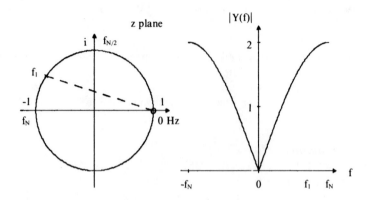

Figure 15.11 The z plane diagram for a single D.C. rejection filter and its transfer function.

Following equation 15.2-3 the recursion equation for such a filter becomes

$$y_n = 0.99505 [x_n - x_{n-1}] + 0.9901 y_{n-1} \qquad 15.4\text{-}5$$

The z plane diagram and transfer function are shown in figure 15.12. The amplitude at any frequency, f_1, is now the ratio of the length of the vector, z_1, from the zero to the frequency, f_1, on the unit circle to the length of the vector from the pole, p_1, to the frequency, f_1.

$$\text{Gain}(f) = G z_1 / p_1 \qquad 15.4\text{-}6$$

The ratio is close to unity at all frequencies except zero. The phase as a function of frequency is given by the difference in phase angle between the zeros to f_1 on the unit circle and the pole to f_1 (see figure 15.12).

$$\phi(f) = \phi_1 - \psi_1 \qquad 15.4\text{-}7$$

If we wish to design a filter which rejects some other frequency such as 50 Hz, the pole and zero are located at the appropriate position on the z plane diagram. The angle, Ω_r, on the z plane, to the pole and zero is determined by the rejection frequency, f_r, and the Nyquist frequency, $f_N = 1/(2\Delta t)$.

Rejection and narrow pass filters

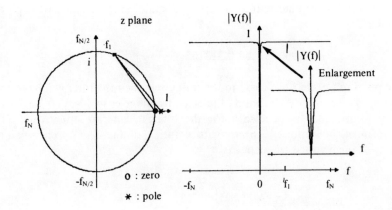

Figure 15.12 The z plane diagram for a D.C. rejection filter with one pole and the transfer function.

$$\Omega_r = \pm 180° \cdot f_r / f_N \qquad 15.4\text{-}8$$

The poles and zeros must be located on both the positive and negative frequency axis because the cosine function is identical for either choice of sign and so it is indistinguishable in the time domain. However a sine function has odd symmetry and if one neglects having poles and zeros in complex conjugate pairs an arbitrary signal input would become complex after passing through such a filter. The zeros should be placed on the unit circle for the simplest form of a rejection filter.

$$z_z = \cos\Omega_r \pm i \sin\Omega_r = R_z \pm iI_z \qquad 15.4\text{-}9$$

The poles are placed just outside the unit circle to insure stability and a sharp cut-off. If the radius of the pole position is r_p then the poles occur at

$$z_p = r_p \cos\Omega_r \pm i r_p \sin\Omega_r = R_p \pm iI_p \qquad 15.4\text{-}10$$

If the Nyquist frequency is 500 Hz (sampling interval of 1 millisecond) and a rejection frequency of 50 Hz, the angle $\Omega_r = 18°$ and the zeros occur at

$$z_z = 0.951057 \pm i\, 0.309017$$

If the two poles are placed at an arbitrary distance such as $r_p = 1.01 = |z|$ then the poles are at

$$z_p = 0.960567 \pm i\, 0.312107$$

If the poles are placed too close to the unit circle then the filter will have a very long impulse response. The band of rejected frequencies will be very narrow but a very long signal is necessary for the filter to operate effectively. The z transform of the impulse response with a static gain factor, G, to insure a gain of unity at the Nyquist frequency is

$$W(z) = \frac{G[z-z_z][z-z_z^*]}{[z-z_p][z-z_p^*]} \qquad 15.4\text{-}11$$

Rewriting this equation with a one as the first term in the denominator and expanding the complex parts into real (R) and imaginary (I) parts yields

$$W(z) = \frac{G[z^2 - 2R_z z + R_z^2 + I_z^2]}{1 + [-2R_p z + z^2]/[R_p^2 + I_p^2]} \qquad 15.4\text{-}12$$

where

$$G = [1 + (2R_p + 1)/(R_p^2 + I_p^2)]/[2 + 2R_z] \qquad 15.4\text{-}13$$

For the example of the 50 Hz rejection filter the z transform of the impulse response is

$$W(z) = \frac{0.990124\,[z^2 - 1.902114\,z + 1]}{1 - 1.883280\,z + 0.980296\,z^2} \qquad 15.4\text{-}14$$

Following equation 15.2-6 the recursion relation is

$$y_n = 0.990124\,[x_n - 1.912114\,x_{n-1} + x_{n-2}]$$
$$- 0.980296\, y_{n-2} + 1.883280\, y_{n-1} \qquad 15.4\text{-}15$$

This equation may be converted to a zero phase shift rejection filter by programming a cascade form as in section 15.3. The z phase diagram and the transfer function are shown in figure 15.13.

Rejection and narrow pass filters

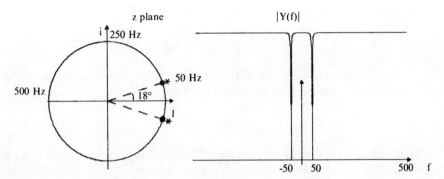

Figure 15.13 The z phase diagram for a 50 cycle rejection filter with two poles and its transfer function.

For a geometrical interpretation of the singularities on a z plane diagram (figure 15.14), let $z_1, z_2, p_1, p_2, p_3, p_4$ be the lengths of the vector from the frequency of interest, f, on the unit circle to the zeros and poles. Then

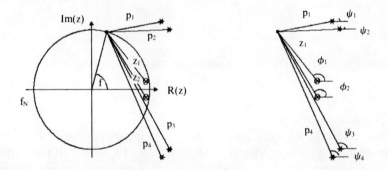

Figure 15.14 Poles and zeros on a z plane diagram. The diagram on the right shows the measurement of the phase lag for each singularity.

the ratio of the lengths of the vectors are related to the normalized transfer function by

$$|Y(f)| = \frac{z_1 z_2}{p_1 p_2 p_3 p_4} \qquad 15.4\text{-}16$$

The phase is given by

$$\phi(f) = \phi_1 + \phi_2 - (\psi_1 + \psi_2 + \psi_3 + \psi_4) \qquad 15.4\text{-}17$$

If the zeros are omitted in 15.4-11 then one obtains a narrow pass filter. The impulse response in the z domain is given by

$$W(z) = G / [(z - z_p)(z - z_p^*)] \qquad 15.4\text{-}18$$

The response of a typical narrow pass filter is shown in figure 15.15. As an exercise, the reader should construct a recursive narrow pass filter with a gain of unity at the Nyquist frequency. More elaborate notch or narrow pass filters are described by various authors such as Ahmed and Peterson (1979) and Hirano, Nishimura and Mitra (1974).

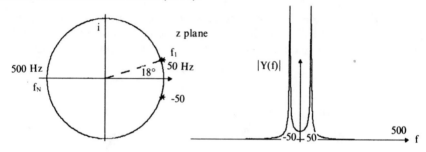

Figure 15.15   The z plane diagram and gain function for a pair of poles at $\pm f_1$.

15.5 Butterworth filters

Gibbs' phenomenon was seen to be a restriction on the practicality of a truncated ideal box-car shaped band pass filter. The square of the response of an ideal low pass filter may be approached in a more satisfactory manner by a general function of the type

$$|Y(W)|^2 = 1 / [1 + A_n(W)] \qquad 15.5\text{-}1$$

where

$$A_n \ll 1 \qquad 0 \leq W \leq 1$$
$$A_n \gg 1 \qquad W > 1$$

Note that the gain (figure 15.16) has been normalized to unity and also the desired cut-off frequency by replacing Ω_L with a normalized frequency, W.

Butterworth filters

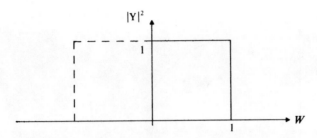

Figure 15.16 Response of a normalized low pass filter given by equation 15.5-1.

$$W = \Omega/\Omega_L \qquad \text{15.5-2}$$

In the following sections on Butterworth filters the symbol Ω will be used for *low pass filters* while a vector, W, will indicate a *normalized frequency*. Butterworth (1930) (see Weinberg, 1962, also) let

$$A_n = W^{2n} \qquad n = 1, 2, 3, ... \qquad \text{15.5-3}$$

to obtain a monotonic function as an approximation to the square response (figure 15.17). The square of a transfer function containing the following expression specifies a Butterworth filter.

$$|Y_L(W)|^2 = 1 / [1 + W^{2n}] \qquad \text{15.5-4}$$

The function $|Y_L|^2$ passes through 1/2 at $W = 1$ for all orders, n. The curve is maximally flat both inside and outside the band pass since it has no tendency to oscillate. A larger value of n gives a greater rate of attenuation. An example of a calculation to determine the value of n necessary is given below.

15.6 Number of poles required for a given attenuation

If the required attenuation, measured in units of power, at $2\Omega_L$ is 48 db down from attenuation at $\Omega = 0$, what value of n is necessary for a Butterworth filter? From a definition of the decibel scale

(Decibel level = $10 \log_{10}$ (power/reference power))

and using 15.5-4 we have that

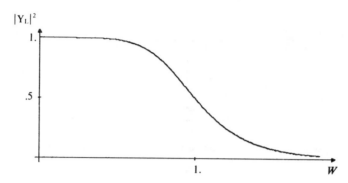

Figure 15.17 The square of the transfer function of a Butterworth filter for n = 3.

$$10 \log_{10} [1/(1 + W^{2n})]_{W=0} / [1/(1 + W^{2n})]_{W=2} = 48 \qquad 15.6\text{-}1$$
$$[1/(1 + 2^{2n})] = 10^{-4.8}$$

Approximately then

$$[1/2^{2n}] \approx 10^{-4.8} \qquad 15.6\text{-}2$$

$$n = 4.8 / 2 \log 2 = 4.8 / 2(.301) = 8 \qquad 15.6\text{-}3$$

Therefore we require n = 8 to obtain the required attenuation. In general the power changes by 6 db for a unit change in n.

15.7 Singularities of a low pass Butterworth filter

The poles and zeros are obtained by examining the Laplace transform of the transfer function. The Laplace transform is obtained from the Fourier transform in equation 15.5-4 by letting $p = \rho + iW$ and if $\rho = 0$.

$$p = iW \qquad 15.7\text{-}1$$

In this chapter p will be used to denote a Laplace transform variable for low pass filters. All other filters can be derived from this basic low pass equation and s will be used for the Laplace transform variable for these other filters.

Singularities of a low pass Butterworth filter

$$|Y_L(p)|^2 = 1 / [1 + (p/i)^{2n}] \qquad 15.7\text{-}2$$

$$|Y_L(p)|^2 = 1 / [1 + (-1)^n p^{2n}] \qquad 15.7\text{-}3$$

Since the numerator is constant there are no zeros in a low pass filter except at infinity. The poles occur at places where the denominator is zero

$$1 + (-1)^n p^{2n} = 0$$

or if we multiply both sides by $(-1)^n$

$$p^{2n} + (-1)^n = 0 \qquad 15.7\text{-}4$$

The poles will be around a unit circle and their location may be obtained from complex number theory by solving for the roots of the equation. In general a complex function may be written in polar form as

$$u = re^{i\phi} \qquad 15.7\text{-}5$$

The mth root, $u^{1/m}$, of a complex number, $u = r \cdot \exp(i\phi)$, is found from

$$u^{1/m} = r^{1/m} e^{i(\phi + 2\pi k)/m} \qquad k = 0, 1, 2, \ldots m-1 \qquad 15.7\text{-}6$$

Example 1

For $n = 1$ the roots of the square of the transfer function of a low pass filter are found from equation 15.7-4.

$$p^2 - 1 = 0$$

This may be solved by factoring

$$(p - 1)(p + 1) = 0$$

$$(iW - 1)(iW + 1) = 0$$

$$iW = \pm 1$$

or by equation 15.7-6 with $\phi = 0$, $u = 1$, $r = 1$ (figure 15.18)

$$p = 1^{1/2} = e^{i2\pi k/2} \qquad k = 0, 1.$$
$$iW = e^0 = 1$$
$$iW = e^{i\pi} = -1$$

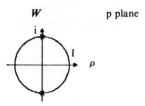

Figure 15.18   Poles of a first order low pass Butterworth filter where $p = \rho + iW$.

Example 2

A second order low pass Butterworth filter is obtained by setting $n = 2$ in equation 15.7-4.

$$p^4 + (-1)^2 = 0$$
$$(p^4 + 1) = 0$$

From 15.7-6 with $u = -1, r = 1$ and $\phi = \pi$

$$p = (-1)^{1/4} = e^{i(\pi + 2\pi k)/4} \qquad k \ 0, 1, 2, 3.$$
$$p = e^{i\pi/4}, \ e^{i3\pi/4}, \ e^{i5\pi/4}, \ e^{i7\pi/4}$$

The poles occur at

$$-(2)^{1/2} \pm i(2)^{1/2}$$
$$+(2)^{1/2} \pm i(2)^{1/2}$$

It is not surprising that the square of the transfer function contains a function, which may be called the Butterworth polynomial, $B_n(p)$, and its complex conjugate. That is, equation 15.5-4 may be written

$$|Y_L(p)|^2 = 1/[1 + W^{2n}] = 1/[B_n(p) \ B_n^*(p)] \qquad 15.7\text{-}7$$

Singularities of a low pass Butterworth filter

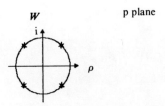

Figure 15.19 Poles of a second order low pass Butterworth filter.

The Laplace transform of the transfer function is then

$$Y_L(p) = 1/B_n(p) \qquad 15.7\text{-}8$$

The roots are on the left side of the p plane (figure 15.19) around the unit circle for the normalized filter. If the frequency is not normalized the poles will lie outside the unit circle at a radius ω_c.

The roots of the transfer function in the p plane are given by

$$p_{2\nu+1} = e^{i\pi(2\nu+1+n)/2n} \qquad \nu = 0, 1, \ldots n\text{-}1 \qquad 15.7\text{-}9$$

or by

$$p_{2\nu+1} = -\sin \pi(2\nu+1)/2n + i \cos \pi(2\nu+1)/2n \qquad 15.7\text{-}10$$

The Butterworth polynomial can be written as

$$B_n(p) = (p - e^{i\pi(1+n)/2n})(p - e^{i\pi(3+n)/2n}) \ldots (p - e^{2\pi(n-1)/2n})$$

or

$$B_n(p) = \prod_{\nu=0}^{n-1} (p - e^{i\pi(2\nu+1+n)/2n}) \qquad 15.7\text{-}11$$

As a polynomial expansion this becomes

$$B_n(p) = \sum_{k=0}^{n} a_k\, p^k \qquad 15.7\text{-}12$$

where

$$a_k = \prod_{\mu=1}^{k} \cos(\mu-1)\gamma / \sin \mu\gamma \qquad 15.7\text{-}13$$

and

$$\gamma = \pi/2n$$
$$a_0 = 1$$

Weinberg (1962, p. 494) supplies a proof for this theorem.

Example

Find the Butterworth polynomial and the square of the transfer function for a 2nd order low pass filter. Letting $n = 2$ equations 15.7-12 and 15.7-13 become

$$B_2(p) = 1 + (2)^{1/2} p + p^2$$
$$B_2(p) \, B_2^*(p) = (1 + (2)^{1/2} p + p^2)(1 + (2)^{1/2} p + p^2)^*$$

Letting $p = iW$ we have

$$B_2(W) \, B_2^*(W) = (1 + (2)^{1/2} iW - W^2)(1 - (2)^{1/2} iW - W^2)$$
$$= 1 + W^4$$
$$|Y_L(W)|^2 = 1 / [1 + W^4]$$

15.8 Frequency transformation for high pass filter

High pass and band pass filters may be easily constructed by an appropriate frequency transformation of a low pass filter. The transformations are carried out in the complex frequency or Laplace transform domain.

Let the frequency for the low pass functions be designated by

$$p = iW \qquad 15.8\text{-}1$$

and let the frequency in high pass functions be given by

$$s = iw \qquad 15.8\text{-}2$$

It is necessary to find a frequency transformation function which will convert one type of filter into another. Let such a function be

Frequency transformation for high pass filter

$$p = F(s) \quad \text{15.8-3}$$

For a *High Pass Filter* the transformation is expected to be naturally the inverse of the low pass frequency transform variable.

$$p = 1/s \quad \text{15.8-4}$$

Substituting 15.5-1 and 15.5-2 one has

$$W = -1/w \quad \text{15.8-5}$$

Example

Convert an nth order low pass Butterworth filter to the equivalent high pass filter. From 15.5-4 the low pass filter is obtained from

$$|Y_L(W)|^2 = 1/[1 + W^{2n}] \quad \text{15.5-4}$$

Applying the transformation in 15.8-5 we have that the square of the transfer function of a high pass filter is given by

$$|Y_H(w)|^2 = \frac{1}{1 + (-1/w)^{2n}} = \frac{w^{2n}}{1 + w^{2n}} \quad \text{15.8-6}$$

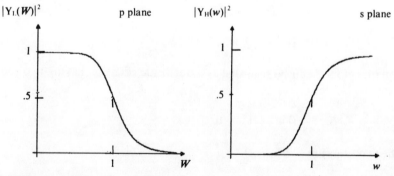

Figure 15.20 The square of the transfer function for an nth order Butterworth filter with low and high pass characteristics.

Notice that a high pass filter is equivalent to subtracting a low pass filter from one in the frequency domain (figure 15.20).

$$1 - |Y_L(W)|^2 \equiv 1 - 1/[1 + W^{2n}] \equiv W^{2n}/[1 + W^{2n}] \qquad 15.8\text{-}7$$

The procedure to be followed to obtain the transform for a high pass filter is as follows.
(1) Translate the specifications of a high pass filter to a normalized form. That is, determine ω_c, the cut off frequency, and the order n necessary for a certain attenuation and normalize the frequency.
(2) Find a low pass filter with the same specification as in (1) and convert this to the high pass equivalent filter.
(3) Calculate the poles and zeros of the filter.
(4) Remove the frequency normalization.
(5) The digital realization of the filter will be described in section 15.10. Synthesis of analog filters of this type are described in many engineering texts.

15.9 Frequency transformation for a band pass filter

It is to be expected that, as in section 15.1, a band pass filter is made up of a combination of a low pass and a high pass filter. The normalized transformation, 15.8-3, is given by
$$p = s + 1/s \qquad 15.9\text{-}1$$

where p is the complex frequency in the low pass function and s is the complex frequency in the band pass function. Substituting 15.8-1 and 15.8-2 we have

$$W = w - 1/w$$

or

$$W = [w^2 - 1]/w \qquad 15.9\text{-}2$$

Equation 15.9-2 can be regarded as a quadratic equation having two roots, w_1 and w_2.

$$w^2 - Ww - 1 = 0 \qquad 15.9\text{-}3$$

The two roots are

$$w_{1,2} = W/2 \pm [W^2/4 + 1]^{1/2} \qquad 15.9\text{-}4$$

Note that $W = 0$ transforms to $w = \pm 1$.
In the low pass domain the band pass is specified by setting $W = 1$.

Frequency transformation for a band pass filter

$$w_{1,2} = 1/2 \pm (5/4)^{1/2}$$
$$= 1/2 \pm 1.118$$
$$w_1 = -0.618$$
$$w_2 = +1.618 \qquad 15.9\text{-}5$$

From $W = -1$ we obtain the conjugate pair of frequencies $w_2 = -1.618$, $w_1 = +0.618$. These four roots specify the band width of the normalized band pass filter (see figure 15.21).

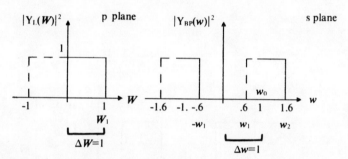

Figure 15.21 Each point in the transfer function of a low pass filter is transformed into two points in the band pass filter.

It is interesting to note that any two arbitrary frequencies in the pass band domain may be obtained by the appropriate transformations from a low pass filter. The normalized band pass width is one and we have the relation that for any frequency

$$w_1 w_2 = 1 \qquad 15.9\text{-}6$$

This is easily verified by substituting w_1 and w_2 from 15.9-4.

The normalized relations must be generalized to the desired band pass filter with unnormalized upper and lower frequency limits, ω_1 and ω_2. The geometric mean, ω_0, which is sometimes imprecisely called the center frequency is

$$\omega_0 = (\omega_1 \omega_2)^{1/2} \qquad 15.9\text{-}7$$

Any other frequency, ω_l, with a particular attenuation is related to a higher frequency, ω_h, by the relation

$$\omega_0 = (\omega_l\omega_h)^{1/2} \qquad 15.9\text{-}8$$

The band width, $\Delta\omega$, is related to the low pass cut-off as seen in figure 15.21.

$$\Delta\omega = \omega_2 - \omega_1 \qquad 15.9\text{-}9$$

The unnormalized low pass and band pass filters are seen in figure 15.22 which may be compared to figure 15.21.

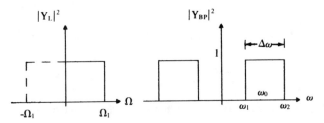

Figure 15.22 Frequency relations between an unnormalized low pass and a band pass filter under the transformation given by equation 15.9-1.

The band width will be rescaled by dividing by the geometric mean.

$$\Omega_1 = [\omega_2 - \omega_1] / \omega_0 \qquad 15.9\text{-}10$$

The scaling procedure affects other frequencies, Ω_h, as follows:

$$\Omega_h = [\omega_h - \omega_l] / \omega_0 \qquad 15.9\text{-}11$$

but from 15.9-7. $\omega_0^2 = \omega_l\omega_h$ so we can eliminate ω_l

$$\Omega_h = \frac{1}{\omega_0}\left(\omega_h - \frac{\omega_0^2}{\omega_h}\right)$$

$$\Omega_h = \frac{\omega_h}{\omega_0} - \frac{\omega_0}{\omega_h} \qquad 15.9\text{-}12$$

Frequency transformation for a band pass filter

This relation is generally applicable to any frequency so it can be written

$$\Omega = \frac{\omega}{\omega_0} - \frac{\omega_0}{\omega} \qquad 15.9\text{-}13$$

The normalized frequency is obtained by dividing 15.9-13 by 15.9-10, the low pass frequency cut-off as in equation 15.5-2.

$$W = \frac{\Omega}{\Omega_1} = \frac{\dfrac{\omega}{\omega_0} - \dfrac{\omega_0}{\omega}}{\dfrac{\omega_2 - \omega_1}{\omega_0}}$$

Then making use of 15.9-7 we have

$$W = [\omega^2 - \omega_1\omega_2] / [\omega(\omega_2 - \omega_1)] \qquad 15.9\text{-}14$$

Note that, as in figure 15.21, frequencies above the geometric mean remain positive in the low pass domain while positive frequencies below ω_0 are mapped into the negative domain.

For a Laplace transformation let $p = iW$ and $s = i\omega$.

$$p/i = \frac{-s^2 - \omega_1\omega_2}{s(\omega_2 - \omega_1)/i}$$

or

$$p = [s^2 + \omega_1\omega_2] / [s(\omega_2 - \omega_1)] \qquad 15.9\text{-}15$$

This equation is then rearranged to form a quadratic in s.

$$s^2 - p(\omega_2 - \omega_1)s + \omega_1\omega_2 = 0$$

The roots of the equation will be required later to locate the poles of the band pass filter and are

$$\left.\begin{array}{c} s \\ \\ s^* \end{array}\right\} = \frac{p(\omega_2 - \omega_1)}{2} \pm \left[\frac{p^2(\omega_2 - \omega_1)^2}{4} - \omega_1\omega_2\right]^{1/2} \qquad 15.9\text{-}16$$

From each of the n poles of a low pass filter in the p plane there are generated

2n poles, in complex conjugate pairs, in the s plane. Equation 15.9-16 may be written

$$\left.\begin{array}{c}s_i\\ \\s_i^*\end{array}\right\} = p_j \frac{(\omega_2 - \omega_1)}{2} \pm [\frac{p_j^2(\omega_2 - \omega_1)^2}{4} - \omega_1\omega_2]^{1/2} \qquad i, j = 1, 2, ...n \qquad 15.9\text{-}17$$

According to equation 15.7-8 the Laplace transform of the low pass transfer function is

$$Y_L(p) = 1 / B_n(p) \qquad 15.7\text{-}8$$

As an example a fourth order Butterworth polynomial will be used. From equation 15.7-11 or from Table 11.2 in Weinberg (1962) this is

$$B_4 = 1 + 2.6131259\ p + 3.4142136\ p^2 + 2.6131259\ p^3 + p^4 \qquad 15.9\text{-}18$$

The poles of the normalized low pass filter (Table 11.1 in Weinberg or equation 15.7-10) occur at

$$p_{1,4} = -0.3826834 \pm i\ 0.9238795$$
$$p_{2,3} = -0.9238795 \pm i\ 0.3826834 \qquad 15.9\text{-}19$$

The transfer function of 15.7-8 has the form

$$Y_L(p) = 1 / [(p - p_1)(p - p_2)(p - p_3)(p - p_4)] \qquad 15.9\text{-}20$$

As required by theory these occur on the unit circle in the p plane. For a band pass filter the transformation is

$$p = [s^2 + 1] / s \qquad 15.9\text{-}1$$

Therefore the Laplace transform for a normalized band pass filter is

$$Y_{NBP}(s) = \frac{1}{\dfrac{(s^2+1-p_1s)}{s} \dfrac{(s^2+1-p_2s)}{s} \dfrac{(s^2+1-p_3s)}{s} \dfrac{(s^2+1-p_4s)}{s}} \qquad 15.9\text{-}2$$

Frequency transformation for a band pass filter

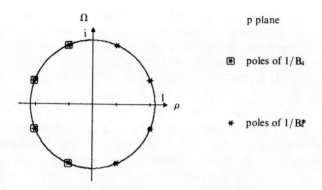

Figure 15.23 Poles of a fourth order low pass Butterworth filter.

or

$$Y_{NBP}(s) = s^4 / [s^8 + g_7 s^7 + g_6 s^6 + g_5 s^5 + g_4 s^4 + g_3 s^3 + g_2 s^2 + g_1 s + g_0] \quad 15.9\text{-}22$$

The factored form of 15.9-21 is more useful but it is the unnormalized transfer function that is required. This may be obtained directly with the pole locations given by equation 15.6-17.

$$\boxed{Y_{BP}(s) = \frac{s}{(s^2+b_1 s+c_1)} \frac{s}{(s^2+b_2 s+c_2)} \frac{s}{(s^2+b_3 s+c_3)} \frac{s}{(s^2+b_4 s+c_4)}} \quad 15.9\text{-}23$$

The coefficients b_j and c_j are obtained from the pole positions, which occur in complex conjugate pairs to make each of the terms in brackets above real.

$$\boxed{(s - s_j)(s - s_j^*) = (s^2 + b_j s + c_j)} \quad 15.9\text{-}24$$

The continuous s plane representation of the transfer function must be converted to a digital or z plane representation. This is accomplished with the standard z transform or a bilinear z transform as discussed in the next section.

In the frequency or Laplace transform domain the *band stop* filter may be obtained by taking the reciprocal of the transformation for the band pass filter.

$$p = s(\omega_2 - \omega_1) / [s^2 + \omega_1 \omega_2] \quad 15.9\text{-}25$$

The transformations presented above are quite general and do not apply only to the Butterworth filters. They may, for instance, be used to design operators with Chebyshev or elliptic functions that allow some ripple to achieve a sharper cut-off (see Guillemin, 1957; Weinberg, 1962; Golden and Kaiser, 1964; Cole and Oakes, 1961).

15.10 The bilinear z transform

Golden and Kaiser (1964) have employed an algebraic transformation which converts a continuous transfer function into one which can be used on sampled data. Furthermore it eliminates aliasing errors which are present when the standard z transform of equation 13.5-1 is applied directly to a transfer function:

$$z = e^{-s(\Delta t)} \qquad 15.10\text{-}1$$

where Δt is the sampling interval. Equation 15.10-1 may be written as

$$\log_e z = -s(\Delta t)$$

which may be expanded in series form (see Peirce No. 841) as

$$\log_e z = -2[(1-z)/(1+z) + (1-z)^3/3(1+z)^3 + \ldots] \quad z > 0$$

Taking the first term one obtains a non-linear warping of the frequency called the bilinear z transform. It preserves the minimum phase property of the filter but introduces a non-linear warping of the frequency scale which must be compensated before applying it. The desired transformation is

$$s_d = \frac{2}{\Delta t} \frac{1-z}{1+z} \qquad 15.10\text{-}2$$

In fact the bilinear z transform is well known in electrical engineering (Jury, 1963; Guillemin, 1949) and in the mathematical theory of complex variables where it is referred to as the linear fractional transformation (Churchill, 1960). This transform is also used in plane wave theory where it is called a reflection coefficient in which the impedance of the uppermost interface is normalized to unity. The transformation

$$s = [1-z]/[1+z] \qquad 15.10\text{-}3$$

The bilinear z transform

maps the right half of the z plane into the interior of the unit circle in the s plane. The inverse function

$$z = [1 - s] / [1 + s] \qquad 15.10\text{-}4$$

has the same form and therefore the entire right half of the s plane maps into the interior of the unit circle in the z plane (see figure 15.24).

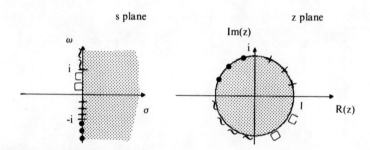

Figure 15.24 Mapping of the bilinear transform. Compare this to figure 13.3 in which an ordinary z transform is shown.

The warping of the frequency domain is found by substituting 15.10-1 into 15.10-2.

$$s_d = \frac{2}{\Delta t} \frac{1 - e^{-s(\Delta t)}}{1 + e^{-s(\Delta t)}}$$

Multiplying numerator and denominator by $\exp(s\Delta t/2)$

$$s_d = \frac{2}{\Delta t} \frac{\exp(s\Delta t/2) - \exp(-s\Delta t/2)}{\exp(s\Delta t/2) + \exp(-s\Delta t/2)}$$

$$s_d = \frac{2}{\Delta t} \tanh\ s\ \Delta t/2 \qquad 15.10\text{-}5$$

the transformation may be converted to the frequency domain by letting $s = i\omega$ and by using the trigonometric identity that $\tanh x = -i \tan ix$

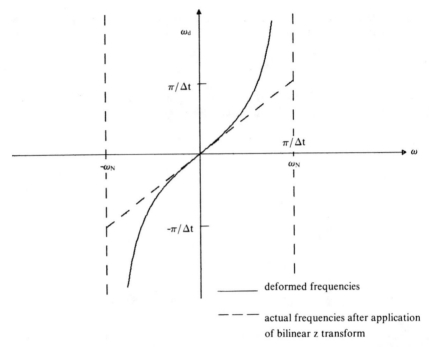

Figure 15.25 Relation between the deformed frequencies for the bilinear z transform and the actual frequencies of the digital filter.

$$\omega_d = \frac{2}{\Delta t} \tan \frac{\omega \Delta t}{2} \qquad -\pi/\Delta t < \omega < \pi/\Delta t \qquad 15.10\text{-}6$$

ω_d = deformed angular frequency used to calculate s_d and bilinear z transform.

ω = angular frequency in the original transfer function.

The right-hand side of equation 15.10-6 is periodic but ω is restricted to the principal values (figure 15.25). The prewarping must be applied *before* the poles of the transfer function in equation 15.9-16 are determined. The entire ω_d axes are mapped uniquely into the frequencies bounded by the Nyquist limits, $\omega = \pm \pi/\Delta t$. Thus the transformation places limits on the frequency band of the

The bilinear z transform

function without introducing aliasing problems. The bilinear z transform is similar to the standard z transform at low frequencies (see figure 13.25). If the digital interval is large the standard z transform is especially susceptible to producing aliasing errors and it is recommended that the bilinear transform be used to avoid this.

Applying the bilinear z transform to the band pass transfer function (equation 15.9-23) it is seen that there are terms in the deformed frequency or s plane such as

$$\frac{s_d}{s_d^2 + b_j s_d + c_j} = \frac{\dfrac{2\Delta t}{(\Delta t)^2}\dfrac{(1-z)(1+z)}{(1+z)^2}}{\dfrac{4}{(\Delta t)^2}\dfrac{(1-z)^2}{(1+z)^2} + \dfrac{2b_j\Delta t}{(\Delta t)^2}\dfrac{(1-z)\cdot(1+z)}{(1+z)^2} + c_j\dfrac{(1+z)^2(\Delta t)^2}{(1+z)^2(\Delta t)^2}}$$

$$= \frac{\dfrac{1}{a_j}(1 - z^2)}{1 + \dfrac{z}{a_j}\left(c_j\Delta t - \dfrac{4}{\Delta t}\right) + \dfrac{z^2}{a_j}\left(\dfrac{2}{\Delta t} - b_j + c_j\dfrac{\Delta t}{2}\right)} \qquad 15.10\text{-}7$$

where

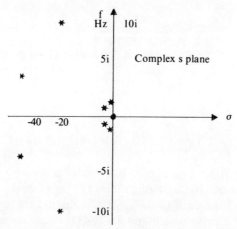

Figure 15.26 Poles and zeros of a band pass filter with cut-off frequencies at 1 and 10 cycles per second. ✶ is the location of a pole of Y in the deformed frequency domain.

$$a_j = 2/\Delta t + b_j + c_j\Delta t/2 \qquad 15.10\text{-}8$$

Equation 15.10-7 may be written more compactly using

$$\frac{S_d}{S_d^2 + b_j S_d + c_j} = \frac{\frac{1}{a_j}(1 - z^2)}{B_j} \qquad 15.10\text{-}9$$

where
$$B_j = 1 - D_{1j}z + D_{2j}z^2 \qquad 15.10\text{-}10$$

D_{2j} and D_{2j} are coefficients of z and z^2 in the denominator of equation 15.10-7. Omitting the static gain factor, $1/(a_1 a_2 a_3 a_4)$, the z transform of a Butterworth band pass filter with 8 poles is given by

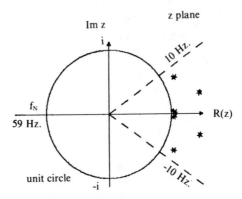

Figure 15.27 Location of poles and zeros in the z plane of a band pass filter with cut-off frequencies at 1 and 10 Hz.

$$W(z) = [(1 - z^2)^4] / [B_1(z) B_2(z)] [B_3(z) B_4(z)] \qquad 15.10\text{-}11$$

As an example, consider a band pass filter with 8 poles with cut-off frequencies at 1 and 10 Hz. Assume the digitizing rate is 118 per second.

From equation 15.10-6 the deformed angular frequencies are 6.28 and 64.36 radians per second or 1.0 and 10.24 Hz. Using equation 15.9-17 the s plane poles occur at the following locations (see also figure 15.26):

σ	$i\omega$	f(Hz)
-47.0526	±i 25.8533	4.03
-20.1726	±i 59.7337	9.51
- 6.6029	±i 3.6280	0.58
- 2.0527	±i 6.0782	0.97

The bilinear z transform

The coefficients of the polynomials in equation 15.9-22 are

$$g_0 = 2.67661 \times 10^{10} \qquad g_4 = 2.16737 \times 10^7$$
$$g_1 = 1.00427 \times 10^{10} \qquad g_5 = 6.96027 \times 10^5$$
$$g_2 = 2.14872 \times 10^9 \qquad g_6 = 1.31337 \times 10^4$$
$$g_2 = 2.81529 \times 10^8 \qquad g_7 = 1.51762 \times 10^2$$

The coefficients for equation 15.9-23 are

$$b_1 = 94.10519 \qquad c_1 = 2882.33984$$
$$b_2 = 40.34528 \qquad c_2 = 3975.04687$$
$$b_3 = 13.20584 \qquad c_3 = 56.76093$$
$$b_4 = 4.10533 \qquad c_4 = 41.15739$$

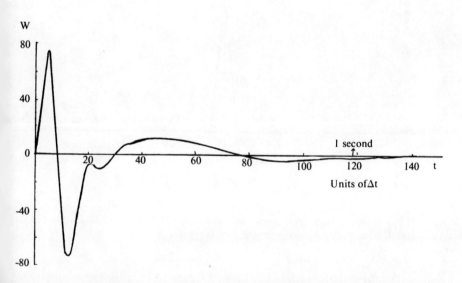

Figure 15.28 Impulse response for a band pass filter with cut-off at 1 and 10 Hz. The digitizing interval is 118 times per second.

The location of the singularities of 15.10-11 (see figure 15.27) are at

$$z_1 = 1.452137 \pm i\ 0.335533$$
$$z_2 = 1.031333 \pm i\ 0.562206$$
$$z_3 = 1.057052 \pm i\ 0.032561$$
$$z_4 = 1.016185 \pm i\ 0.052410$$

The terms in the denominator of equation 15.10-11 are

$$B_1 = 1 - 1.307476z + 0.450190z^2$$
$$B_2 = 1 - 1.494886z + 0.724783z^2$$
$$B_3 = 1 - 1.890261z + 0.894119z^2$$
$$B_4 = 1 - 1.962924z + 0.965829z^2$$

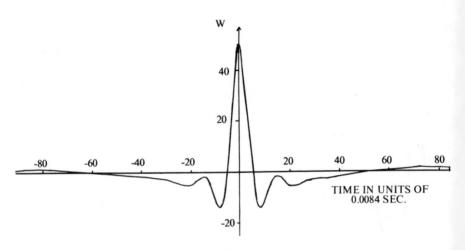

Figure 15.29 Impulse responses for a band pass zero phase shift filter with cut-off a 1 and 10 Hz. The abscissa is in units of the digitizing interval, $\Delta t = 1/118$ second.

A sketch of the impulse response is given if figure 15.28 while the modulus and phase of the transfer function are shown in figure 15.30. It is seen that approximately 130 points are necessary to define the impulse response adequately. The impulse response should be checked in all cases to see that this condition is satisfied. The impulse response of a zero phase shift 1-10 Hz band pass filter with 8 poles is given in figure 15.29. The parameters are identical to the Butterworth operator illustrated in figure 15.28. The transfer function of the cascade form is dotted in figure 15.30 where it is seen that a steeper roll-off has been obtained because the filter operates twice in cascade on the same data.

A Fortran subroutine for zero phase shift, Butterworth function, band

The bilinear z transform

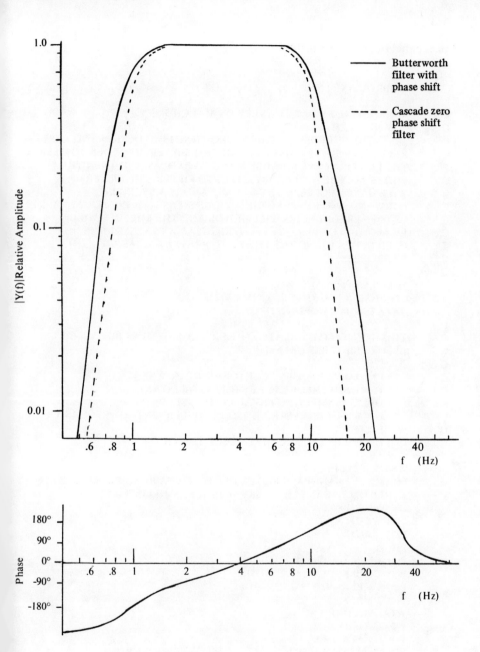

Figure 15.30 Gain and phase response of a band pass Butterworth filter with 8 poles.

pass panel filter is listed below:

```
      SUBROUTINE BNDPAS(F1,F2,DELT,D,G)
C
C     SUBROUTINE BY DAVE GANLEY ON MARCH 5, 1977.
C
C            THE PURPOSE OF THIS SUBROUTINE IS TO DESIGN AND APPLY A
C     RECURSIVE BUTTERWORTH BAND PASS FILTER. IN ORDER TO DESIGN
C     THE FILTER A CALL MUST BE MADE TO BNDPAS AND THEN THE
C     FILTER MAY BE APPLIED BY CALLS TO FILTER. THE FILTER
C     WILL HAVE 8 POLES IN THE S PLANE AND IS APPLIED IN
C     FORWARD AND REVERSE DIRECTIONS SO AS TO HAVE ZERO PHASE
C     CUTOFF FREQUENCIES WILL BE -6DB AND THE ROLLOFF WILL
C     BE ABOUT 96 DB PER OCTAVE. A BILINEAR Z TRANSFORM IS
C     USED IN DESIGNING THE FILTER TO PREVENT ALIASING
C     PROBLEMS.
C
      COMPLEX P(4),S(8),Z1,Z2
      DIMENSION D(8),X(1),XC(3),XD(3),XE(3)
      DATA ISW/0/,TWOPI/6.2831853/
C
C     THIS SECTION CALCULATES THE FILTER AND MUST BE CALLED
C     BEFORE FILTER IS CALLED
C
C        F1 = LOW FREQUENCY CUTOFF (6 DB DOWN)
C        F2 = HIGH FREQUENCY CUTOFF (6 DB DOWN)
C        DELT = SAMPLE INTERVAL IN MILLISECONDS
C        D = WILL CONTAIN 8 Z DOMAIN COEFFICIENTS OF RECURSIVE
C        FILTER
C        G = WILL CONTAIN THE GAIN OF THE FILTER
C
      WRITE (6,1) F1,F2,DELT
    1 FORMAT ('1 BANDPASS FILTER DESIGN FOR A BAND FROM ',F8.3,' TO',F8
     ..3,' HERTZ.',//' SAMPLE INTERVAL IS ',F5.2,' MILLISECONDS.')
      DT=DELT/1000.0
      TDT=2.0/DT
      FDT=4.0/DT
      ISW=1
      P(1)=CMPLX(-.3826834,.9238795)
      P(2)=CMPLX(-.3826834,-.9238795)
      P(3)=CMPLX(-.9238795,.3826834)
      P(4)=CMPLX(-.9238795,-.3826834)
      W1=TWOPI*F1
      W2=TWOPI*F2
      W1=TDT*TAN(W1/TDT)
      W2=TDT*TAN(W2/TDT)
      HWID=(W2-W1)/2.0
      WW=W1*W2
      DO 19 I=1,4
      Z1=P(I)*HWID
      Z2=Z1*Z1-WW
```

The bilinear z transform

```
      Z2=CSQRT(Z2)
      S(I)=Z1+Z2
   19 S(I+4)=Z1-Z2
      WRITE (6,2) S
    2 FORMAT ('-S PLANE POLES ARE AT:',/' ',8(/' ',E12.6,' + I ',E12.6))
      G=.5/HWID
      G=G*G
      G=G*G
      DO 29 I=1,7,2
      B=-2.0*REAL(S(I))
      Z1=S(I)*S(I+1)
      C=REAL(Z1)
      A=TDT+B+C/TDT
      G=G*A
      D(I)=(C*DT-FDT)/A
   29 D(I+1)=(A-2.0*B)/A
      G=G*G
      WRITE (6,3)
    3 FORMAT ('-FILTER IS (1-Z** 2)**4 / B1*B2*B3*B4')
      WRITE (6,4) D
    4 FORMAT (4(/' B(I) = 1 + ',E12.6,' Z + ',E12.6,' Z**2'))
      WRITE (6,5) G
    5 FORMAT ('-FILTER GAIN IS ',E12.6)
      RETURN
C
C
C
      ENTRY FILTER(X,N,D,G,IG)
C
C     X = DATA VECTOR OF LENGTH N CONTAINING DATA TO BE FILTERED
C     D = FILTER COEFFICIENTS CALCULATED BY BNDPAS
C     G = FILTER GAIN
C     IG = 1 MEANS TO REMOVE THE FILTER GAIN SO THAT THE GAIN IS
C     UNITY
C
      IF (ISW.EQ.1) GO TO 31
      WRITE (6,6)
    6 FORMAT ('1BNDPAS MUST BE CALLED BEFORE FILTER')
      CALL EXIT
C
C     APPLY FILTER IN FORWARD DIRECTION
C
   31 XM2=X(1)
      XM1=X(2)
      XM=X(3)
      XC(1)=XM2
      XC(2)=XM1-D(1)*XC(1)
      XC(3)=XM-XM2-D(1)*XC(2)-D(2)*XC(1)
      XD(1)=XC(1)
      XD(2)=XC(2)-D(3)*XD(1)
      XD(3)=XC(3)-XC(1)-D(3)*XD(2)-D(4)*XD(1)
      XE(1)=XD(1)
```

```
      XE(2)=XD(2)-D(5)*XE(1)
      XE(3)=XD(3)-XD(1)-D(5)*XE(2)-D(6)*XE(1)
      X(1)=XE(1)
      X(2)=XE(2)-D(7)*X(1)
      X(3)=XE(3)-XE(1)-D(7)*X(2)-D(8)*X(1)
      DO 39 I=4,N
      XM2=XM1
      XM1=XM
      XM=X(I)
      K=I-((I-1)/3)*3
      GO TO (34,35,36),K
   34 M=1
      M1=3
      M2=2
      GO TO 37
   35 M=2
      M1=1
      M2=3
      GO TO 37
   36 M=3
      M1=2
      M2=1
   37 XC(M)=XM-XM2-D(1)*XC(M1)-D(2)*XC(M2)
      XD(M)=XC(M)-XC(M2)-D(3)*XD(M1)-D(4)*XD(M2)
      XE(M)=XD(M)-XD(M2)-D(5)*XE(M1)-D(6)*XE(M2)
   39 X(I)=XE(M)-XE(M2)-D(7)*X(I-1)-D(8)*X(I-2)
C
C     FILTER IN REVERSE DIRECTION
      XM2=X(N)
      XM1=X(N-1)
      XM=X(N-2)
      XC(1)=XM2
      XC(2)=XM1-D(1)*XC(1)
      XC(3)=XM-XM2-D(1)*XC(2)-D(2)*XC(1)
      XD(1)=XC(1)
      XD(2)=XC(2)-D(3)*XD(1)
      XD(3)=XC(3)-XC(1)-D(3)*XD(2)-D(4)*XD(1)
      XE(1)=XD(1)
      XE(2)=XD(2)-D(5)*XE(1)
      XE(3)=XD(3)-XD(1)-D(5)*XE(2)-D(6)*XE(1)
      X(N)=XE(1)
      X(N-1)=XE(2)-D(7)*X(1)
      X(N-2)=XE(3)-XE(1)-D(7)*X(2)-D(8)*X(1)
      DO 49 I=4,N
      XM2=XM1
      XM1=XM
      J=N-I+1
      XM=X(J)
      K=I-((I-1)/3)*3
      GO TO (44,45,46),K
```

The bilinear z transform

```
   44 M=1
      M1=3
      M2=2
      GO TO 47
   45 M=2
      M1=1
      M2=3
      GO TO 47
   46 M=3
      M1=2
      M2=1
   47 XC(M)=XM-XM2-D(1)*XC(M1)-D(2)*XC(M2)
      XD(M)=XC(M)-XC(M2)-D(3)*XD(M1)-D(4)*XD(M2)
      XE(M)=XD(M)-XD(M2)-D(5)*XE(M1)-D(6)*XE(M2)
   49 X(J)=XE(M)-XE(M2)-D(7)*X(J+1)-D(8)*X(J+2)
      IF (IG.NE.1) RETURN
      DO 59 I=1,N
   59 X(I)=X(I)/G
      RETURN
      END
```

15.11 Other digital filters

In conclusion, digital filters have many advantages because of their flexibility and economy, particularly at low frequencies. If very sharp cut-offs and great economy are required it is easiest to obtain these using a recursive design. The only disadvantage of this form is that they require a long data set for greatest effect since they have an infinite memory. For band pass filtering one may use functions other than the Butterworth such as the Bessel, Chebyshev or elliptic. Thus the Chebyshev filter of the first kind has ripple within the pass and decays monotonically in the stop band (figure 15.31). The nth order Chebyshev polynomial with ripple in the pass band specified by ϵ is given by

$$A_n = \epsilon^2 \, T_n^2(W) \qquad n=0,1,2,... \qquad 15.11\text{-}1$$

in equation 15.5-1. The function T_n is

$$T_n = \begin{cases} \cos(n \cos^{-1}W) & |W| \leq 1 \\ \cosh(n \cosh^{-1}W) & |W| > 1 \end{cases} \qquad 15.11\text{-}2$$

If the ripple is to be confined to the reject band then the filter of the second kind has the form

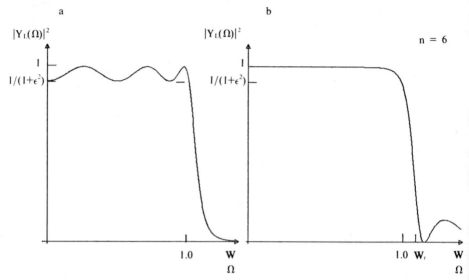

Figure 15.31 Amplitude-squared response of a Chebyshev filter for even n. (a) Filter of the first kind with ripple in the pass band. (b) Filter of the second kind with ripple in the stop band.

$$A_n = \epsilon^2 [T_n(W_r) / T_n(W_r / W)]^2 \qquad 15.11\text{-}3$$

where W_r is the frequency at which the reject band first reaches its desired cut-off value. This second type of Chebyshev filter is preferred to the first since the phase is less variable in the pass band. Note that in equation 15.11-2 the cosine function becomes imaginary for W greater than unity so it becomes a hyperbolic function which increases monotonically with frequency. The poles of the filter in the p plane are on an ellipse (Guillemin, 1957; Gold and Radar, 1969; Oppenheim and Schafer, 1975).

A filter with an even more rapid rate of attenuation for a given number of poles and zeros is based on an example using a Jacobian elliptic function. The ripple occurs in both the pass and stop band (figure 15.32). The elliptic filter is the lowest order filter which can be used when the signal and the noise are separated by the smallest transition region. The phase distortion is relatively severe and the location of the poles and zeros are computed by more complex equations (see Daniels, 1974 or Gold and Radar, 1969). It is recommended that the Butterworth function, which yields an optimum filter when the signal and noise are clearly separated in bands, be used whenever simple low pass, high pass, or band pass filtering is required. The Chebyshev or elliptic filters should only be used in special circumstances where a very sharp

Other digital filters

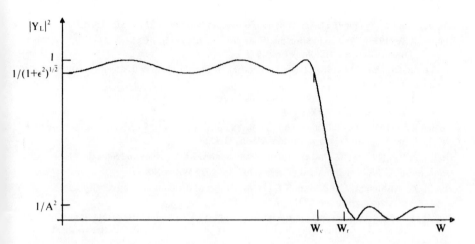

Figure 15.32   Amplitude-squared response of an elliptic filter for even n.

cut-off is required and where a long impulse response does not create difficulties. Many other forms of differentiating and band pass filtering are available (Daniels, 1974; Constantinides, 1969; Golden, 1968; Rabiner and Gold, 1975; Chan and Rabiner, 1973; Chan and Chai, 1979). For most geophysical applications the rejection or notch filter and the Butterworth band pass filter with zero phase shift are the most useful and least likely to be misapplied.

References

Ahmed,I., Abu-el-Haija and Peterson,A.M.(1979), A structure for digital notch filters. *Transactions IEEE on Acoustics, Speech and Signal Processing* **ASSP-27**, 193-195.

Butterworth,S.(1930), On the theory of filter amplifiers *Wireless Engr.* **1**, 536-541.

Chan,A.K. and Chai,C.K.(1979), Non-recursive digital filter designs by means of Korovkin Kernals. *Transactions IEEE on Acoustics, Speech and Signal Processing* **ASSP-27**, 218-222.

Chan,D.S.K. and Rabiner,L.R.(1973), Analysis of quantization errors in the direct form for finite impulse response digital filters. *Transactions IEEE on Audio and Electroacoustics* **AV-21**, 354-366.

Churchill,R.V.(1960), Complex variables and applications. McGraw-Hill Book Co., Inc., New York, 2nd Ed., 73-77, 286.

Cole,A.J. and Oakes,J.B.(1961), Linear vacuum tube and transfer circuits. McGraw-Hill Book Co., 164-177.

Constantinides,A.G.(1969), Design of bandpass digital filters, *Proc. IEEE* **57**, 1229-1231.

Daniels,R.W.(1974), Approximation methods for the designing of passive, active, and digital filters. McGraw-Hill Book Co., 1974.

Gold,B. and Radar,C.M.(1969), Digital processing of signals. Chapter 3, McGraw-Hill Book Co., New-York.

Golden,R.M. and Kaiser,J.F.(1964), Design of a side-band sampled data filter. *The Bell Telephone Technical Journal, 1533-1547*.

Golden,R.M.(1968), Digital filter synthesis by sampled-data transformation, *Transactions IEEE on Audio and Electroacoustics* **AV-16**, 321-329.

Guillemin,E.A.(1949), The Mathematics of circuit analysis. John Wiley and Sons Inc., New York.

Guillemin,E.A.(1957), Synthesis of passive networks. John Wiley, 588-591.

Hirano,K., Nishimura,S. and Mitra,S.K.(1974), Design of digital notch filters. *Transactions IEEE on Communications* **COM-22**, 964-970.

Jury,E.I.(1963), Recent advances in the field of sampled data and digital control systems. In theory of control systems using discrete information, Ed. J. F. Coales, Butterworths, London, 1-6.

Kaiser,J.F.(1966), Digital Filters. Ch. 7 in *Systems Analysis by Digital Computer*. Editors: F.F. Kuo and J.F. Kaiser, 218-285, John Wiley and Sons, New York.

Martin,M.A.(1959), Frequency domain applications to data processing. *IRE Transcations on Space Electronics and Telemetry* **Vol.SET-5**, 33-41.

Oppenheim,A.V. and Schafer,R.W.(1975), Digital signal processing. Chapter 5, Prentice-Hall, Inc., Englewood Cliffs, New Jersey.

Ormsby,J.F.A.(1961), Design of numerical filters with applications to missle data processing *Jour. Acm* **8**, 440-466.

Peirce,B.O.(1956), A short table of intergrals. Ginn and Co., Boston.

Rabiner,L.R. and Gold,B.(1975), Theory and application of digital signal processing. Chapter 4, Prentice-Hall, Inc., Englewood Cliffs, New Jersey.

Radar,C.M. and Gold,B.(1967), Digital filter disign techniques in the frequency domain *Proc. IEEE* **55**, 149-171.

Robinson,E.A.(1966), Fortran II programs for filtering and spectral analysis of single channel time series. *Geophys. Pros. XIV*.

Shanks,J.L.(1967), Recursion filters for digital processing *Geophysics* **32**, 33-51.

Tattersall,G.D. and Carey,M.J.(1979), Digital band limiting filters for P.C.M. transmission systems. *Transactions IEEE on Communications* **COM-27**, 240-246.

Tooley,R.D.(1955), Tables of special linear operators. *Part 8 of geophysical Analysis Group Report No. 9*. Editor: S.M. Simpson, Jr., Department of Geology and Geophysics, M.I.T., Cambridge, Mass.

Truxal,J.G.(1955), Automatic feedback control system syntheses. McGraw-Hill, 531.

Weinberg,L.(1962), Network analysis and synthesis. McGraw-Hill Book Co., Inc., Toronto, 485-498.

16 Wave propagation in layered media in terms of filter theory

16.1 Introduction
There has been considerable success in restating the problem of plane wave propagation in layered media by means of filter theory (Baranov and Kunetz, 1960; Wuenschel, 1960; Goupillaud, 1961; Kunetz and D'Erceville, 1962; Trorey, 1962; Kunetz, 1964; Sherwood and Trorey 1965; Robinson and Treitel, 1965; Trietel and Robinson, 1966). The method was originally used in optics by Abels (1946) and Crook (1948). The technique has had its most immediate success in attacking problems of generating synthetic seismograms and removing objectionable reverberations from field seismograms. As an introduction to the subject the article by Robinson and Treitel (1965) is recommended.

16.2 Reflection and transmission at an interface
Consider a system with two semi-infinite elastic layers in welded contact. Assume that a plane compressional wave has been generated and is propagating downward in the positive y direction at normal incidence to the plane separating the two media. Since no shear waves are generated at the boundary under conditions of normal incidence one needs to consider only a reflected and a transmitted compressional wave.

Let the displacement due to the incident wave, V_i, in the first media be given by the real part of

$$V_i = A_i e^{i\omega(y/\alpha_1 - t)} \qquad 16.2\text{-}1$$

$$\alpha = [(\lambda_1 + 2\mu_1)/\rho_1]^{1/2} \qquad 16.2\text{-}2$$

The compressional wave velocity is α; λ and μ are Lame's elastic constants and ρ is the density; t is the time and ω is the angular frequency of the incident sinusoidal plane wave. The reflected and transmitted waves are

$$V_r = A_r e^{-i\omega(y/\alpha_1 + t)} \qquad 16.2\text{-}3$$

$$V_t = A_t e^{i\omega(y/\alpha_2 - t)} \qquad 16.2\text{-}4$$

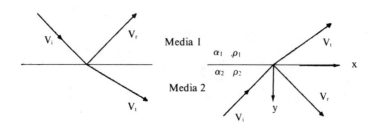

Figure 16.1 Co-ordinate system and symbols used.

At the boundary two conditions must be satisfied and these will determine the amplitudes of the reflected, A_r, and transmitted, A_t, waves. The first is continuity of displacement

$$V_i + V_r = V_t \qquad 16.2\text{-}5$$

or setting $y = 0$ one has

$$A_i + A_r = A_t \qquad 16.2\text{-}6$$

The second condition is continuity of normal stress. If u, v and w are displacements in the x, y and z directions then normal stress in the y direction is

$$P_{yy} = \lambda\left(\frac{\partial u}{\partial x} + \frac{\partial v}{\partial y} + \frac{\partial w}{\partial z}\right) + 2\mu\left(\frac{\partial v}{\partial y}\right)$$

or since all derivatives with respect to x and z are zero for plane waves propagating in the y direction one has that

$$P_{yy} = (\lambda + 2\mu)\frac{\partial v}{\partial y} \qquad 16.2\text{-}7$$

Reflection and transmission at an interface

Substituting the appropriate terms for incident, reflected and transmitted waves one has

$$(\lambda_1 + 2\mu_1) [+ \frac{\omega}{\alpha_1} V_i - \frac{\omega}{\alpha_1} V_r] = +(\lambda_2 + 2\mu_2) \frac{\omega}{\alpha_2} V_t \qquad 16.2\text{-}8$$

or at $y = 0$

$$\frac{\lambda_1 + 2\mu_1}{\alpha_1} (A_i - A_r) = \frac{(\lambda_2 + 2\mu_2)}{\alpha_2} A_t \qquad 16.2\text{-}9$$

The acoustic impedance, Z, is defined as the product of density and velocity.

$$Z_1 = \rho_1 \alpha_1 = [(\lambda_1 + 2\mu_1) / \alpha_1^2] \cdot \alpha_1 = (\lambda_1 + 2\mu_1) / \alpha_1 \qquad 16.2\text{-}10$$

With this substitution equation 16.2-9 becomes

$$A_i - A_r = \frac{Z_2}{Z_1} A_t \qquad 16.2\text{-}11$$

The amplitude of the reflected and transmitted curves may be solved from equations 16.2-6 and 16.2-11. Adding 16.2-6 and 16.2-11

$$A_t = \left[\frac{2}{1 + \frac{Z_2}{Z_1}} \right] A_i \qquad 16.2\text{-}12$$

The reflection amplitude is

$$A_r = -A_i + \left[\frac{2}{1 + \frac{Z_2}{Z_1}} \right] A_i \qquad 16.2\text{-}13$$

or

$$A_r = -A_i [1 - \frac{2}{\frac{Z_2 + Z_2}{Z_1}}] \qquad 16.2\text{-}14$$

$$= -A_i \left[\frac{Z_1 + Z_2 - 2Z_1}{Z_1 + Z_2} \right]$$

$$= A_i \left[\frac{Z_1 - Z_2}{Z_1 + Z_2} \right]$$

$$A_t = \frac{2Z_1}{Z_1 + Z_2} A_i \qquad 16.2\text{-}15$$

The reflection and transmission coefficient are then defined as

$$r = \frac{Z_1 - Z_2}{Z_1 + Z_2} = \frac{\rho_1 \alpha_1 - \rho_2 \alpha_2}{\rho_1 \alpha_1 + \rho_2 \alpha_2} \qquad 16.2\text{-}16$$

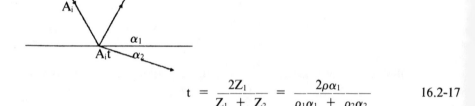

$$t = \frac{2Z_1}{Z_1 + Z_2} = \frac{2\rho\alpha_1}{\rho_1 \alpha_1 + \rho_2 \alpha_2} \qquad 16.2\text{-}17$$

Similarly the coefficient for waves in the upward direction are

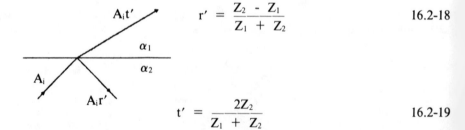

$$r' = \frac{Z_2 - Z_1}{Z_1 + Z_2} \qquad 16.2\text{-}18$$

$$t' = \frac{2Z_2}{Z_1 + Z_2} \qquad 16.2\text{-}19$$

16.3 Reflection in a multi-layered media

The stratified sedimentary section of the earth will be modelled by n horizontal layers in welded contact. Plane waves are assumed to propagate at normal incidence but in the diagram a horizontal displacement is given to the rays to simulate time. Let d(k,k+2m-1) be the displacement amplitude of the

Reflection in a multi-layered media

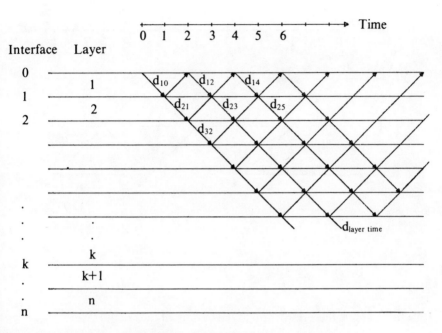

Figure 16.2 Indexing used in describing an n layered media.

downward going wave at the top layer k at time k+2m-1 where m is a time index running from 0 to infinity. Noting that from 13.2-1

$$z = e^{-i\omega \Delta t} \qquad 16.3\text{-}1$$

represents a time delay of one unit of the digitizing interval, ($\Delta t = 1$), one can form a z transform by summing all the displacements due to the down-going wavelets with an appropriate time delay.

$$D_k(z) = \sum_{m=0}^{\infty} d_{k,k+2m-1} \, z^{k+2m-1} \qquad 16.3\text{-}2$$

Thus for example at the top of the second layer the z transform is

$$D_2 = d_{21}z^1 + d_{23}z^3 + d_{25}z^5 + \dots$$

Let d'(k,k+2m) indicate the displacement of down-going waves at the bottom of the k layer at time k+2m. The sum of displacements in the down-going wave at the bottom of the layer is

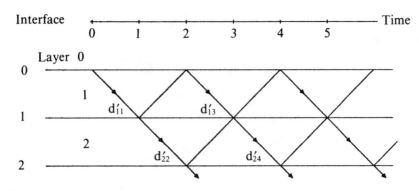

Figure 16.3 The displacement wavelets for the down-going wave.

$$D'_k(z) = \sum_{m=0}^{\infty} d'_{k,k+2m} z^{k+2m} \qquad 16.3\text{-}3$$

Thus for example the z transform at the bottom of the second layer is

$$D'_2(z) = d'_{22}z^2 + d'_{24}z^4 + \ldots$$

The waves in equations 16.3-2 and 16.3-3 are related by the transmission time through the layer which is equal to the unit delay time, z, and an attenuation coefficient, A(z).

$$D'_k(z) = zA_k(z)\, D_k(z) \qquad 16.3\text{-}4$$

If $A_k(z) = 1$ there is no attenuation.

A similar analysis for up-going waves may be made and leads to the conclusion that the z transform for up-going waves at the top of the layer is

$$U_k(z) = zA_k(z)\, U'_k(z) \qquad 16.3\text{-}5$$

where the z transform of the up-going wave at the bottom of the layer is

$$U'_k = \sum_{m=0}^{\infty} u'_{k,k+2m} z^{k+2m} \qquad 16.3\text{-}6$$

As illustrated in figure 16.5, the coefficient u'(k,j) is made up of a reflection of

Reflection in a multi-layered media

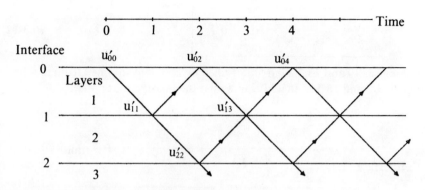

Figure 16.4 The displacement wavelets for up-going waves.

$d'(k,j)$ and the transmitted part $u(k+1,j)$

$$u'_{k,j} = r_k d'_{k,j} + t'_k u_{k+1,j} \qquad 16.3\text{-}7$$

The k which is a subscript on r and t denotes the interface. The coefficient $d(k+1,j)$ is made up of the transmitted part of $d'(k,j)$ and the reflected part of $u(k+1,j)$

$$d_{k+1,j} = t_k d'_{k,j} + r'_k u_{k+1,j} \qquad 16.3\text{-}8$$

or solving for $d'(k,j)$

$$\boxed{d'_{k,j} = (d_{k+1,j} - r'_k u_{k+1,j}) / t_k} \qquad 16.3\text{-}9$$

Substitute into 16.3-7

$$u'_{k,j} = \frac{r_k}{t_k}(d_{k+1,j} - r'_k u_{k+1,j}) + \frac{t_k}{t_k} t'_k u_{k+1,j} \qquad 16.3\text{-}10$$

The reflection and transmission coefficient 16.2-16 to 16.2-19 satisfy the relation

$$t_k t'_k - r_k r'_k = 1 \qquad 16.3\text{-}11$$

so 16.3-10 becomes

$$u'_{k,j} = \frac{r_k}{t_k} d_{k+1,j} + \frac{1}{t_k} u^{at\,o}_{k+1,j} \qquad 16.3\text{-}12$$

Equations 16.3-9 and 16.3-12 are multiplied by z^{k+2m} and summed from $m = 0$ to ∞ as in 16.3-3 and 16.3-6. In addition we use the relation that $r = -r'$.

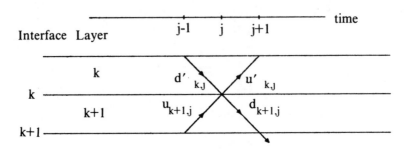

Figure 16.5 Division of amplitude at the k interface.

$$D'_k = \sum_{m=0}^{\infty} d'_{k,k+2m} z^{k+2m} = \frac{1}{t_k} [\sum_{m=0}^{\infty} d_{k+1,k+2m} z^{k+2m} + r_k \sum_{m=0}^{\infty} u_{k+1,k+2m} z^{k+2m}]$$
$$16.3\text{-}13$$

$$U'_k = \sum_{m=0}^{\infty} u'_{k,k+2m} z^{k+2m} = \frac{1}{t_k} [r_k \sum_{m=0}^{\infty} d_{k+1,k+2m} z^{k+2m} + \sum_{m=0}^{\infty} u_{k+1,k+2m} z^{k+2m}]$$
$$16.3\text{-}14$$

Comparing 16.3-3 and 16.3-6 these may be written as

$$D'_k(z) = [D_{k+1}(z) + r_k U_{k+1}(z)] / t_k \qquad 16.3\text{-}15$$

and

$$U'_k(z) = [r_k D_{k+1}(z) + U_{k+1}(z)] / t_k \qquad 16.3\text{-}16$$

Reflection in a multi-layered media

If there is no absorption then $D'_k = zD_k$ and $U_k = zU'_k$ in which case the equations for the wave motion become

$$D_k(z) = \frac{z^{-1}}{t_k} D_{k+1}(z) + \frac{r_k z^{-1}}{t_k} U_{k+1}(z) \qquad 16.3\text{-}17$$

$$U_k(z) = \frac{zr_k}{t_k} D_{k+1}(z) + \frac{z}{t_k} U_{k+1}(z) \qquad 16.3\text{-}18$$

$$\begin{bmatrix} D_k(z) \\ U_k(z) \end{bmatrix} = \frac{1}{t_k} \begin{bmatrix} z^{-1} & r_k z^{-1} \\ r_k z & z \end{bmatrix} \begin{bmatrix} D_{k+1}(z) \\ U_{k+1}(z) \end{bmatrix}$$

or

$$\begin{bmatrix} D_k \\ U_k \end{bmatrix} = \mathbf{M}_k \begin{bmatrix} D_{k+1} \\ U_{k+1} \end{bmatrix} \qquad 16.3\text{-}19$$

where $\mathbf{M}_k$ is the *communication matrix*. At the surface, using equations 16.3-15 and 16.3-16

$$\begin{bmatrix} D'_0(z) \\ U'_0(z) \end{bmatrix} \frac{1}{t_0} \begin{bmatrix} 1 & r_0 \\ r_0 & 1 \end{bmatrix} \begin{bmatrix} D_1(z) \\ U_1(z) \end{bmatrix} = \mathbf{M}_0 \begin{bmatrix} D_1 \\ U_1 \end{bmatrix} \qquad 16.3\text{-}20$$

Successive substitution for D_{k+1} and U_{k+1} leads to the following equation (with $D'_0(z) = D_0(z)$ and $U'_0 = U_0(z)$ since the zeroth layer at the surface has no thickness and so no delay).

$$\begin{bmatrix} D_0(z) \\ U_0(z) \end{bmatrix} = \mathbf{M}_0 \mathbf{M}_1 \mathbf{M} \ldots \mathbf{M}_n \begin{bmatrix} D_{n+1}(z) \\ U_{n+1}(z) \end{bmatrix}$$

or

$$\begin{bmatrix} D_0(z) \\ \\ U_0(z) \end{bmatrix} = \prod_{k=0}^{n} \mathbf{M}_k \begin{bmatrix} D_{n+1} \\ \\ U_{n+1} \end{bmatrix} \qquad 16.3\text{-}21$$

Normally $U_{n+1} = 0$ since no energy is reflected back up into the system. The down-going wave is usually simulated by a unit spike at time 0 so

$$D_0(z) = d_{00} + d_{02}z^2 + d_{04}z^4 + \ldots = 1 \qquad 16.3\text{-}22$$

Therefore equation 16.3-21 becomes

$$\begin{bmatrix} 1 \\ \\ U_0(z) \end{bmatrix} = \prod_{k=0}^{n} \mathbf{M}_k \begin{bmatrix} D_{n+1}(z) \\ \\ 0 \end{bmatrix} \qquad 16.3\text{-}23$$

The physical model is specified by the reflection and transmission coefficients while a more realistic seismic wavelet may be simulated by a final convolution of $U_0(z)$ with the z transform of the wavelet.

The matrix, $\mathbf{M}_k$, is related to one developed by Thomson (1950) for body waves at any angle of incidence. Baranov and Kunetz (1960) illustrate a synthetic seismogram calculated on the basis of a similar theoretical development. Sherwood and Trorey (1965) insert absorption coefficients in the layers as illustrated below and obtain an appropriate synthetic seismogram which includes frequency dependent dissipation of energy, but omits dispersion.

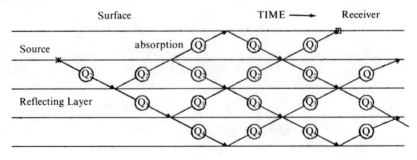

Figure 16.6 Absorption in multi-layer media.

16.4 Synthetic seismogram including dissipation

Approximate methods of introducing attenuation into reflection seismograms were attempted by Trorey (1962). These did not include the dispersive nature of dissipative materials. A rigorous treatment of the subject is introduced by Lockett (1962). Application of the general theory of linear viscoelasticity may be found in Borcherdt (1973), Buchen (1971), Cooper and Reiss (1966), and Silva (1976). Because the wave velocity in the media is frequency dependent, it is necessary to solve a matrix equation at several discrete frequencies to obtain a transfer function for the earth at all frequencies by interpolation. An inverse Fourier transform is taken to obtain a synthetic seismogram for an earth model with dissipation. The spatial attenuation factor, α, or its related value, Q, must be specified for each layer as in figure 16.6. The specific attenuation factor, Q, is defined as the ratio of peak energy, E, stored in a cycle to the energy dissipated per cycle, $\Delta E / 2\pi$. That is

$$Q = 2\pi E/\Delta E = |\omega| / (2c\alpha) \qquad 16.4\text{-}1$$

where ω is the angular frequency and c is the phase velocity. The quantity Q is useful in describing solids because it is found experimentally to be nearly independent of frequency and, for most earth materials, is numerically between 10 and 1000. When attenuation is present the phase velocity depends upon the frequency (see Futterman, 1962 and Savage, 1976). Therefore a z transform method is not applicable, strictly speaking. A general matrix technique is possible as was mentioned above provided that a fine enough frequency resolution is used to avoid aliasing in the time domain. An example of reflection synthetic seismograms with and without dissipation is shown in figure 16.7.

16.5 Reverberation in a water layer

Either a z transform or a matrix method are also useful in designing a filter to remove the reverberations caused by near surface layers. Backus (1959) has developed techniques for removing reverberations in a water layer where water-air interface has a reflection coefficient close to -1 and the water bottom interface has a value, r_1, which may be high also. If the shot is near the surface and causes a unit spike in pressure then the water layer generates an infinite series of spikes each delayed by L, the two-way travel time through the water layer. The z transform of the water reverberation generated by the source is

$$W_1(z) = 1 - r_1 z^L + r_1^2 z^{2L} - z_1^3 z^{3L} + \dots \qquad 16.5\text{-}1$$

The water layer will also act as a reverberation filter to the pulse travelling up from the reflecting layer. The resulting z transform for water reverberation is

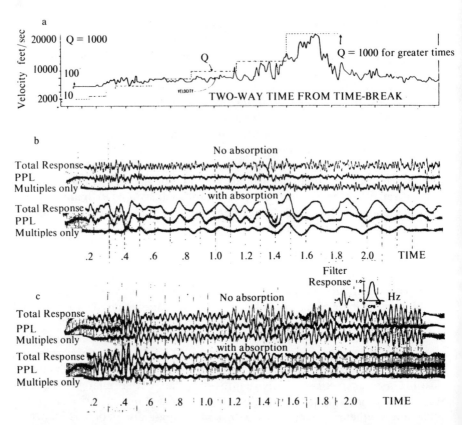

Figure 16.7 Hypothetical velocity and Q logs plotted as a function of two-way time. The Q is assumed to be constant at 1,000 for times greater than 1.870 s; the shot is located just beneath the weathered layer and has an uphole time of 46 ms. For the calculation of theoretical seismograms, the velocity was averaged and digitized over 4 ms. intervals of two-way time. Density is assumed to be constant and the surface reflection coefficient is assumed to be unity. (b) Unfiltered impulse responses, both with and without absorption for the model in (a). The responses are particle velocity at the shot depth. The primaries plus peg-leg multiples (PPL) were computed with a transmission width of 50 ms. The upgoing energy from the shot is assumed to be zero (that is, no ghost reflections are included in the responses). The leading edge of the sharp pulse at the beginning of each trace is the time-break (zero time, or the instant of shot detonation). The heavy timing lines are 100 ms. apart. (c) The responses of (b) passed through a filter 6 db down at 18 and 36 Hz. and possessing a delay of 0.06 s. The filter waveform and its spectrum are shown at the top of the figure. Note the decay in amplitude of the absorptive response when the low frequencies are filtered out. (Trorey, 1962, courtesy of the Society of Exploration Geophysicists.)

Reverberation in a water layer

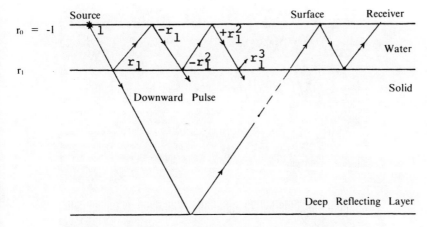

Figure 16.8 Reverberations set up in an elastic solid overlain by a layer of water.

$$W(z) = (1 - r_1z^L + r_1^2z^{2L} - r_1^3z^{3L} + ...)^2 \qquad 16.5\text{-}2$$

Since r_1 is less than 1, this is a convergent geometric series.

$$W(z) = \frac{1}{(1 + r_1z^L)^2} \qquad 16.5\text{-}3$$

To eliminate the train of waves the inverse filter is needed.

$$\begin{aligned} W^{-1}(z) &= (1 + r_1z^L)^2 \\ &= 1 + 2r_1z^L + r_1^2z^{2L} \end{aligned} \qquad 16.5\text{-}4$$

Suppose the depth of water is 50 feet and the velocity in water is 5,000 ft/sec., then the two-way travel time is 20 milliseconds. For a 4 millisecond digitizing interval L = 5 and the impulse response function according to 16.5-4 is

$$\begin{aligned} W^{-1}(z) &= (1, 0, 0, 0, 0, 2r_1, 0, 0, 0, 0, r_1^2) \\ \text{Time} &= (0, 1, 2, 3, 4, 5, 6, 7, 8, 9, 10) \end{aligned} \qquad 16.5\text{-}5$$

A convolution of the marine seismic record with this impulse response will help remove the reverberations. Backus (1959) has applied a similar type of inverse to remove the reverberations present in marine records from the Persian Gulf and Lake Maracaibo (figure 16.9).

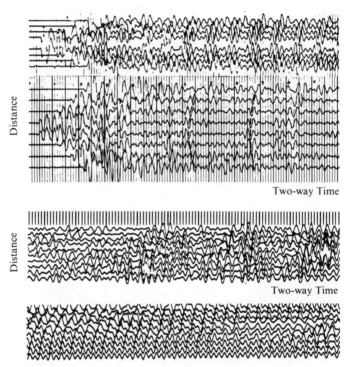

Figure 16.9 Upper diagram: An illustration of approximate inverse filtering applied to a Persian Gulf record. The upper record is a conventional playback (75-21 cps pass band) of a reverberation record from the Persian Gulf. The lower record is a playback of the same record with approximate inverse filtering applied. This illustrates the effectiveness of inverse filtering on conventionally recorded data taken in deep water (200 ft.) for a hard bottom. *Lower diagram*: The lower record is a conventional playback of an end-on profile from Lake Maracaibo. Water depth was about 80 ft. and the two dominant frequencies on the record are 30 cps and 58 cps indicating that the effective trap bottom is acoustically soft. The upper record is a playback of the same data with inverse filtering applied. (Backus, 1959 Courtesy of the Soc. of Exploration Geophysics.)

16.6 Removal of ghost reflections

Another example of reverberation filtering involves the removal of ghosts from seismic records. A weathered layer overlies most regions of the solid earth. This consists of surface rocks which are aerated, leached, altered chemically and are not saturated with water. The base of the weathered layer does not usually correspond to any geological formation but is often coincident with the water table. The velocity in this layer is highly irregular from place to place and has typically low values between 400 and 1,600 m/s. This layer alters the spectral character of the record and, because of its low variable velocity, delays arrivals in an erratic and significant amount. The delays can be easily corrected if the velocities are known but the spectral alterations can only be carried out by an inverse filter. The surface or any velocity discontinuity with the weathered layer

Removal of ghost reflections

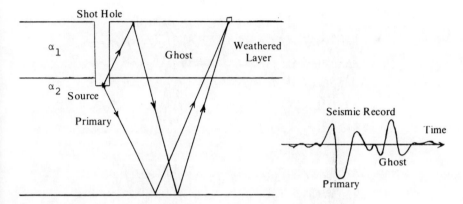

Figure 16.10 Ghost reflection produced by wave reflected from the surface.

produces a ghost which is similar to the direct or primary reflector.

A method of removing ghosts by least squares optimum filters has been given by Foster, Sengbush and Watson (1964). The principle is described more simply by Robinson (1966, p. 482) and his example is used. Hammond (1962) has presented some excellent examples of ghosts (figure 16.11).

Let the extra two-way travel time for the ghost be L seconds. We will assume the ghost and primary have identical shape but the ghost has a smaller amplitude and the ratio of ghost to primary is r_1.

If the primary is represented by a unit pulse, the z transform of the downgoing wave at the level of the shot is

$$W(z) = 1 + r_1 z^L \qquad 16.6\text{-}1$$

since the ghost is delayed by L units of digital time. The inverse filter then has the following z transform:

$$W^{-1}(z) = \frac{1}{1 + r_1 z^L} \qquad 16.6\text{-}2$$

This is an infinite length filter which acts similar to a feedback amplifier with unit gain. The output from the amplifier is fed to a negative feedback loop which attenuates the signal by r_1, delays it by L unit of time, inverts it and mixes it with the input.

$$Y(z) = X(z) \frac{1}{1 + r_1 z^L} \qquad 16.6\text{-}3$$

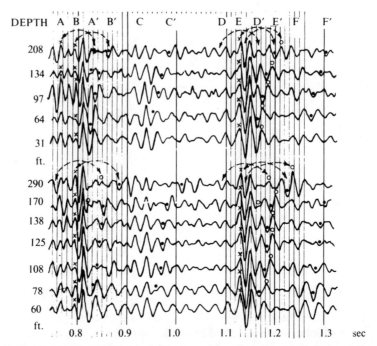

Figure 16.11 Composite records composed of recordings from a series of shots at different hole depths. The upper five traces are from one hole while the bottom seven traces are from another hole 50 feet away. The symbols o and o indicate the peaks and troughs of the ghosts. The symbols x and ' indicate peak and trough of the true reflections. (From Hammond, 1962, courtesy of the Society of Exploration Geophysicists.)

or

$$Y(z) = X(z) - r_1 z^L Y(z) \qquad 16.6\text{-}4$$

Foster, Sengbush and Watson (1964) give a derivation of a finite length operator which deghosts records effectively and Schneider, Larner, Burg and Backus (1964) have designed a multi-channel ghost eliminating filter.

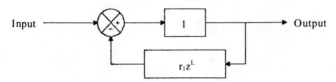

Figure 16.12 Analog feedback system which acts as an inverse filter.

References

Abels,F.(1946), Nouvelles formules relative a la lumiere réflechie et transmise par un empilement de almes a faces paralleles. *Comptes Rendue de L'Academie des Sciences* **223**, 891-893.

Backus,M.M.(1959), Water reverberations: Their Nature and Elimination. *Geophysics* **24**, 233,261.

Baranov,V. and Kunetz,G.(1960), Film synthétique avec réflexions multiples. Théorie et calcul pratique. *Geophys. Prosp.* **8**, 315-325.

Borcherdt,R.D.(1973), Energy and plane waves in linear viscoelastic media. *Journal of Geophysical Research* **78**, 2442-2453.

Buchen,P.W.(1971), Plane waves in linear viscoelastic media. *Geophysical Journal* **23**, 531-542.

Claerbout,J.F. and Robinson,E.A.(1964), The error in least-squares inverse filtering. *Geophysics* **29**, 48-120.

Cooper,H.F.J. and Reiss,E.L.(1966), Reflection of plane viscoelastic waves from plane boundaries. *Journal of Acoustical Society of America* **39**, 1133-1138.

Crook,A.W.(1948), The reflection and transmission of light by any system of parallel isotropic films. *Journal of the Optical Society of America* **38**, 954-964.

Foster,M.R., Sengbush,R.L. and Watson,R.J.(1964), Suboptimum filter systems for seismic data processing. *Geophys. Prosp.* **12**, 173-191.

Futterman,W,I,(1962), Dispersive body waves. *Journal of geophysical Research* **67**, 5279-5291.

Ganley,D.C.(1979), The seismic measurement of absorption and dispersion. Phd Thesis, University of Alberta, Edmonton, Alberta.

Goupillaud,P.L.(1961), An approach to inverse filtering of near-surface layer effects from seismic records. *Geophysics* **26**, 754-760.

Hammond,J.W.(1962), Ghost elimination from reflection records. *Geophysics* **27**, 48-60.

Koehler,F. and Taner,M.T.(1977), Direct and inverse problems relating reflection coefficients and reflection response for horizontally layered media. *Geophysics* **42**, 1199-1206.

Kunetz,G. and D'Erceville,I.(1962), Sur certaines propriétés d'une onde acoustique plane de compression dans un milieu stratifié. *Annales de Geophysique* **18**, 351-359.

Kunetz,G.(1964), Généralisation des opérateurs d'anti-résonance a un nombre quelconque de réflecteures. *Geophys. Prosp.* **12**, 283-289.

Lockett,F.J.(1962), The reflection and refraction of waves at an interface between viscoelastic media. *Journal of Mechanics and Physical Solids* **10**, 53-64.

Mendel,J.M.,Nahi,N.E. and Chan,M.(1979), Synthetic seismograms using the state space approach. *Geophysics* **44**, 880-895.

Robinson,E.A. and Treitel,S.(1965), Dispersive digital filters *Review of Geophysics* **3**, 433-461.

Robinson,E.A.(1966), Multichannel z-transforms and minimum delay. *Geophysics* **31**, 482-500.

Robinson,E.A. and Treitel,S.(1976), Net downgoing energy and the resulting minimum phase property of downgoing waves. *Geophysics* **41**, 1394-1396.

Savage,J.C.(1976), Anelastic degradation of acoustic pulses in rocks - Comments. *Physics of the Earth and Planetary Interiors* **11**, 284-285.

Schneider,W.A., Larner,K.L., Burg,J.P. and Backus,M.M.(1964), A new data-processing technique for the elimination of ghost-arrivals on reflection seismograms. *Geophysics* **29**, 783-805.

Sherwood,J.W.C. and Trorey,A.W.(1965), Minimum phase and related properties of a horizontal stratified absorptive eath due to plane acoustic waves. *Geophysics* **30**, 191-197.

Silva, W.(1976), Body waves in a layered anelastic solid. *Bulletin of the seismological Society of America* **66**, 1539-1554.

Thomson,W.T.(1950), Transmission of elastic waves through a stratified medium. *Journal of Applied Physics* **21**, 89-93.

Treitel,S. and Robinson,E.A.(1966), Seismic wave propagation in layered media in terms of communication theory. *Geophysics* **31**, 17-32.

Trorey,A.W.(1962), Theoretical seismograms with frequency and depth dependent absorption. *Geophysics* **27**, 766-785.

Wuenschel,P.C.(1960), Seismogram synthesis including multiples and transmission coefficients. *Geophysics* **25**, 106-129.

17 Velocity filters

17.1 Introduction

A class of two-dimensional linear operators which have been highly successful in detecting and extracting desired signals from a background of similar coherent noise are velocity filters. These operate in the frequency and wave number domain and so they require a spatial array of sensing devices. In the simplest variety it is assumed that the signal wave propagation is from one direction which is known. This is the usual case in exploration seismology but if the signals are due to natural sources such as earthquakes or storm generated waves in oceans then more elaborate arrays are required.

Time sequence analysis for the group and phase velocity of dispersive events such as seismic surface waves will not be discussed. However, some recent techniques are given in the references under Dziewonski et al. (1969), Landisman et al. (1969), Godforth and Herrin (1979) and Seneff (1978). Matched filters as discussed by Turin (1960) may also be of value in this analysis.

The term *velocity filter* was used first by Embree, Burg and Backus (1963) and it should be noted that Fail and Grau (1963) refer to them as *filtre en eventail*. Wiggins (1966) calls these two-dimensional operators ω-k filters. When applied to two-dimensional arrays of seismometers such operators are usually called *beam forming* filters. The *pie slice* process is an industrial trade name applied to a commonly used form of this class. The term *velocity filter* will be used here because it seems to be the simplest descriptive phrase whereas *fan* and *pie* are rather obscure when discussing wave phenomena. Velocity operators given by researchers above are empirical in nature and others which are based on the Wiener-Hopf optimum filter theory have been derived by Foster, Sengbush and Watson (1964) and Schneider, Prince and Giles (1965). These are generally more elaborate and are used to de-ghost (rf. Chapter 16) seismic records and to reduce multiple reflections. Velocity filters may operate in the digital domain as above, or, equally effectively, in the analog domain with the use of laser sources and optical processing (Dobrin, Ingalls and Long, 1965).

It is assumed that plane waves of the signal have different velocities from those of coherent noise. The filter is designed to pass the signal without distortion but reject the noise. Several traces of data recorded at different locations are used in the analysis. Suppose that the detectors are equally spaced to form a linear array and that the signal arrives at times τ_A apart on adjacent detectors (figure 17.1). τ is called the signal moveout. From seismic records,

300

Velocity filters

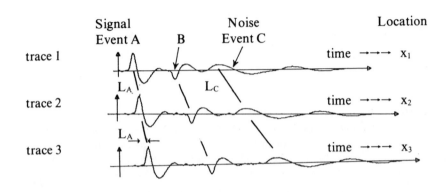

Figure 17.1 Signal recorded along a linear array of seismometers.

signal and noise spectra are obtained and displayed on a graph of frequency versus wave number or wavelength. A typical example is shown in figure 17.2. Non-dispersive events travelling with a constant velocity are described by straight lines through the origin. The signal consists of energy reflected at nearly vertical incidence from great depths. Therefore the reflected signals are received at nearly the same time by seismometers at a variety of distances. The apparent wavelength is then very large or can even be infinite. The noise will usually consist of waves travelling close to the surface or scattered from the sides and will have much lower velocities and shorter apparent wavelengths.

Velocity filter may be designed to pass only events such as A in figure 17.1 and 17.2 with a particular apparent velocity. The transfer function of such a filter will be designed to approximate that shown in figure 17.3. Rather than design a special filter for each band of velocities the nth channel is delayed by $n\tau_A$ so that events such as A appear to arrive at the same time (i.e., with infinite apparent velocity) at all detectors. A general velocity filter, with transfer function shown in figure 17.4, is used to enhance the desired events and attenuate others with different phase velocity. The time shift can be removed as a final step in presenting the filtered output.

17.2 Linear velocity filters with box-car shaped pass bands

Let us find the impulse response of a filter whose two-dimensional transfer function $Y(f,k)$ is unity within the pass band and zero outside.

$$Y(f,k) = \begin{cases} 1 & -|f|/V < k < |f|/V \\ 0 & \text{otherwise} \end{cases} \qquad 17.2\text{-}1$$

Linear velocity filters with box-car shaped pass bands

In this equation k is spatial frequency defined as

$$k = |f|/V = K/2\pi = 1/\lambda \qquad 17.2\text{-}2$$

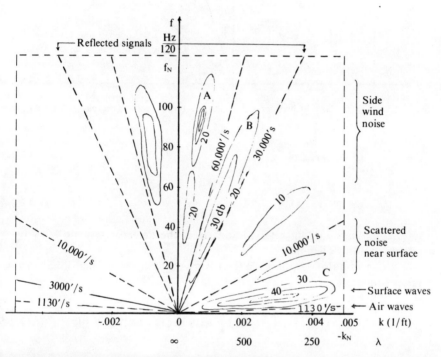

Figure 17.2 Spectrum of coherent seismic signals in a typical reflection recording. The data for the experiment was generated by an impulsive source and the elastic waves were recorded on a linear array of seismometers on the surface of the earth. The sampling rate is assumed to be 4 milliseconds and the detector spacing is 100 feet. Events A, B, and C are illustrated in figure 17.1.

K is the wave number and λ the apparent wavelength. The impulse response was given by Embree, Burg and Backus (1963) and by Fail and Grau (1963) who also derived the response for both trapezoidal and box-car shaped functions. The upper frequency limit is given by the Nyquist limits

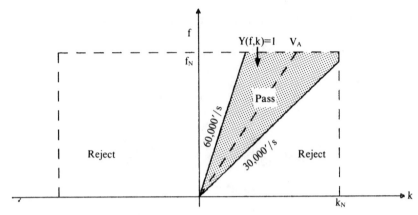

Figure 17.3 A particular velocity filter designed to pass events centered on velocity $V_A = \Delta x / \tau_A$.

$$f_N = 1/2\Delta t \qquad 17.2\text{-}3$$

$$k_N = 1/2\Delta x \qquad 17.2\text{-}4$$

where Δt is the sampling interval and Δx is the detector spacing of a linear array of recording elements.

The impulse response, $W(t,x)$ is given by a two-dimensional inverse Fourier transform.

$$W(t,x) = \int_{-\infty}^{\infty} \int_{-\infty}^{\infty} Y(f,k) \, exp(2\pi i [ft - kx]) \, dk \, df \qquad 17.2\text{-}5$$

The transfer function in 17.2-1 can be substituted and the integral multiplied by $\Delta t \cdot \Delta x$ because of the discrete sampling of a continuous function. Also let

$$T_n = n\Delta t$$

be the nth sample point and

$$X_m = (m + 1/2) \, \Delta x \qquad m = \text{-L-1, L, ... 0, ... L}$$

be the mth detector.

$$W(T_n, X_m) = \Delta t \, \Delta x \int_{-f_N}^{f_N} \int_{-k}^{k} e^{2\pi i (fT_n - kX_m)} \, dk \, df \qquad 17.2\text{-}6$$

Linear velocity filters with box-car shaped pass bands

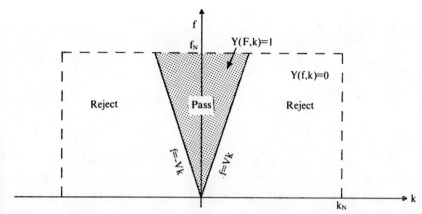

Figure 17.4 A general velocity filter designed to pass events with apparent velocity $|V| > f/k$. By applying time shifting of $n\tau_A$ to the data this can simulate the filter shown in figure 17.3.

Applying 17.2-2 to the limits and integrating with respect to k yields

$$\frac{1}{-2\pi i X_m} \int_{-|f|/V}^{|f|/V} (-2\pi i X_m) e^{-2\pi k X_m} dk = \frac{e^{2\pi i k X_m} - e^{-2\pi i k X_m}}{2\pi i X_m}$$

$$= \frac{\sin 2\pi i k X_m}{\pi X_m} \quad 17.2\text{-}7$$

Then substituting into 17.2-6 yields

$$W(T_n, X_m) = \Delta x \, \Delta t \int_{-f_N}^{f_N} \frac{\sin 2\pi X_m |f|/V}{\pi X_m} e^{2\pi i f T_n} df \quad 17.2\text{-}8$$

By the symmetry properties of the integrand or the necessity of a real impulse response it is clear that the imaginary part of the integral must vanish. Therefore only a cosine Fourier transform need be evaluated.

$$W(T_n, X_m) = \frac{2 \, \Delta t \, \Delta x}{\pi X_m} \int_0^{1/2\Delta t} \cos(2\pi f n \Delta t) \cdot \sin(2\pi f \, X_m \, \Delta t / \Delta x) \, df \quad 17.2\text{-}9$$

From Pierce's table of integrals, number 371

$$\int \sin(mx) \cos(nx) \, dx = \frac{\cos(mx+nx)}{2(m+n)} - \frac{\cos(mx-nx)}{2(m-n)} \qquad 17.2\text{-}10$$

$$W(T_n, X_m) = \frac{2 \, \Delta t \, \Delta x}{\pi X_m} \left[-\frac{\cos[2\pi \Delta t(X_m/\Delta x + n)/2\Delta t]}{4\pi \Delta t \, (X_m/\Delta x + n)} - \frac{\cos[2\pi \Delta t(X_m/\Delta x - n)/(2\Delta t)]}{4\pi \Delta t \, (X_m/\Delta x - n)} \right.$$

$$\left. - \frac{1}{4\pi \Delta t} \left(\frac{1}{X_m/\Delta x + n} + \frac{1}{X_m/\Delta x - n} \right) \right] \qquad 17.2\text{-}11$$

The two cosine terms cancel each other since $X_m/\Delta x$ and n are integers. Therefore

$$\boxed{W(T_n, X_m) = \frac{1}{\pi^2[(X_m/\Delta x)^2 - n^2]}} \qquad 17.2\text{-}12$$

This impulse response is symmetric in both space and time and has zero phase shift. It may be convolved directly with the field data to produce a velocity filtered record section. Figure 17.5 illustrates the operation of the filter on 12 traces of reflection data and is set up to pass event dipping at +11.7 km/sec. A 2D Fourier transform may be taken of the impulse response for the 12 channels to obtain the actual two-dimensional transfer function as shown in figure 17.6.

17.3 Recursive relation for a velocity filter

Treitel, Shanks and Frasier (1967) have shown how to transform the velocity filter of Embree, Burg and Backus into the z domain and obtain a recursive relation for computation. The time consuming two-dimensional convolution can then be implemented much more economically through a recursive relation. For m traces, each output trace requires m/2 convolutions with an ordinary convolution technique whereas the recursive relation allows one to perform the operation with only one convolution regardless of the value of m.

Several changes in the indexing must be made to simplify the computations. Since $X_m/(\Delta x/2)$ is always an odd integer, define a new spatial index, μ.

$$\mu = 2m \mp 1 = \frac{X_m}{\Delta x/2} \qquad 17.3\text{-}1$$

The impulse response in 17.2-12 is then

Recursive relation for a velocity filter

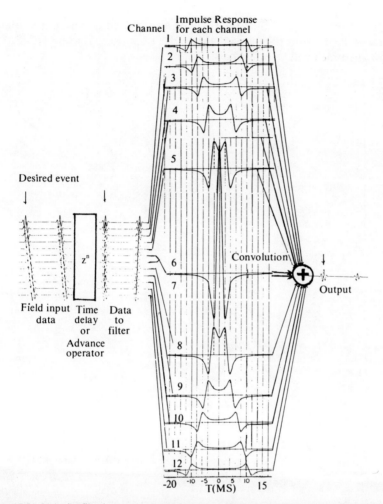

Figure 17.5 Velocity filtering as a convolution. The transfer function for the 12 channels is shown in the central diagram. Twelve input traces are time shifted so the desired event arrives at the same time on all channels. They are then convolved and summed to produce the output trace. Note that the second event with a slightly different velocity has been attenuated. (From Embree et al., 1963, courtesy of the Society of Exploration Geophysicists.)

$$W_n(\mu) = \frac{1}{\pi^2 (\mu^2/4 - n^2)} \qquad 17.3\text{-}2$$

$\mu = \pm 1, \pm 3 \ldots \pm L$ ——— spatial index
$n = 0, \pm 1, \ldots \pm N$ ——— time index.

The output from the space-time filter will be a sum of convolution over all the input traces, $x_n(\mu)$, with the individual impulse responses given by 17.3-2. The output at any time t will be called $y_{n'}$.

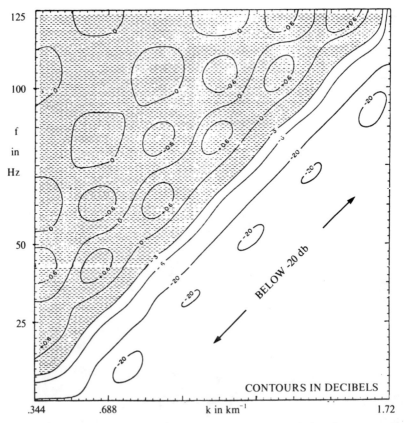

Figure 17.6 Transfer function of a velocity filter as approximated by the impulse response shown in figure 17.5 using 12 traces of data. (From Embree et al., 1963, courtesy of the Society of Exploration Geophysicists.)

$$y_n = \sum_{\mu=-L}^{L} x_n(\mu * W_n(\mu))$$ 17.3-3

Furthermore, the impulse response can be expressed as the sum of two partial functions to allow us to design a parallel form of a digital filter (see figure 15.9). The form can be seen in the last term of equation 17.2-11. Thus

Recursive relation for a velocity filter

$$W_n(\mu) = \frac{1}{\pi^2 \mu}\left[\frac{1}{\mu/2 - n} + \frac{1}{\mu/2 + n}\right] \qquad 17.3\text{-}4$$

or

$$W_n(\mu) = \frac{1}{\pi^2 \mu}[r_n(\mu) + q_n(\mu)] \qquad 17.3\text{-}5$$

where the operators r_n and q_n are obtained from equation 17.3-4. A plot of r_n and q_n (figure 17.7) shows that they are both antisymmetric in time about the points $n = \mu/2$ and $n = -\mu/2$ respectively. Values of $r_n(\mu)$ and $q_n(\mu)$ for which $\mu \neq 1$ have the same shape as $r_n(1)$ and $q_n(1)$ but are displaced in time. The operator q_n is related to r_n by a negative sign and a shift of μ units of time.

$$q_n(\mu) = -r_{n+\mu}(\mu) = \frac{1}{\mu/2 + n} \qquad 17.3\text{-}6$$

From the symmetry properties

$$r_n(\mu) = \frac{1}{\mu/2 - n} = \frac{1}{1/2 - [n - \mu/2 + 1/2]} = r_{n-\mu/2+1/2}(1) \qquad 17.3\text{-}7$$

The impulse response can then be written in terms of the response of the central detectors.

$$W_n(\mu) = \frac{1}{\pi^2 \mu}[r_{n-\mu/2+1/2}(1) - r_{n+\mu/2+1/2}(1)] \qquad 17.3\text{-}8$$

The output signal in 17.3-3 can then be written as

$$y_n = 1/\pi^2 \sum_{\mu=-L}^{L} 1/\mu \left[[x_n(\mu) * r_{n-(\mu-1)/2}(1)] - [x_n(\mu) * r_{n+(\mu+1)/2}(1)]\right] \qquad 17.3\text{-}9$$

The delay factor in the convolution can be associated with the input data rather than the impulse response so 17.3-9 becomes

$$y_n = \frac{1}{\pi^2} r_n(1) * \sum_{\mu=-L}^{L} \frac{1}{\mu}[x_{n-(\mu-1)/2}(\mu) - x_{n+(\mu+1)/2}(\mu)] \qquad 17.3\text{-}10$$

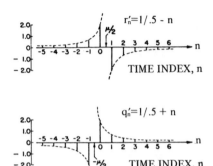

Figure 17.7 A plot of the operators r_n and q_n for $\mu = 1$. (From Treitel et al., 1967, courtesy of the Society of Exploration Geophysicists.)

where

$$r_n(1) = 1/(0.5 - n) \qquad 17.3\text{-}11$$

In terms of z transforms the convolution (*) above becomes a multiplication with the z transform of the output being

$$Y(z) = \frac{1}{\pi^2} R_1(z) \sum_{\mu=-L}^{L} \frac{X_\mu(z)}{\mu} [z^{(\mu-1)/2} - z^{-(\mu+1)/2}] \qquad 17.3\text{-}12$$

where

$$R_1(z) = \lim_{\nu \to \infty} \sum_{n=-\nu+1}^{\nu} \frac{z^n}{(1/2) - n} \qquad 17.3\text{-}13$$

or

$$\begin{aligned} R_1(z) &= \lim_{\nu \to \infty} \left[\sum_{n=-\nu+1}^{0} \frac{1}{1/2 - n} z^n + \sum_{n=1}^{\nu} \frac{1}{1/2 - n} z^n \right] \\ &= \lim_{\nu \to \infty} \left[\sum_{n=0}^{\nu-1} \frac{1}{1/2 + n} z^{-n} + \sum_{n=1}^{\nu} \frac{1}{1/2 - n} z^n \right] \\ &= \lim_{\nu \to \infty} \left[\sum_{n=0}^{\nu-1} \frac{1}{1/2 + n} z^{-n} - z \sum_{n=0}^{\nu-1} \frac{1}{1/2 + n} z^n \right] \qquad 17.3\text{-}14 \end{aligned}$$

Recursive relation for a velocity filter

Terms such as z^n are pure time shift operators and $R_1(z)$ can be approximated sufficiently closely by letting $\nu = 11$ according to Treitel, Shanks and Frasier. Shanks (1967) has shown how to convert operators like 17.3-14 into a recursive filter. A close approximation is

$$R_1(z) \simeq \frac{2(1 - 0.65465z^{-1})}{1 - 0.98612z^{-1} + 0.13091z^{-2}} - \frac{2z(1 - 0.65465z)}{1 - 0.98612z + 0.13091z^2} \qquad 17.3\text{-}15$$

Figure 17.8 is a block diagram illustrating the steps necessary in a computer algorithm to realize a velocity filtering operation. The recursive relation in 17.3-15 can be evaluated as a feedback system as shown in section 15.8. Notice that both positive time shifts (z^n) and negative time shifts (z^{-n}) are required.

A theoretical example of velocity filtering is illustrated in figure 17.9. Seven reflection events having an apparent velocity ranging from +11.7 km/sec. to infinity and then to -11.7 km/sec. are shown in the top diagram. The lower three diagrams illustrate the filtering accomplished by passing three different velocity bands. An eight-channel velocity filter was used so that the 12 input channels are compressed into five output traces. Events with apparent velocity or dip within the pass band are not modified while those on the cut-off velocity are attenuated by one half. All reflections whose apparent velocities are outside the pass band are severely attenuated.

Some actual reflection data from southern Alberta are illustrated in figure 17.10 (Clowes and Kanasewich, 1972). The upper diagram shows the original field data recorded with channels separated 0.29 kilometers along a linear array about 3.2 kilometers long. The lower three diagrams show the same data velocity filtered with three different pass bands. The reflections labelled R and M are probably the Riel (Conrad) and Mohorovicic discontinuities in the lower part of the crust. It is clear that significant improvement in signal-to-noise ratio can be obtained with these two-dimensional operators.

17.4 Non-linear velocity filters

The simplest form of velocity filtering is a procedure known as delay and sum. Even with some form of weighting of the individual channels, the technique is not very satisfactory as the acceptance lobe on the response curve is very broad. Figure 17.11 shows some original field seismic data recorded by the University of Alberta 1128 to 1472 kilometers from an explosive source in Lake Superior. In figure 17.12 the same data is shown after very narrow Butterworth zero-phase band-pass filtering. If the data in figure 17.11 is band-pass filtered from 1 to 5 Hz and passed through a delay and sum velocity filter the results are as in figure 17.13. The signal-to-noise ratio is still quite poor and it is easier to determine the first arrival of coherent energy in figure 17.12 than in figure 17.13. Also this simple linear velocity filter does not discriminate between arrivals at 8, 10, 12, or 14 kilometers per second so no new information is

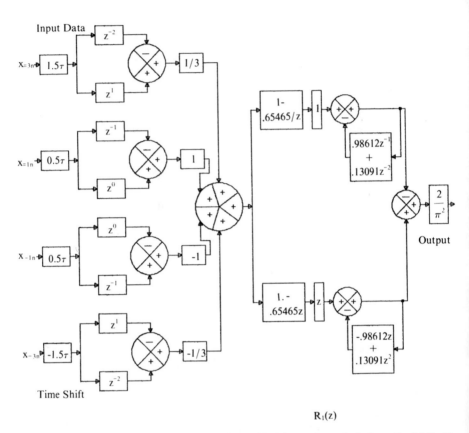

Figure 17.8 Block diagram for a velocity filter with 4 input channels (l=3, μ=-3, -1,1,3). The relevant equations are 17.3-10 and 17.3-15. An initial time shift of τ per trace is incorporated to rotate the filter to the velocity which is desired to be transmitted.

obtained by this procedure.

A non-linear method of velocity filtering was suggested by Muirhead (1968) and Kanasewich, Hemmings and Alpaslan (1973) which we call the Nth root stack. This consists of forming a delay of the various traces to align a group of arrivals with a particular velocity, taking the Nth (usually the 4th or 8th root) root of the signal, summing, and then raising the result to the Nth power. Usually the sign is preserved in the process. Let x_{ij} represent the jth channel at the ith time for the case with K channels and M samples. Let the jth channel with gain G_j be normalized to some common gain, G.

Non-linear velocity filters

$$X_{ij} = \frac{G}{G_j} x_{ij} \quad 1 \leq i \leq M \quad 17.4\text{-}1$$
$$1 \leq j \leq K$$

Assuming a velocity V for the incoming wave, a lag L_j is determined for channel j which is a distance D_j from the center of the array.

$$L_j = \frac{D_j}{V}$$

For an nth root beam forming filter output is

$$y_i = R_{iV} |R_{iV}|^{N-1}$$

where

$$R_{iV} = \frac{1}{K} \sum_{j=1}^{K} \frac{X_{i+L_j,j}}{|X_{i+L_j,j}|} |X_{i+L_j,j}|^{1/N} \quad 17.4\text{-}2$$

The effectiveness and operation is most easily illustrated for the case where N = 2 and there are only two channels. Consider a noise spike and a signal on one channel and only a signal on the second. Figure 17.14 illustrates the operation of the second root beam forming filter. Note that the signal is passed without distortion. Figure 17.15 illustrates the effectiveness of the operator on the field data shown in figure 17.11. Note that the signals are distorted by the nonlinear operation. However, random noise before the first arrival of coherent energy is highly attenuated. Furthermore the amplitude of the arrivals at 8 to 10 kilometers per second (2nd and 3rd traces) is higher indicating that the velocity can be determined even for a small array of 6 detectors spaced over 2.4 kilometers.

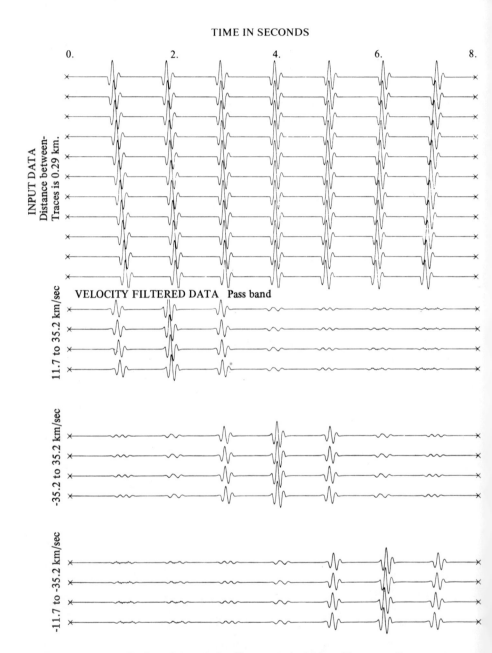

Figure 17.9 Application of the velocity filter to synthetic data. The upper diagram represents 12 input traces with coherent events at various apparent velocities from +11.7 to ∞ and -∞ to -11.7 km/sec. The lower three diagrams have pass bands from +35.2 to 11.7 km/sec., <-35.2 to >35.2 km/sec. and -11.7 to -35.2 km/sec. An eight channel operator was used.

Non-linear velocity filters

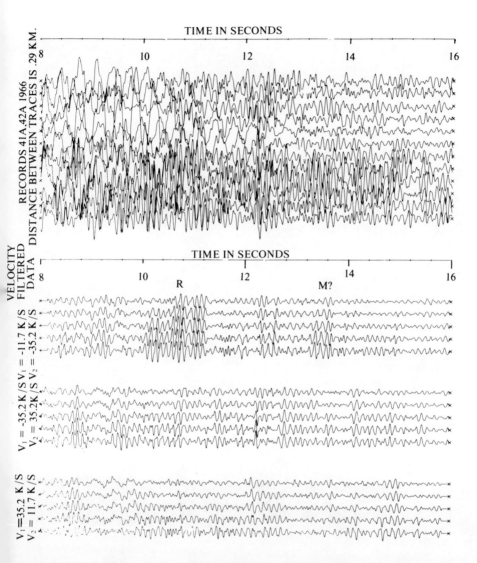

Figure 17.10 Application of velocity filtering to reflection field data. The upper diagram shows the unfiltered data; the other three diagrams show filtered data with the same pass bands as in figure 17.9.

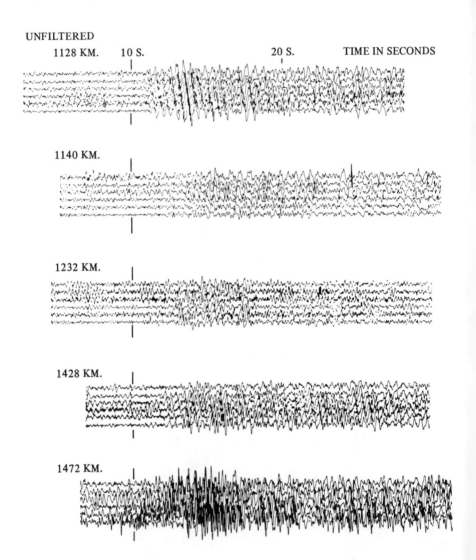

Figure 17.11 Five seismic records from the Early Rise Experiment recorded in Western Canada by seismologists from the University of Alberta. The source was a 5 ton chemical explosive in Lake Superior.

Non-linear velocity filters

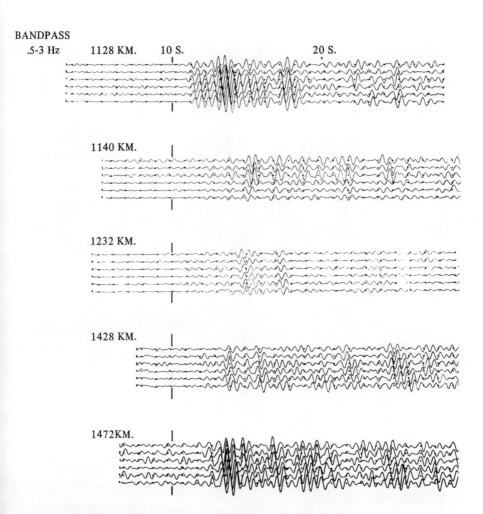

Figure 17.12 The five seismic records from figure 17.11 after band-pass filtering by a zero phase shift, 0.5 to 3.0 Hz Butterworth filter.

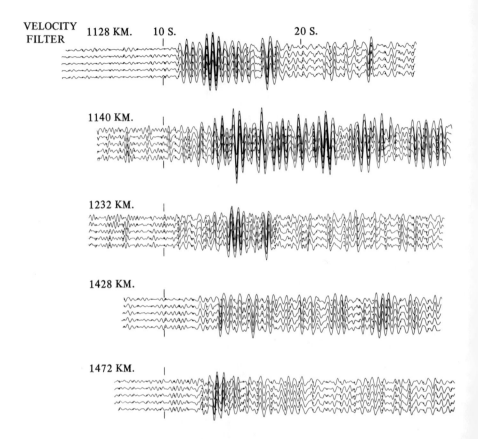

Figure 17.13 A delay and sum velocity filtering of the records in figure 17.11 after an initial filtering with a zero phase shift 1 to 5 Hz Butterworth filter. The 5 traces represent a velocity filtered output at 6, 8, 10, 12 and 14 km/sec. respectively.

Non-linear velocity filters

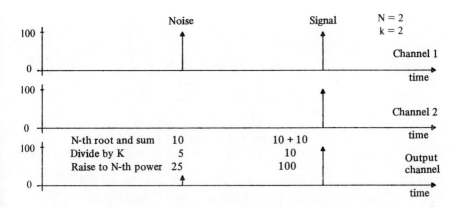

Figure 17.14 Simplified operation of an Nth root beam forming filter.

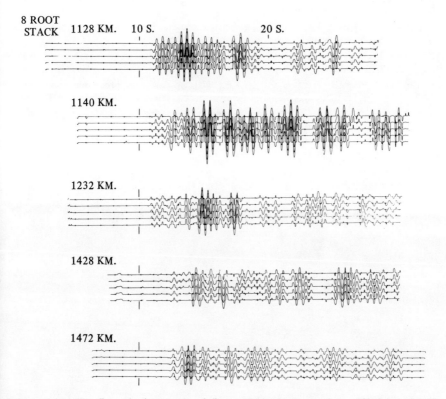

Figure 17.15 The five seismic records of figure 17.11 after zero phase shift 0.5 to 3.0 Hz Butterworth filtering and an 8th root beam forming operation. The 5 traces on each record represent velocity filtered

References

Clowes,R.M. and Kanasewich,E.R.(1972), Digital filtering of deep crustal seismic reflection. *Can. J. Earth Sci.* **9**, 434-451.

Dobrin,M.B., Ingalls,A.L. and Long,J.A.(1965), Velocity and frequency filtering of seismic data using laser light. *Geophysics* **30**, 1144-1178.

Dziewonski,A.M., Bloch,S. and Landisman,M.(1969), A technique for the analysis of surface waves. *Bulletin Seismological Society of America* **59**, 427-444.

Embree,P., Burg,J.P. and Backus,M.M.(1963), Wide band velocity filtering - The Pie Slice Process. *Geophysics* **28**, 948-974.

Fail,J.P. and Grau,G.(1963), Les Filtres en Eventail. *Geophysical Prospecting* **11**, 131-163.

Foster,M.R., Sengbush,R.L. and Watson,R.J.(1964), Design of suboptimum filter systems for multi-trace seismic data-processing. *Geophysical Prospecting* **12**, 173-191.

Godforth,T. and Herrin,E.(1979), Phase matched filters. Application to the study of Love waves. *Bulletin Seismological Society of America* **69**, 27-44.

Kanasewich,E.R., Hemmings,C.D. and Alpaslan,T.(1973), Nth root stack non-linear multichannel filter. *Geophysics* **27**, 927-938.

Landisman,M., Dziewonski,A. and Sato,Y.(1969), Recent improvements in the analysis of surface wave observations. *Geophysical Journal* **17**, 369-403.

Muirhead,K.J.(1968), Eliminating false alarms when detecting seismic events automically. *Nature* **217**, 533-534.

Schneider,W.A.,Prince,Jr.,E.R., and Giles,B.F.(1965), A new data processing technique for multiple attenuation exploiting differential normal move-out. *Geophysics* **30**, 348-362.

Seneff,S.(1978), A fast new method for frequency filter analysis of surface waves. *Bulletin Seismological Society of America* **68**, 1031-1048.

Shanks,J.L.(1967), Recursion filters for digital processing. *Geophysics* **32**, 33-51.

Treitel,S., Shanks,J.L. and Frasier,C.W.(1967), Some aspects of fan filtering. *Geophysics* **32**, 789-800.

Turin,G.L.(1960), An introduction to matched filters. *IRE Transactions on Information Theory*. 311-329.

Wiggins,R.A.(1966), ω-k filter design. *Geophysical Prospecting* **14**, 427-440.

18 Velocity spectra

18.1 Common depth point and correlations

Since the introduction by Mayne (1962) of the common depth point method (CDP) of seismic reflection exploration, the determination of compressional wave velocities has been intensively explored. The ray paths from source to receiver in the subsurface were shown in figure 1.6. A velocity spectrum is a three-dimensional graph showing the power of signals arriving at different two-way vertical travel times over a spectrum of velocities. The technique was pioneered by Schneider and Backus (1968) and by Taner and Koehler (1969). A useful review was given by Montalbetti (1971). Apart from obtaining subsurface root mean square velocities it is also possible to identify primary and multiple reflections and to obtain interval velocities and so detect charges in stratigraphy.

In obtaining the velocity spectra some method must be used to find the coherency. There are many different approaches and only a brief summary will be given. Let the amplitude of the digitized seismic trace at distance i and time t be $f(i,t(i))$. The summation is made only for signals with a common reflection depth point (see figure 1.6, Chapter 1) assuming that the layering is plane. The time distance relation is approximated first by a hyperbolic relation involving the root mean square velocity, V, to a reflector and the vertical incidence time, T.

$$t^2 = T^2 + x^2/V^2$$

Higher order equations are also used and are discussed by May and Straley (1979). A velocity stack over M receivers in which $t(i)$ corresponds to a trajectory with a particular velocity is given by

$$s_T = \sum_{i=1}^{M} f_{i,t(i)} \qquad 18.1\text{-}1$$

It is more usual to use a cross correlation as a coherency measure in which the summation is over all channel combinations, k,i. The correlation is centered at t_0 and taken over a window of length W centered at time T.

$$C_T = \frac{2}{M(M-1)W} \sum_{t=-W/2}^{W/2} \sum_{k=1}^{M} \sum_{i=1}^{M} f_{i\,t}\, f_{i+k\,t(i+k)} / [\sum_t f_{i\,t}^2 \sum_t f_{i+k\,t(i+k)}^2]^{1/2}$$

$$18.1\text{-}2$$

This is called the *statistically normalized cross correlation*. The normalization is to ensure that C(T) varies between ±1 and the denominator has the geometric mean of the energy in the channels over the time gate. An expression which requires fewer multiplications is the *energy normalized* cross correlation introduced by Neidell and Taner (1971). It uses the average trace energy in the denominator.

$$C_E = \frac{\sum_t [s_T^2 - \sum_{i=1}^{M} f_{i,t}^2]}{2(M-1) \sum_t \sum_i f_{i,t}^2} \qquad 18.1\text{-}3$$

Another measure introduced by Taner and Koehler (1969) for velocity spectra computation is the *semblance*. This is a ratio of stacked trace energy to input energy.

$$S_T = \frac{\sum_t s_T^2}{M \sum_t \sum_{i=1}^{M} f_{i\,t(i)}^2} \qquad 18.1\text{-}4$$

Semblance may be described as a multichannel filter which measures the common signal power over the channels according to a specified lag pattern. It has a possible range of 0 to 1 and is a measure of coherency which is independent of the joint power level of the traces. The statistics of semblance are discussed by Douze and Laster (1979).

Figure 18.1 shows a velocity spectra using the semblance measure as computed from the 24 traces shown on the right. The velocity spectra displays power as a function of velocity and two-way reflection time. Note the multiple energy between 3.2 and 3.8 seconds with velocities of 7000 to 8000 ft/sec. Another example was given in figures 1.7 and 1.8 of Chapter 1 where the statistically normalized cross correlation was used on marine seismic data. With velocity spectra data it is possible to classify signals, extract common signals and to separate multiple reflected energy from primary energy. Figure 18.2 from Schneider and Backus (1968) illustrates how it is possible to enhance either the primary or the multiple energy using a velocity operator derived from multichannel optimum filter theory (Schneider, Prince and Giles, 1965). Note the large amount of multiple energy between 1.8 and 5.0 seconds. It is particularly notable that one can establish that there is a primary dipping reflector at 3.2 seconds even with abundant but non-dipping multiples at the same time.

The determination of seismic wave velocity is difficult to carry out accurately because of the effect of structural dip and lateral variations in the

Common depth point and correlations

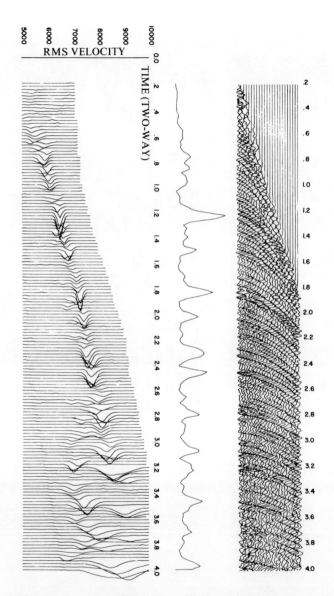

Figure 18.1 Typical velocity spectra display. The original seismic field records are shown on the right diagram. The central graph shows peak power and the left diagram is the velocity spectra. (From Taner and Koehler, 1969, courtesy of the Society of Exploration Geophysicists.)

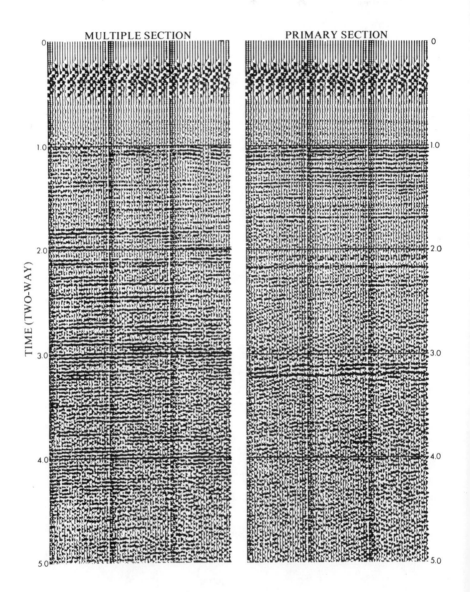

Figure 18.2 Three seismic profiles from a Florida marine line. The right half of the figure is a 6-fold stock made by applying continuously time varying filters prior to stock to enhance multiples and attenuate primaries. (From Schneider and Backus, 1968, courtesy of the Society of Exploration Geophysicists.)

velocity. This may be done more effectively by migrating seismic data. Migration is a technical process by which the subsurface reflecting horizons or diffracting points are placed in their correct horizontal position. The velocity function and the structural section are determined simultaneously in such a manner that the observed data is satisfied according to the differential equations governing the propagation of the elastic waves. Commonly used approximate techniques of wave equation migration are the finite difference method of Claerbout and Doherty (1972), the Kirchoff summation method of Schneider (1978), and the frequency-wavenumber domain method of Stolt (1978). Lateral velocity variations are not effectively treated by these techniques at the present time. Larner et al. (1978), Judson et al. (1980) and Schultz and Sherwood (1980) discuss partial solutions to the problem. Other aspects are discussed by Berkhout (1979), Bolondi et al. (1978), Hood (1978), Hubral (1980), Gazdag (1978, 1980) and Krey (1976).

18.2 Vespa and Covespa

Velocity spectra analysis has been used in studying teleseismic data with medium and large arrays of seismometers. The graph that is obtained usually plots seismic energy as a function of arrival time and wave slowness, $dT/d\Delta$, rather than velocity. The plot has been called a *vespagram*. The velocity spectral analysis (Vespa) used by Davies, Kelly and Filson (1971) involves the formation of a beam by delay and summation of the traces of an array and the calculation of the power in the beam over a specified time window which is stepped down the record. This is then repeated for different slownesses, or inverse velocities, keeping the beam at a constant azimuth, until a two-dimensional contour plot (vespagram) of power in slowness and in time is generated. In practice the array is pointed in a constant azimuth toward the source and steered over a range of wave slownesses so that the power in the beam of arriving waves in time may be examined.

Doornbos and Husebye (1972) pointed out that the proper interpretation of the vespagram requires knowledge not only of the response of the array as a function of slowness but also as a function of time. The factors which determine the time response are essentially the signal variation across the array, the duration of the signal, and the symmetry of the signal. As they have discovered, it may be difficult to properly attribute energy to a phase which follows very closely in time to a dominant arrival and is separated only in slowness due to leakage of energy into sidelobes.

The process incorporated in the Covespa technique as introduced by Gutowski and Kanasewich (1973) involves a zero-lag cross correlation in which only the high coherency measures are accepted so that the problem with sidelobe leakage is minimized and is less likely to be misinterpreted. The equation is a generalization to a two-dimension array of a one-dimensional form used by Montalbetti (1971).

$$cc(\phi,S,t) = \frac{2}{m(m-1) \cdot T} \underset{t\ k\ i}{\Sigma\Sigma\Sigma} \frac{f_{i,t}(\phi,S) \cdot f_{i+k,t}(\phi,S)}{(\Sigma_t f^2_{i,t} \cdot \Sigma_t f^2_{i+k,t})^{1/2}} \qquad 18.2\text{-}1$$

where
    cc is the coherency,
    m is the number of channels,
    k is an incremental integer on channel i ($i \neq k$),
    T is the length of the time window, and
    f(i,t) is the amplitude of ith channel at time t.

The computation begins by inserting appropriate delays into the traces corresponding to a certain slowness, S, and azimuth, ϕ. For each time along the records the zero-lag cross correlations of all combinations of two stations within a specified time window (1s for P waves, 4s for S waves) are computed, normalized to unity and summed. Thus for five stations the summation would involve 10 cross correlation functions. Since the coherency, cc, is normalized to unity, it will give a value of unity at a certain time and slowness if the phases and shapes of the signal within the window at all sensors are the same. The range of acceptable coherencies ($0.5 \leqslant cc \leqslant 1.0$) assures us that only similar signals irrespective of their power are plotted. The Covespa process is successful with as few as 3 stations. If the signal is modified too much from station to station by near-surface reverberations, the coherency tends to be very low in the coda of each phase.

Figure 18.3 shows an earthquake recorded by the variable aperture seismic array near Edmonton. The coherency and azimuth is nearly constant for the first 30 seconds. The slowness and azimuth then exhibit variations at 40 and 75 seconds. The coherency contours of the covespagram and the wave train of the beams suggest that energy arrives in groups of about 25s in length repeating every 40s. The repetitive energy groupings often seen on covespagrams could be due to reflection, single or multiple, from velocity transitions zones. A discontinuity in the low velocity layer at 150 km would generate a time difference of about 40s. However, the energy pattern which is nearly constant within each group must be interpreted on the basis of the source function, pP, sP, and any reverberations within the crust. From the time duration of such groups it may be possible to derive some information concerning the size of these sources. Assuming a shallow seismic disturbance propagating at a velocity of 2 or 3 km s^{-1} and a coherent group, typically 20 - 30s in duration, then the source length should be less than 40 - 60 km for magnitudes of 5.1 - 6.0. This is compatible with observations of surface rupture length of many shallow earthquakes (Bonilla, 1970, figure 3.16).

Vespa and Covespa

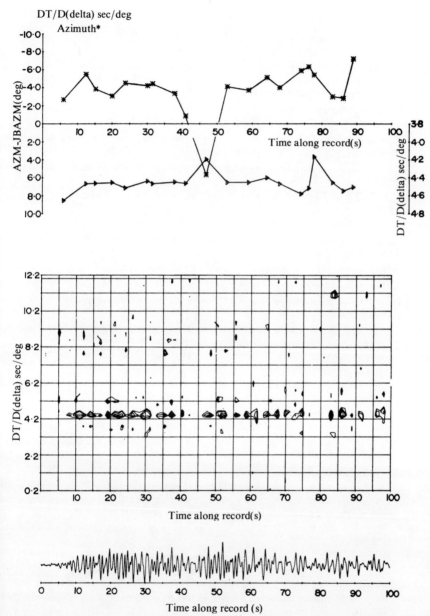

Figure 18.3 Covespa analysis of a New Ireland earthquake (h=20 km, Δ=95.8 degrees, magnitude=5.5) recorded on the VASA array in Alberta. The slowness and azimuth from the Jeffreys Bullen tables is 4.56 sec/deg. and 274.1 degrees. The top graph shows the variation of azimuth and slowness with time. The second graph shows the variation of slowness with time. The lower graph is a sum of seismic recordings aligned at the observed velocity and azimuth (23.95 km/sec. and 271.5 degrees). (From Gutowski and Kanasewich, 1973).

References

Berkhout,A.J.(1979), Steep dip dinite difference migration. *Geophysical Prospecting* **27**, 196-213.
Bolondi,G., Rocca,F. and Savelli,S.(1978), A frequency domain approach to two dimensional migration. *Gephysical Prospecting* **26**, 750-772.
Bonilla,M.G.(1970), Surface faulting and related effects. Ch. 3, 47-74 in *Earthquake Engineering*. Ed. R.L. Wiegel. Prentice-Hall, Inc., Englewood Cliffs, N.J., 518.
Cassano,E. and Rocca,F.(1973), Multichannel linear filters for optimal rejection of multiple reflections. *Geophysics* **38**, 1053-1061.
Claerbout,J. and Doherty,S.(1972), Downward continuation of moveout corrected seismograms. *Geophysics* **37**, 741-768.
Davies,D., Kelly,E.J. and Filson,J.R.(1971), Vespa process for analysis of seismic signals. *Nature* **232**, 8-13.
Dohr, G.P. and Stiller, P.K.(1975), Migration velocity determination. *Geophysics* **40**, 6-16.
Doornbos,D.J. and Husebye,E.S.(1972), Array analysis of PKP phases and their precursors. *Phys. Earth Planet. Int.* **5**, 387.
Douze,E.J. and Laster,S.J.(1979), Statistics of semblance. *Geophysics* **44**, 1999-2003.
Gazdag,J.(1978), Wave migration with the phase shift method. *Geophysics* **43**, 1342-1351.
Gazdag,J.(1980), Wave equation migration with the accurate space derivative method. *Geophysical Prospecting* **28**, 60-70.
Gutowski,P.R. and Kanasewich,E.R.(1973), Velocity spectral evidence of upper mantle discontinuities. *Geophys. J.R. Astr. Soc.* **36**, 21-32.
Havskov,J. and Kanasewich,E.R.(1978), Determination of the dip and strike of the Moho fram Array studies. *Bulletin Seismological Society of America* **68**, 1415-1419.
Hood,P.(1978), Finite difference and wave number migration. *Geophysical Prospecting* **26**, 773-789.
Hubral,P.(1980), Computation of the normal moveout velocity in 3D laterally inhomogeneous media with curved interfaces. *Geophysical Prospecting* **28**, 221-239.
Judson,D.R., Lin,J., Schultz,P.S. and Sherwood,J.W.C.(1980), Depth migration after stack. *Geophysics* **45**, 361-375.
Krey,T. and Toth,F.(1973), Remarks on wavenumber filtering in the field. *Geophysics* **38**, 959-970.
Krey,T.(1976), Computation of interval velocities from common reflection point moveout times for n layers with arbitrary dips and curvations in three dimensions when assuming small shot-geophone distances. *Geophysical Prospecting* **24**, 91-111.
Larner,K., Hatton,L. and Gibson,B.(1978), Depth migration of complex offshore profiles. Proc. OTC, Houston.
May,B.T. and Straley,D.K.(1979), Higher order moveout spectra. *Geophysics* **44**, 1193-1207.
Mayne,W.H.(1962), Common reflection point horizontal data stacking techniques. *Geophysics* **27**, 927-938.
Mercado, E.(1978), Maximum likelihood filtering of reflection seismograph data. *Geophysics* **43**, 407-513.

Montalbetti,J.F.(1971), Computer determination of seismic velocities - a review. *Journal of the Canadian Society of Exploration Geophysicists* **7**, 32-45.
Neidell,N.S. and Taner,M.T.(1971), Semblance and other coherency measures for multichannel data. *Geophysics* **36**, 482-497.

References

Sattlegger,J.W.(1975), Migration velocity determination. *Geophysics* **40**, 1-5.

Schneider,W.A., Prince,E.R., and Giles,B.F.(1965), A new data processing technique for multiple attenuation exploiting differential normal moveout. *Geophysics* **33**, 105-126.

Schneider,W.A. and Backus,M.M.(1969), Dynamic correlation analysis. *Geophysics* **32**, 33-51.

Schneider,W.(1978), Integral formulation for migration in two or three dimensions. *Geophysics* **43**, 49-76.

Schultz,P.S. and Sherwood,J.W.C.(1980), Depth migration before stack. *Geophysics* **45**, 376-393.

Shen,Wen-Wa(1979), A constrained minimum power adaptive beam former with time varying adaptation rate. *Geophysics* **44**, 1088-1096.

Stolt,R.(1978), Migration by Fourier transform. *Geophysics* **43**, 23-48.

Taner,M.T. and Koehler,F.(1969), Velocity spectra-digital computer derivation and application of velocity function. *Geophysics* **34**, 859-881.

Taylor,H.L., Banks,S.C. and McCoy,J.F.(1979), Deconvolution with l1 norm. *Geophysics* **44**, 39-52.

Ward,R.W. and Reining,J.B.(1979), Cubic spline approximation of inaccurate RMS velocity data. *Geophysical Prospecting* **27**, 443-457.

Woods,J.W. and Lintz,R.R.(1973), Plane waves at small arrays. *Geophysics* **38**, 1023-1041.

19 Polarization analysis

19.1 Introduction

The polarization properties of electromagnetic fields in optics and radio transmission have been the subject of considerable study and much of the theory is derived from this work (Wolf, 1959; Landau and Lifshitz, 1962; Born and Wolf, 1965). The nature and the source of geomagnetic micropulsation are not well understood and, since the introduction of broad-band, high gain, three-component systems for measuring the time varying natural occurring magnetic and electric fields, polarization studies have been undertaken to test various physical models. Early examples which include the necessary background mathematical exposition include the studies of Paulson, Egeland and Eleman (1965) and Fowler, Kotick and Elliott (1967).

In seismology the type of elastic waves which are produced by an earthquake or an explosion are well known from theoretical and laboratory model studies. However, the seismic recordings are always contaminated by noise which makes the detection and interpretation of small seismic events difficult. Therefore polarization analysis has been used to devise filters which will separate elastic body waves into compressional (P) and shear (S) phases and also enhance or attenuate surface Rayleigh and Love waves as desired (Flinn, 1965; Archambeau and Flinn, 1965; Montalbetti and Kanasewich, 1970).

19.2 Theory of polarization

Consider a plane quasi-monochromatic wave propagating in the z direction. A quasi-monochromatic wave is one on which most of the energy is confined over a small bandwidth, Δf, about the mean frequency, f.

$$\Delta f / f \ll 1 \qquad \qquad 19.2\text{-}1$$

To observe the polarization properties at any one point it is required that the complex amplitude and phase are relatively constant over a time, T, called the coherence interval determined by

$$1/f < T < 1/\Delta f \qquad \qquad 19.2\text{-}2$$

The recorded components are real functions of time but may be written as

Theory of polarization

$$E_x(t) = A_x(t)\ e^{i(\phi_x(t) - 2\pi ft)}$$
$$E_y(t) = A_y(t)\ e^{i(\phi_y(t) - 2\pi ft)}$$
19.2-3

where the amplitude, A, and phase, ϕ, are functions of time that vary slowly relative to the period of the sinusoidal oscillation. It will be assumed that the time average of both components is zero.

If the wave is monochromatic and invariant with time the field is perfectly polarized. Letting $-2\pi ft = \tau$ and taking the real part of 19.2-3, we have

$$E_x = A_x \cos(\tau + \phi_x)$$
$$E_y = A_y \cos(\tau + \phi_y)$$
19.2-4

To eliminate τ we make use of the following two trigonometric identities:

$$\cos(\tau \pm \phi) = \cos \tau \cos \phi \mp \sin \tau \sin \phi$$
$$\sin(\tau - \phi) = \sin \tau \cos \phi - \cos \tau \sin \phi$$
19.2-5

Using the first of these on 19.2-4 gives

$$E_x/A_x = \cos \tau \cos \phi_x - \sin \tau \sin \phi_x$$
$$E_y/A_y = \cos \tau \cos \phi_y - \sin \tau \sin \phi_y$$
19.2-6

Multiply the first by $\sin \phi_y$ and the second by $\sin \phi_y$ and subtract the two. For a second equation multiply the first by $\cos \phi_y$ and the second by $\cos \phi_y$ and subtract the two.

$$\frac{E_x}{A_x} \sin \phi_y - \frac{E_y}{A_y} \sin \phi_x = \cos \tau \cos(\phi_y - \phi_x)$$
$$\frac{E_x}{A_y} \sin \phi_y - \frac{E_y}{A_y} \cos \phi_x = \sin \tau \sin(\phi_y\ \phi_x)$$
19.2-7

When equations 19.2-7 are squared and added we obtain an equation for an ellipse which is described by the end points of the electric vector at a point in space.

$$\frac{E_x^2}{A_x^2} + \frac{E_y^2}{A_y^2} - 2\frac{E_x E_y}{A_x A_y} \cos(\phi_y - \phi_x) = \sin^2(\phi_y - \phi_x) \qquad 19.2\text{-}8$$

The ellipse (figure 19.1) is contained within a rectangle of sides $2A_x$ by $2A_y$.

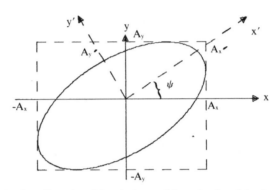

Figure 19.1   Polarization ellipse describing the locus of the end point of the electric vector.

The equation in 19.2-8 may be written in the matrix form as

$$[E_y E_x] \begin{bmatrix} A_x^2 & -A_x A_y \cos(\phi_y - \phi_x) \\ -A_x A_y \cos(\phi_y - \phi_x) & A_y^2 \end{bmatrix} \begin{bmatrix} E_y \\ E_x \end{bmatrix} = A_x^2 A_y^2 \sin^2(\phi_y - \phi_x) \qquad 19.2\text{-}9$$

or

$$\mathbf{E S E}^T = A_x^2 A_y^2 \sin^2(\phi_y - \phi_x) \qquad 19.2\text{-}10$$

where $\mathbf{E}^T$ is the transpose of $\mathbf{E}$. An orthogonal transformation may be made to rotate by an angle ψ to a new coordinate axis, x', y', in which the major axis of the ellipse coincides with the x' axis. The rotation matrix, $\mathbf{T}$,

$$\mathbf{T} = \begin{bmatrix} \cos \psi & \sin \psi \\ -\sin \psi & \cos \psi \end{bmatrix} \qquad 19.2\text{-}11$$

will convert $\mathbf{S}$ to the diagonal form, $\mathbf{S}'$.

$$\mathbf{T}^T \mathbf{S} \mathbf{T} = \mathbf{S}' \qquad 19.2\text{-}12$$

Theory of polarization

The equation for the ellipse is then

$$[E_y E_x] \begin{bmatrix} A_x^2 & 0 \\ 0 & A_y^2 \end{bmatrix} \begin{bmatrix} E_y \\ E_x \end{bmatrix} = A_x^2 A_y^2 \qquad 19.2\text{-}13$$

In the rotated co-ordinate system, the field components are

$$E_{x'} = A_{x'} \cos(L + \phi)$$
$$E_{y'} = A_{y'} \sin(L + \phi) \qquad 19.2\text{-}14$$

Born and Wolf (1965, pp. 24-27) derive the polarization parameters. The angle ψ is given by

$$\tan 2\psi = \frac{2 A_x A_y}{A_x^2 - A_y^2} \cos(\phi_y - \phi_x) \qquad 19.2\text{-}15$$

The ratio of the minor axis to the major axis, or ellipticity, is given by

$$\tan \beta = \frac{A_{y'}}{A_{x'}} \qquad (-\pi/4 < \beta \leq \pi/4) \qquad 19.2\text{-}16$$

where β is given by

$$\sin 2\beta = \frac{2 A_x A_y}{A_x^2 + A_y^2} \sin(\phi_y - \phi_x) \qquad 19.2\text{-}17$$

When looking, face-on, into the propagating wave the polarization is said to be right handed when the rotation is clockwise ($\beta \geq 0$) and left handed when the rotation is counter clockwise ($\beta < 0$). This is opposite to the usual terminology for a right- or left-handed screw.

Waves which have time varying amplitudes and phases as in equations 19.2-3 are not analyzed easily by their field vectors. It is advisable to form the auto and cross covariances and obtain a coherency matrix, **C**, as a function of lag, L.

$$\mathbf{C}(L) = \begin{bmatrix} C_{xx}(L) & C_{xy}(L) \\ C_{yx}(L) & C_{yy}(L) \end{bmatrix} \qquad 19.2\text{-}18$$

The cross covariance is given by the expectation over components such as E_x and the complex conjugate of E_y.

$$C_{xy}(L) = \lim_{T \to \infty} 1/T \int_{-T}^{T} E_x(t) \, E_y^*(t + L) \, dt \qquad 19.2\text{-}19$$

The matrix, **C**, is Hermitian since $C_{yx}(L) = C_{xy}^*(L)$ and it is non-negative definite. At zero lag the intensity of the wave as measured by the energy flux density is given by the trace of the matrix.

$$\text{tr } \mathbf{C}(0) = C_{xx}(0) + C_{yy}(0) \qquad 19.2\text{-}20$$

The coefficient of correlation is given by

$$\mu_{xy} = \frac{|C_{xy}(0)|}{(C_{xx}(0) \, C_{yy}(0))^{1/2}} \qquad 19.2\text{-}21$$

This gives the degree of coherence between the quasi-periodic components, E_x and E_y. Note that it depends upon the non-diagonal elements. For *perfect coherence*, $\mu_{xy} = 1$ and the determinant of the coherency matrix is zero.

$$\det \mathbf{C} \equiv 0 \qquad 19.2\text{-}22$$

If there is no polarization then E_x and E_y are independent and the non-diagonal terms of the coherency matrix are zero.

Born and Wolf (1965) showed that if several independent signals propagate in the same direction, the coherency matrix of the resultant signal is the sum of the coherency matrices of the individual signals. A quasi-monochromatic wave may be taken as the sum of a totally polarized signal and one that is unpolarized. Both Paulson et al. (1965) and Fowler et al. (1967) have analyzed such a situation. Let the coherency matrix for data with no polarization be

$$\mathbf{C}_u(0) = \begin{bmatrix} U^2 & 0 \\ 0 & U^2 \end{bmatrix} \qquad 19.2\text{-}23$$

where $C_{xx}(0) = C_{yy}(0) = U^2$. A perfectly coherent signal is given by a matrix that satisfies equation 19.2-22.

$$\mathbf{C}_p(0) = \begin{bmatrix} A^2 & B - iC \\ B + iC & (B^2 + C^2)/A^2 \end{bmatrix} = \begin{bmatrix} \rho_{xx} & \rho_{xy} \\ \rho_{xy}^* & \rho_{yy} \end{bmatrix} \qquad 19.2\text{-}24$$

A partially polarized wave is given by the sum

Theory of polarization

$$\mathbf{C}(0) = \begin{bmatrix} A^2+U^2 & B-iC \\ B+iC & (B^2+C^2)/A^2 + U^2 \end{bmatrix} \qquad 19.2\text{-}25$$

Equating the elements in 19.2-18 for zero lag with those in 19.2-25 we find that

$$\begin{aligned} C_{xx}(0) &= A^2 + U^2 \\ C_{yy}(0) &= (B^2+C^2)/A^2 + U^2 \\ |C_{xy}(0)|^2 &= B^2 + C^2 \end{aligned} \qquad 19.2\text{-}26$$

Solving these equations and letting

$$S = ([C_{xx}(0) - C_{yy}(0)]^2 + 4|C_{xy}(0)|^2)^{1/2} \qquad 19.2\text{-}27$$

we obtain

$$\begin{aligned} A^2 &= \frac{1}{2} (C_{xx}(0) - C_{yy}(0) + S) \\ U^2 &= \frac{1}{2} (C_{xx}(0) + C_{yy}(0) - S) \\ \frac{B^2+C^2}{A^2} &= \frac{1}{2} (-C_{xx}(0) + C_{yy}(0) + S) \end{aligned} \qquad 19.2\text{-}28$$

The degree of polarization in percent is obtained from the ratio of the trace for a perfectly coherent signal to the trace for a partially coherent signal.

$$P = 100 \frac{A^2 + (B^2+C^2)/A^2}{A^2 + (B^2+C^2)/A^2 + 2U^2} \qquad 19.2\text{-}29$$

or using 19.2-27 and 19.2-28

$$P = \frac{100\, S}{C_{xx}(0) + C_{yy}(0)} \qquad 19.2\text{-}30$$

The coherency matrix, $\mathbf{C}_p(0)$, may be incorporated into an equation for an ellipse.

$$[y\ x] \begin{bmatrix} \rho_{xx} & \rho_{xy} \\ \rho_{xy}^* & \rho_{yy} \end{bmatrix} \begin{bmatrix} y \\ x \end{bmatrix} = \rho_{xx}\ \rho_{yy} \qquad 19.2\text{-}31$$

An orthogonal rotation by ψ will make the major axis coincide with the new axis x'.

$$\tan 2\psi = \frac{2\,\Re[\rho_{xy}]}{\rho_{xx}(0) - \rho_{yy}(0)} \qquad 19.2\text{-}32$$

The ratio of the minor axis to the major axis, or ellipticity, is given by

$$\tan \beta = (\rho_{\acute{y}\acute{y}}/\rho_{\acute{x}\acute{x}})^{0.5} \qquad 19.2\text{-}33$$

where

$$\sin 2\beta = \frac{i(\rho_{yx} - \rho_{xy})}{[(\rho_{xx} - \rho_{yy})^2 + 4\rho_{yx}\rho_{xy}]^{1/2}} \qquad 19.2\text{-}34$$

The sense of polarization is given by the sign of β and is right handed or clockwise when β is positive and counter-clockwise when β is negative.

For computational ease and incorporation of a lag window it is advantageous to compute the auto and cross covariances through the Fourier transform of the power and cross-power spectral estimates, P_{ij}.

$$c_{ij}(0) = \int_{-W_N}^{W_N} P_{ij}(w)\,dw \qquad 19.2\text{-}35$$

19.3 Time domain polarization filters

Digital computer techniques allow one to take advantage of the polarization properties of seismic data for the enhancement of the signal to noise ratio. When the signal and noise exhibit similar spectral characteristics band-pass filters are not effective. However, the availability of three component seismograms allows one to determine filter functions which take advantage of the polarization properties of both body waves and surface waves to favor the detection of desired phases.

Elastic body waves, which may be generally separated into P (compressional) and S (shear) phases, can be considered as non-dispersive group arrivals with maximum power in the 0.3- to 10-second period range. Surface Rayleigh and Love waves may be described as dispersive group arrivals with maximum power in the 2- to 100-s period range for earthquakes of moderate magnitude, the observable periods extending up to 57 minutes for larger teleseismic events. Superimposed on these signals is microseismic background noise as well as signal generated noise. Signal generated noise is the result of multiple reflections and refractions of P and S body waves at crustal interfaces and in homogeneities and local conversion of body waves to surface waves; this type of noise originates primarily under the recording station. Microseismic noise, which is considered to consist mainly of fundamental and higher mode Rayleigh waves, has been shown to exhibit a sharp peak in the 5- to 8-s period range (Brune and Oliver, 1959). As a result of this sharp peak in the microseismic noise spectrum, frequency band-pass filtering is often employed to improve the signal to noise ratio. Although this type of filtering is very effective in removing microseismic background from both long and short period data, it often cannot distinguish between signal and signal generated noise. Difficulty may still arise in the attempted identification of phases whose frequency characteristics are similar.

Both compressional and shear waves exhibit a high degree of linear polarization. Particle motion coincides with the azimuth of propagation for the transverse (S) phases. Surface waves of the Rayleigh type are generally elliptically polarized in the vertical-radial plane, the fundamental modes displaying retrograde particle motion and the higher modes prograde or tetrograde ellipticity. Surface Love waves are also found to be rectilinearly polarized, but in a horizontal plane orthogonal to the direction of wave propagation. Since most microseismic background is of the Rayleigh type, it will exhibit elliptical polarization, but with little preferred directionality. Signal generated noise may also be polarized, but again the direction of polarization is random in nature. By using these various characteristics of polarized particle trajectories, filters may be designed which preserve motion if it satisfies specified conditions of polarization in a particular direction and which attenuate motion that does not satisfy the desired criteria.

Shimshoni and Smith (1964) suggest that the time averaged cross product of vertical and radial components of ground motion may be used as a measure of rectilinearity. The computed cross product is multiplied by the original signal so that a function of ground motion which enhances rectilinearly polarized signal is obtained. Another process described by the same workers computes the parameters of an equivalent ellipse at each instant in time from the Fourier components of the vertical and radial motions. Eccentricity, major axis and angle of inclination of the ellipse from the vertical are displayed for the frequency at which maximum power is arriving and used to provide criterion for the identification of P and SV type motion.

Various workers have applied polarization filtering techniques to recorded seismic data for improvement of the signal to noise ratio. Lewis and Meyer (1968) apply a phase filter of the REMODE type as described by Archambeau and Flinn (1965) and originally developed by Mims and Sax (1965) with subsequent work by Griffin (1966a,b) to data recorded during the Early Rise experiment in the summer of 1966. Archambeau, Flinn and Lambert (1969) used the same filter to study multiple P phases from NTS explosions. In another application, Basham and Ellis (1969) used a REMODE filter designated as a P-Detection (P-D) filter to process P-wave codes of numerous seismic events recorded in western Alberta. The polarization filter, applied to 25s of record following the P onset, aids in identification of numerous compressional wave arrivals including P, pP and sP. PcP and PKP phases for events at appropriate epicentral distances are also detected.

The design of the polarization filter used by Montalbetti and Kanasewich (1970) is a variation on one described by Flinn (1965) in which both rectilinearity and direction of particle motion is considered. In order to obtain measures of these two quantities, the covariance matrix for a set of N points taken over each of the three orthogonal components of ground motion, R (Radial), T (Transverse) and Z (Vertical), is computed. For a three-component digital seismogram then, a specified time window of length $N\Delta t$, where Δt is the sampling interval, is thus considered. To determine the covariance matrix for this set of observations, the means, variances and covariances must be calculated for the three variables R, T and Z.

We define the mean or expected, E, value of N observations of the random variable $X_{1i}(i=1,2,...,N)$ as

$$\mu_1 = \frac{1}{N} \sum_{i=1}^{N} X_{1i} = E(X_1) \qquad 19.3\text{-}1$$

The covariance between N observations of two variables X_1 and X_2 is given by

$$Cov[X_1,X_2] = \frac{1}{N} \sum_{i=1}^{N} (X_{1i} - \mu_1)(X_{2i} - \mu_2) \qquad 19.3\text{-}2$$

where μ_1 and μ_2 are computed as in equation 19.3-1. It is evident that

$$Cov[X_1,X_2] = Cov[X_2,X_1]$$

The quantity $Cov[X_1,X_2]$ is defined as the autocovariance or simply the variance of X_1. The matrix with $Cov[X_r,X_s]$ in its rth row and sth column (r,s=1,2,...,n) is the covariance matrix for the set of n random variables X_j,j=1,2,...,n. If **X** is the vector of the random variables and μ the vector of means for each of these variables, the covariance matrix **V** is defined by

$$\mathbf{V} = E[(\mathbf{X} - \mu)(\mathbf{X} - \mu)^t] \qquad 19.3\text{-}3$$

The superscript t indicates the column transpose of the vector. For our case of three variables R, T and Z considered over the time window $N\Delta t$, equation 19.3-3 is represented by

Time domain polarization filters

$$V = \begin{bmatrix} Var[R] & Cov[R,T] & Cov[R,Z] \\ Cov[R,T] & Var[T] & Cov[T,Z] \\ Cov[R,Z] & Cov[T,Z] & Var[Z] \end{bmatrix} \qquad 19.3\text{-}4$$

where the covariances and variances are defined in equation 19.3-2.

If the covariance matrix given by equation 19.3-4 is diagonalized, an estimate of the rectilinearity of particle motion trajectory over the specified time window can be obtained from the ratio of the principal axis of this matrix. The direction of polarization may be measured by considering the eigenvector of the largest principal axis. If λ_1 is the largest eigenvalue and λ_2 the next largest eigenvalue of the covariance matrix, then a function of the form

$$F(\lambda_1, \lambda_2) = 1 - (\lambda_2/\lambda_1)^n \qquad 19.3\text{-}5$$

would be close to unity when rectilinearity is high ($\lambda_1 \gg \lambda_2$) and close to zero when the two principal axes approach one another in magnitude (low rectilinearity). The direction of polarization can be determined by considering the components of the eigenvector associated with the largest eigenvalue with respect to the coordinate directions R, T and Z.

To illustrate this, figures 19.2 (a-d) show some computations applied to sets of data in two dimensions. These points were computed for an ellipse and then perturbed by the addition of random noise. In figures 19.2(b) and (d), the data points were generated from a 45° rotation of the original ellipses determined for figures 19.2 (a) and (b) respectively. The computed covariance matrix, correlation coefficient ρ, $F(\lambda_1, \lambda_2)$ for n = 1, and the eigenvector or the principal axis, E, are shown for each case. The correlation coefficient is determined from the usual definition.

$$\rho_{12}^2 = \frac{(Cov[X_1, X_2])^2}{Var[X_1]\, Var[X_2]} \qquad 19.3\text{-}6$$

Comparing figures 19.2 (a) and (b) we see that in the first case $Cov[X_1, X_2]$ is small with respect to the variance terms on the main diagonal, and the eigenvector E indicates a preferred direction along the X_2-axis. In the second case, figure 19.2 (b), $Cov[S_1, S_2]$ is larger and $E = (0.726, 0.689)$ shows no preferred orientation along either the X_1 or X_2 co-ordinate axis. In both instances $F(\lambda_1, \lambda_2)$, the rectilinearity function, is approximately the same. Similar properties of the covariance matrix and the eigenvector E are indicated by figures 19.2 (c) and (d), but here the rectilinearity, $F(\lambda_1, \lambda_2)$, is low in both cases. We see then that when the off-diagonal terms of the covariance matrix are significant the diagonalization will introduce a rotation. After rotation, the orientation in space of the principal axis of the covariance matrix will be given by the components of its eigenvector relative to the original co-ordiante system. $F(\lambda_1,\lambda_2)$ will give an estimate of the degree of polarization along the major axis. In each of figures 19.2 (a)-(d), the data points could be considered as representing particle motion trajectory over a specified time in one of the orthogonal planes of the R, T, Z co-ordinate system.

Suppose now that in the matrix of equation 19.3-4 the covariance terms are small with respect to those on the main diagonal, on which $Var[R] > Var[T] > Var[Z]$. The length of the two principal axes would thus correspond very closely to $Var[R]$ and $Var[T]$, and the eigenvector associated with the major axis, $Var[R]$, would have its largest component in the R co-ordinate direction. If $Var[R] > Var[T]$, then $F(\lambda_1,\lambda_2)$ in equation 19.3-5 would be close to unity and we would have high rectilinearity in the R direction. If $Var[R] \sim Var[T]$, then the direction of polarization would still be predominantly along the R co-ordinate axis, but the rectilinearily would be low. If the off-diagonal terms of the matrix were significant, the diagonalization would introduce a rotation, and the orientation in space of the major axis would be given by the components of its eigenvector relative to the original R, T, Z co-ordinate system. By combining $F(\lambda_1,\lambda_2)$ and the appropriate eigenvector, we see that measures of rectilinearity and directionality can be obtained. Determination of suitable filter function is then possible.

Applying this analysis to three-component digital seismograms, the covariance matrix is computed for a specified time window of length $N\Delta t$ centred about t_0, where t_0 is allowed to range over the entire record length of interest. Eigenvalues and the corresponding eigenvectors of this matrix are then determined. The measure of rectilinearity for the time t_0 is given by

$$RL(t_0) = F(\lambda_1,\lambda_2)]^J \qquad\qquad 19.3\text{-}7$$

where $F(\lambda_1,\lambda_2)$ is defined in equation 19.3-5. If we present the eigenvector of the principal axis with respect to the R, T, Z co-ordinate system by $\mathbf{E} = (e_1, e_2, e_3)$ then the direction functions at time t_0 are given by

$$D_i(t_0) = (e_i)^K \qquad i = R, T, Z \equiv 1, 2, 3. \qquad\qquad 19.3\text{-}8$$

Since the eigenvector is normalized ($|\mathbf{E}|=1$), we see that for each of these functions D_i

$$0 < D_i < 1$$

The exponents J, K and n (equation 19.3-5) are determined empirically. Once these functions are computed, the window is moved down the record one sample interval and the calculations are repeated.

Time domain polarization filters

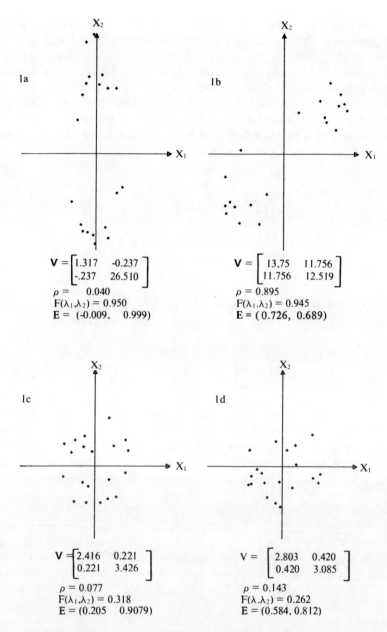

Figure 19.2 Examples of the covariance matrix, correlation coefficient, ρ, the rectilinearity function $F(\lambda_1,\lambda_2)$ for $n=1$, and the eigenvector, **E**, of the principal axis for sets of 20 points in two dimensions. (From Montalbetti and Kanasewich, 1970, courtesy of the Geophys. J. R. Astr. Soc.)

Weighting functions of this form are used by Flinn (1965) as point by point gain controls and applied to the appropriate component of ground motion. In the course of the present study, it was found that an additional computation was advantageous. When the end of the desired record is reached, the quantities defined by equation 19.3-7 and 19.3-8 are then averaged over a window equal to about half the original window length. This has the effect of *smoothing* these operators so that contributions due to any anomalous spikes are subdued. If this time window consists of M points, the gain functions are given by

$$RL^*(t_0) = \frac{1}{M}\sum_{t=-L}^{L} RL(t_0+L)$$

$$D_i^*(t_0) = \frac{1}{M}\sum_{t=-L}^{L} D_i(t_0+L), \quad i=R,T,Z \qquad 19.3\text{-}9$$

$$L = (M-1)/2$$

These operators are then used as a point gain control to modulate the rotated records so that any time, t, the filtered seismograms are given by

$$\begin{aligned} R_f(t) &= R(t) \cdot RL^*(t) \cdot D_R^*(t) \\ T_f(t) &= T(t) \cdot RL^*(t) \cdot D_T^*(t) \\ Z_f(t) &= Z(t) \cdot RL^*(t) \cdot D_Z^*(t) \end{aligned} \qquad 19.3\text{-}10$$

In most cases, the data were band-pass filtered with a zero phase shift digital filter before rotation so that the spectral content of the record was restricted. The window length could be specified so as to be consistent with one or two cycles of the dominant period. Values of J and K which appeared quite adequate were 1 and 2 respectively. For rectilinearity, $n = 0.5$ or 1 was used.

Figure 19.3 shows an event recorded at the University of Alberta's Edmonton Geophysical Observatory processed by this method. The earthquake, of magnitude 5.9, occurred in the Philippines. Traces 4-7 are the filter functions as given by equation 19.3-9. The last three represent the records after applying these operators as in equation 19.3-10. The filter functions and the processed records are displayed from the point in time where calculations began. We see a good separation of events labelled P_1, P_2 and P_3 in the P, PP and PPP codas of the processed seismogram. The good correlation of these events with their respective PP times suggests a multiplicity of sources at the same location rather than a single release of energy. Attenuation of the radial and transverse traces, especially during the P and PP events, illustrates how the filter enhances motion which exhibits a preferred direction of polarization, in this case, the Z direction.

Time domain polarization filters

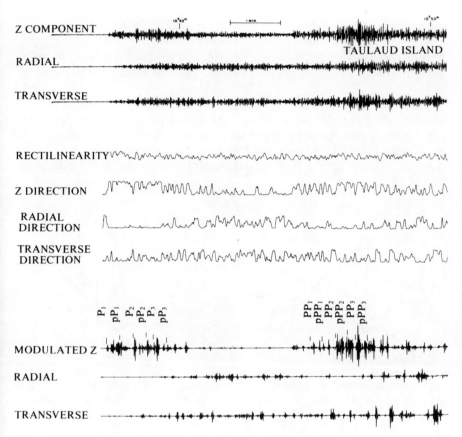

Figure 19.3 Example of a magnitude 5.9 earthquake which occurred in the Phillippines at 10h 29m 40.4s U.T. on January 30, 1969 processed by the polarization filter with J=1, K=L=M=2, n=1/2. Universal time is indicated along the top of the figure. The epicentral distance was 103° and depth was 70km. The first three traces are the vertical, radial and transverse components of ground motion respectively; the next four represent the computed filter operators; the last three are the filtered seismograms. The data were band-pass filtered 0.3 to 4 Hz with a zero phase shift operator before the operation shown here. (From Montalbetti and Kanasewich, 1970, courtesy of the Geophys. J.R. Astr. Soc.)

19.4 The remode filter

A particular form of polarization filter which considers only the rectilinearity in the radial and vertical directions instead of in three-dimensional space has been called *REMODE* for Rectilinear Motion detector. It was originated by Mims and Sax (1965) in the time domain and by Archambeau and Flinn (1965) in the frequency domain. Additional subsequent development was by Griffin (1966 a,b).

The vertical (Z) and radial (R) components of seismic signals are rotated so that the expected direction of the incident body waves bisect the angle between the two orthogonal components (**Z,R** in figure 19.4). That is

$$Z = Z \cos(\pi/4 - \Theta) + N \cos \alpha \cdot \sin(\pi/4 - \Theta) + E \sin \alpha \cdot \sin(\pi/4 - \Theta)$$
$$R = Z \sin(\pi/4 - \Theta) - N \cos \alpha \cdot \cos(\pi/4 - \Theta) - E \sin \alpha \cdot \cos(\pi/4 - \Theta)$$
$$19.4\text{-}1$$

where Θ is the angle of incidence, α is the azimuth of a great circle path, and Z,N,E are the vertical, north-south and east-west components of motion. All the P and SV motion is represented by these two rotated components so that the similar shape of **Z** and **R** signal waveforms contracts with a dissimilar shape for any noise.

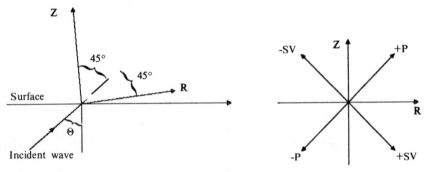

Figure 19.4 Rotation to obtain the **Z** and **R** components for a REMODE filter. On the right: particle motion for either an incident compressional (P) or a vertically polarized shear (SV) wave.

The filter operator is obtained from a cross correlation function, C(T) or **Z**(t) and R(t) over a window, W, centered at some time, t, on the record.

$$C(+T) = \sum_{t-W/2}^{t+W/2} Z(t) \ R(t+T) \qquad 19.4\text{-}2$$

To insure that the operator is an even function and introduces no phase distortion, the negative lags are generated from the positive lags.

$$C(-T) = C(+T) \qquad 19.4\text{-}3$$

If the polarization in the **R-Z** plane is predominantly rectilinear, C(T) will be large. If the motion is elliptical or random, it will be small. By convolving C(T) with the original time series, motion of high rectilinearity is enhanced and

The remode filter

elliptically polarized motion is attenuated. The output from the filter is given by

$$Y_R(t) = P_p K \sum_{T=-L}^{L} R(t-T)\, C(T)$$

$$Y_Z(t) = P_p K \sum_{T=-L}^{L} Z(t-T)\, C(T)$$

19.4-4

where P_p is a polarization operator and K is a normalizing factor. The operator, $C(T)$, is different for every data point, t, so the processing is expensive computationally. It is recommended that the maximum lag, T_{max}, be half the interval, L. A normalizing factor, K, determined from autocorrelation of the input may be used to enhance weak signals with high rectilinearity.

$$K(t) = [\sum_{t-W/2}^{t+W/2} R^2(t) \sum_{t-W/2}^{t+W/2} Z^2(t)]^{-1/2}$$

19.4-5

The REMODE operator, as described above, will enhance both linearly polarized P and SV phases. Following a suggestion of Griffin (1966a), Basham and Ellis (1969) have tuned the filter to accept either P or SV waves. Note that for P or compressional motion in the insert diagram of figure 19.4

$$R(t)Z(t) > 0$$

19.4-6

while for vertically polarized shear waves

$$R(t)Z(t) < 0$$

19.4-7

The polarization operator, P_p, is defined as

$$P_p = \begin{bmatrix} 1 & R(t)Z(t) > 0 \\ 0 & R(t)Z(t) \leq 0 \end{bmatrix}$$

19.4-8

for P phases and

$$P_s = \begin{bmatrix} 1 & R(t)Z(t) < 0 \\ 0 & R(t)Z(t) > 0 \end{bmatrix}$$

19.4-9

for SV phases.

Archambeau and Flinn (1965) have shown that a REMODE filter may be approximated in the frequency domain by a cosine function

$$Y(\omega) = \cos[\phi_R(\omega) - \phi_Z(\omega)] = \cos\phi \qquad 19.4\text{-}10$$

where ϕ_R and ϕ_Z are the phase angles obtained from a Fourier transform, $r(\omega)$, $z(\omega)$, of a short segment of $R(t)$ and $Z(t)$. In practice the function $Y(\omega)$ is obtained from a cross-power spectral estimate, $P_{ZR}(\omega)$, using the relation

$$Y(\omega) = \frac{\Re[P_{ZR}(\omega)]}{ABS[P_{ZR}(\omega)]} \qquad 19.4\text{-}11$$

The argument, ϕ, will be 0° or 180° for rectilinear motion of 90° for pure Rayleigh waves. The sensitivity of the phase response can be increased by raising the cosine to a higher power.

$$Y(\omega) = \cos^n\phi \qquad 19.4\text{-}12$$

Lewis and Meyer (1968) recommend letting n = 7. The transfer function is multiplied by $r(\omega)$ and $z(\omega)$.

$$r'(\omega) = Y(\omega)r(\omega)$$
$$z'(\omega) = Y(\omega)z(\omega0 \qquad 19.4\text{-}13$$

An inverse Fourier transform is taken to recover the filtered segment in the time domain. The window is moved along in short steps, a new transfer function is determined and, by repeated application, the entire trace is filtered.

19.5 Surface wave discrimination filter

Rayleigh and Love waves may be enhanced by designing operators which pass elliptical particle motion in the vertical-radial plane and linearly polarized motion in the transverse horizontal direction. The filter must be designed in the frequency domain since surface wave trains are dispersive. Following Simons (1968) the discrete Fourier series of three components, vertical, Z, radial, R, and transverse, T, of ground motion are obtained. The amplitude coefficients at each frequency are weighted according to how closely the three-dimensional particle motion trajectory at the frequency corresponds to theoretical Love or Rayleigh wave patterns arriving from a specified direction. These modified coefficients are then used to reconstruct the trace in time domain.

Consider a time segment of length $N\Delta t$, where Δt is the sampling interval, then for each component of ground motion, the amplitude and phase

Surface wave discrimination filter

of each harmonic is given in terms of the discrete Fourier coefficients a(nf) and b(nf) by

$$A_i(nf) = [a_i^2(nf) + b_i^2(nf)]^{1/2} \qquad 19.5\text{-}1$$

$$\phi(nf) = \arctan\ [b_i(nf)]/a_i(nf) \qquad n=0,1,...N/2 \qquad 19.5\text{-}2$$

where i = Z, R, T corresponds to the vertical, radial and transverse components of motion. A measure of the apparent horizontal azimuth (figure 19.5) can be obtained from the function

$$\beta(nf) = \arctan\ [A_T(nf)/A_R(nf)] \qquad 19.5\text{-}3$$

and an estimate of eccentricity of the particle motion ellipse is represented by

$$\psi(nf) = \arctan\ [A(nf)/A_z(nf)] \qquad 19.5\text{-}4$$

where

$$A(nf) = [A_R^2(nf) + A_T^2(nf)]^{1/2}$$

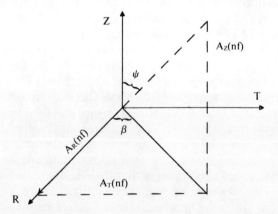

Figure 19.5 The relation between the apparent horizontal azimuth, β, the eccentricity, ψ, and the three orthogonal components of ground motion.

The phase difference between vertical and radial components is also determined as

$$\alpha(nf) = \phi_R(nf) - \phi_z(nf) \qquad 19.5\text{-}5$$

Functions of α, β and ψ are then used to weight the amplitude coefficients according to the following equations:

$$A'_Z(nf) = A_Z(nf) \cdot \cos^M[\beta(nf)] \cdot \cos^K[\psi(nf)\text{-}\Theta] \cdot \sin^N[\alpha(nf)]$$
$$A'_R(nf) = A_R(nf) \cdot \cos^M[\beta(nf)] \cdot \cos^K[\psi(nf)\text{-}\Theta] \cdot \sin^N[\alpha(nf)] \qquad 19.5\text{-}6$$
$$A'_T(nf) = A_T(nf) \cdot \sin^M[\beta(nf)] \cdot \sin^K[\psi(nf)]$$

The exponents M, K and N are determined empirically. We see that the Z and R components receive identical weights and that each function of α, β and ψ lies in the range (0,1).

If at a particular frequency motion in the horizontal plane is purely radial ($\beta(nf)=0$), $A_Z(nf)$ and $A_R(nf)$ are preserved and $A_T(nf)$ is attenuated. Motion which is purely transverse in nature ($\beta(nf)=\pi/2$) will, on the other hand, be attenuated in the vertical-radial plane. The parameter Θ in the equations for $A'_Z(nf)$ and $A'_R(nf)$ can be chosen so that a particular horizontal/vertical displacement ratio for the particle motion trajectory is preserved. If $\psi(nf)=\Theta$ then the radial and vertical amplitudes receive unit weighting from the function $\cos^K[\psi(nf)\text{-}\Theta]$. For motion which is purely horizontal $\psi(nf)=\pi/2$, and the function $\sin^K[\psi(nf)]$ applies unit weight to the transverse amplitude. The function $\sin^N[\alpha(nf)]$ will attenuate the radial and vertical amplitudes by an amount which varies from 1 to 0 according to how closely the phase difference between radial and vertical components departs from the theoretical $\pi/2$ value for fundamental retrograde Rayleigh waves.

The combined effect of these weighting factors is thus seen to enhance pure Rayleigh or pure Love waves of some particular period arriving at the recording station. If, for example, motion in the horizontal plane is predominantly radial, the horizontal/vertical displacement ratio corresponds to the specified Θ, and the phase difference between radial and vertical components is $\pi/2$, then the Z and R coefficients will be preserved and the transverse component attenuated. This corresponds to the case of pure Rayleigh motion. Conversely, a dominant arrival at some particular period on the transverse component along with little or no amplitude on the vertical record corresponds to a Love phase so that the amplitude coefficient for the transverse trace will be preserved and the Z-R motion will be attenuated. Particle trajectories which lie between these limits will be subject to varying degrees of attenuation. Prograde Rayleigh type motion will be attenuated if we specify that

Surface wave discrimination filter

$$\sin[\alpha(nf)] \equiv 0 \quad , \quad \pi \leqslant \alpha(nf) \leqslant 2\pi \qquad 19.5\text{-}7$$

In practice, the Fourier coefficients are determined over the specified window and then modified using the weighting functions as in equation 19.5-6. As no modification is applied to the phase angles of each Fourier component, these are used along with the modified amplitude coefficients to reconstruct the trace in the time domain. The window is then moved down the record some fraction of the original window length and calculations are repeated. The final output is the arithmetic average of all overlapping time segments at any particular point on the seismogram. A cosine taper applied to about the first and last 10 percent of the data points in each window is required to reduce slight phasing effects introduced by the discontinuities in the original time series at the beginning and end of each window.

19.6 Polarization states and unitary matrices

The previous sections have shown that a matrix formulation is of great value in the analysis of the polarization state of a waveform. In quantum mechanics, Hermitian matrices are used in determining observable quantities. In other types of wave fields the exact nature of the waveform is not understood because of the complex effects of anisotropy, attenuation and dispersion. Furthermore, they have not always been modelled properly because it is not certain which terms in the differential equations are significant and because the initial and boundary conditions are not known or are very complicated. It is desirable to make a more formal, if brief, presentation of matrix algebra so that the current literature may be read. The following section will present some definitions and theorems concerning adjoint, unitary and Hermitian matrices and their use in the transformation of a cross-power spectral matrix. A more formal exposition on the subject including the proofs is found in books by Pease (1965) and Mal'cev (1963) and in the papers by Samson (1973, 1977).

The *unitary inner product* of two column vectors, **x** and **y**, produces a scalar and is defined as follows:

$$< \mathbf{x}, \mathbf{y} > = \sum_i x_i^* y_i = \mathbf{x}^{*T} \mathbf{y} = \mathbf{x}^H \mathbf{y} \qquad 19.6\text{-}1$$

The asterisk, *, indicates a complex conjugate, T is the transpose and H indicates a *Hermitian transpose* of a vector or matrix. The data vectors, **x** and **y**, may be real or complex valued numbers.

$$\mathbf{x}^H = \mathbf{x}^{*T} = [x_1^*, x_2^* \ldots x_n^*] \qquad 19.6\text{-}2$$

Let **S** be a matrix operator which acts on **y**, one of the vectors. The *adjoint* of matrix **S** is S^A if the following inner product holds for all **x** and **y**:

$$< x, Sy > \; = \; < S^A x, y > \qquad 19.6\text{-}3$$

For the unitary inner product in 19.6-3 the adjoint of a matrix is equal to its Hermitian conjugate.

$$S^A = S^H \qquad 19.6\text{-}4$$

If we find that $S^A = S$ then for complex numbers the type of matrix is *Hermitian*. For real numbers it is symmetric and the adjoint is equal to the transpose ($S^A = S^T$).

U is a *unitary matrix* if it consists of complex numbers, is non-singular and its adjoint is equal to the reciprocal of **U**.

$$U^A = U^{-1} \qquad 19.6\text{-}5$$

From this it follows that the product of a unitary matrix and its adjoint is equal to the identity matrix, **I**.

$$UU^A = I \qquad 19.6\text{-}6$$

The eigenvalues, λ_i, of a unitary matrix have unit modulus.

$$|\lambda_i|^2 = 1 \qquad 19.6\text{-}7$$

Given any matrix, **S**, we can find two distinct Hermitian matrices, S_1 and S_2.

$$\begin{aligned} S_1 &= (S + S^H) / 2 \\ S_2 &= \text{-}(S - S^H) \, i \, / \, 2 \end{aligned} \qquad 19.6\text{-}8$$

If we let $S = S_1 U_1 = U_2 S_2$ where U_1 and U_2 are unitary matrices then $S^H = U_1^H S_1^H = U_1^{-1} S_1 = S_2^H U_2^H = S_2 U_2^{-1}$. It follows that $S_1^2 = S_1 U_1 U_1^{-1} S_1 = SS^H$ and similarly $S_2^2 = S^H S$. A solution is obtained from the squares.

$$S_1^2 = SS^H \qquad S_2^2 = S^H S \qquad 19.6\text{-}9$$

Each Hermitian matrix in equation 19.6-9 is positive definite and it represents a

Polarization states and unitary matrices

stochastic process. For S_1 let the eigenvalues and eigenvectors be respectively λ_j and u_j.

$$S_1^2 u_j = \lambda_j^2 u_j \qquad j = 1, 2, \ldots n \qquad 19.6\text{-}10$$

The eigenvectors are real. Let U_1 be a square matrix with columns that are eigenvectors of S_1^2. U_1 is a unitary matrix and its inverse is the Hermitian conjugate of U.

$$U_1^{-1} = U_1^H \qquad 19.6\text{-}11$$

U_1 is the matrix that will diagonalize S_1^2 by a similarity transformation.

$$S_1^2 = U_1 \, [diag \, (\lambda_1^2, \lambda_2^2 \ldots \lambda_n^2)] \, U_1^{-1} \qquad 19.6\text{-}12$$

The symbol *diag* (λ_j^2) indicates a square matrix with λ_j^2 on the diagonal and zero everywhere else. If we choose the positive sign for λ_i we can obtain an S_1 that is positive definite.

$$S_1 = U_1 \, [diag \, (\lambda_1, \lambda_2 \ldots \lambda_n)] \, U_1^{-1} \qquad 19.6\text{-}13$$

A similar analysis will yield S_2.

Note that from 19.6-8 S_1 and S_2 can be considered to be components of a complex matrix S.

$$S = S_1 + i S_2 \qquad 19.6\text{-}14$$

The matrix S can also be formed from a product of a Hermitian matrix, S_j, and a unitary matrix, V_j.

$$S = S_1 V_1 = V_2 S_2 \qquad 19.6\text{-}15$$

If S is non-singular the unitary matrices are obtained as follows:

$$V_1 = S_1^{-1} S \qquad V_2 = S S_2^{-1} \qquad 19.6\text{-}16$$

The two unitary matrices can be written in polar form.

$$V_1 = exp \, (iS_1) \qquad V_2 = exp \, (iS_2) \qquad 19.6\text{-}17$$

The definition of the exponential is as follows:

$$\mathbf{V}_1 = \mathbf{U}_1 \; [diag \; (exp[i\lambda_1] \ldots exp \; [i\lambda_n])] \; \mathbf{U}_1^{-1} \qquad 19.6\text{-}18$$

The matrix **S** in 19.6-15 can be interpreted geometrically in a manner similar to the polar representation of a complex number. By 19.6-7 the eigenvalues of a unitary matrix have unit modulus. They may be viewed as points on a unit circle and the λ_j can be any multiple of 2π without changing $\mathbf{V}_j$ or **S**.

Consider the geophysical case of designing a filter when there is a measurement of the polarization matrix from an array consisting of n sensors. The signals in the time domain will be **x**(t).

$$\mathbf{x}^T(t) = [x_1(t), x_2(t), \ldots x_n(t)]$$

The cross-power spectral matrix is obtained from the following set of equations. The Fourier transform, **z**(k), is usually weighted by a Daniell window in which the coefficients, a_k, are all equal to unity. The bandwidth, designated by Δf, must not be too narrow. The H denotes a *Hermitian adjoint*.

$$\mathbf{S}[f(k, \Delta f)] = (k_2 - k_1)^{-1} \sum_{k=k_1}^{k_2} a_k \, \mathbf{z}(k) \, \mathbf{z}^H(k) \qquad 19.6\text{-}19$$

$$\mathbf{z}(k) = \sum_{t=0}^{N-1} \mathbf{x}(t) \, exp(-2\pi i k t) \qquad k = 0, 1, \ldots (N/2) - 1$$

$$\Delta f = (k_2 - k_1) / (n \, \Delta t)$$
$$f(k) = (k_1 + k_2) / (2N \, \Delta t)$$

The matrix **S** is non-negative and Hermitian.

Following Samson (1977) a transformation of **S** by an appropriate unitary matrix, **U**, will diagonalize the matrix. The H denotes a Hermitian adjoint and the eigenvalues are λ_j.

$$\mathbf{U}^H \mathbf{S} \mathbf{U} = diag \; (\lambda_1, \lambda_2, \ldots \lambda_n) \qquad 19.6\text{-}20$$

The cross-power spectral matrix can be expanded as a set of outer products of the eigenvectors, $\mathbf{u}_j$ of **S**, and their Hermitian adjoints.

$$\mathbf{S} = \sum_{j=1}^{n} \lambda_j \, \mathbf{u}_j \, \mathbf{u}_j^H \qquad 19.6\text{-}21$$

Polarization states and unitary matrices

Samson (1973) defines the degree of polarization, P, by the following relation:

$$P^2 = [2(n-1)(tr\mathbf{S})^2]^{-1} \sum (\lambda_j - \lambda_k)^2 \qquad j,k = 1,..n \qquad 19.6\text{-}22$$

The sum is taken over all j and k and the trace of the matrix ($tr\mathbf{S} = \Sigma \lambda_j$) is denoted by tr. This definition is favored over other possible ones because it yields values that are interpreted easily and it will be found that it can be computed without having to diagonalize $\mathbf{S}$. The polarization is zero if all the eigenvalues are equal. This indicates that $\mathbf{S}$ is invariant under all unitary transformations and the data has maximum randomness or is completely unpolarized. If all the eigenvalues are zero except for one then $p = 1.0$ and the data is *purely polarized*. The advantage of this analysis is that if one is seeking pure polarized sections in the time domain then they occur where there is only one non-zero eigenvalue.

The axis of purely polarized spectral data can be rotated so that the direction of the major axis is r_1 in real space. The minor axis, r_2, in real space, is determined within a sign by the condition that it be orthogonal to r.

$$r_1^T r_2 = 0 \qquad 19.6\text{-}23$$

The rotation is in the sense of r_2 moving toward r_1.

$$Im\,\mathbf{S} = r_2\,r_1^T - r_1\,r_2^T \qquad 19.6\text{-}24$$

The eigenvectors, except for a phase factor, ϕ, are then specified in complex space.

$$\mathbf{u}_1 = exp\,(-i\phi)\,[r_1 + i\,r_2] \qquad 19.6\text{-}25$$

The spectral matrix of a pure state is given by the outer product of $\mathbf{u}_1$ and its Hermitian adjoint.

$$\mathbf{S} = \lambda_1\,\mathbf{u}_1\,\mathbf{u}_1^H \qquad 19.6\text{-}26$$

If the complex cross spectral matrix is purely polarized in unitary space it has only one real eigenvalue. The *real part* of $\mathbf{S}$ has one or two non-zero eigenvalues in geometric space. Let us make an orthogonal rotation, $\mathbf{R}$, containing the vectors $r_1^{-1}r_1$ and $r_2^{-1}r_2$ to obtain a 2 square matrix $\mathbf{J}$ and an n-2 square null matrix.

$$\mathbf{R}^T \mathbf{S} \mathbf{R} = diag[\mathbf{J}, \mathbf{O}_{n-2}] \qquad 19.6\text{-}27$$

where

$$\mathbf{J} = \begin{bmatrix} J_{11} & i J_{12} \\ -i J_{12} & J_{22} \end{bmatrix} \qquad |\mathbf{J}| = 0 \qquad 19.6\text{-}28$$

The matrix $\mathbf{J}$ is similar to that in 19.2-18 and specifies the polarization ellipse as shown in figure 19.7.

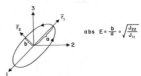

Figure 19.6 The polarization ellipse in a three-dimensional space (Samson, 1977).

The major axis is in the direction r_1 and the ellipticity, E, is given by the ratio of the minor to the major axis.

$$abs\ E = (J_{22} / J_{11})^{1/2} \qquad 19.6\text{-}29$$

Samson (1977) derives the following expressions from the scalar invariants of $\mathbf{J}$:

$$J_{11} = [tr\ \mathbf{S} + ((tr\mathbf{S})^2 + 2\,tr(Im\mathbf{S})^2)]^{1/2} / 2 \qquad 19.6\text{-}30$$

$$J_{22} = [tr\mathbf{S} - ((tr\mathbf{S})^2 + 2\,tr(Im\mathbf{S})^2)^{1/2} / 2 \qquad 19.6\text{-}31$$

$$J_{12}^2 = -tr\ (Im\mathbf{S})^2 / 2 \qquad 19.6\text{-}32$$

In real vector space the degree of polarization is as follows:

$$P_R^2 = [n\,tr\ (Re\mathbf{S})^2 - (tr\mathbf{S})^2] / [(n-1)(tr\mathbf{S})^2] \qquad 19.6\text{-}33$$

Note that all the quantities required to derive a polarization filter are obtained with very little computation and without the necessity of diagonalizing a matrix. This situation only holds if one is interested in pure states. For more complex polarizations with 2 or more eigenvalues it is necessary to diagonalize the cross-power spectral matrix.

Pure states with specified ellipticity, E $(0 < E < 1.0)$, are obtained with

Polarization states and unitary matrices

a function similar to that used by Archambeau et al., 1965.

$$D_1 = P^a (1 - abs(abs (J_{12} / J_{11}) - E))^b \qquad 19.6\text{-}34$$

The numbers a and b are determined empirically but we should choose $a > b$ since J_{12} / J_{11} is a measure of ellipticity only when $P = 1.0$. The polarization filter introduced here is constructed in the frequency domain and appears to be effective when used in array data to suppress noise and enhance waveforms of arbitary shape. Alternative formulations of the equations and their application is discussed at length in a series of papers by Samson and Olson.

References

Archambeau,C.B. and Flinn,E.A.(1965), Automated analysis of seismic radiations for source characteristics. *Proc. I.E.E.E.* **53**, 1876-1884.

Archambeau,C.B., Flinn,E.A. and Lambert,D.G.(1969), Fine structure of the upper mantle. *J. Geophys. Res.* **74**, 5825-5865.

Arthur,C.W., McPherron,R.L. and Means,J.D.(1976), A comparative study of three techniques for using the spectral matrix in wave analysis. *Radio Science* **11**, 833-845.

Basham,P.W. and Ellis,R.M.(1969), The composition of P codas using magnetic tape seismograms. *Bull. Seism. Soc. Am.* **59**, 473-486.

Born,M. and Wolf,E.(1965), Principles of optics. 3rd ed. Pergamon Press, New York, 544-555.

Brune,J.N. and Oliver,J.(1959), The seismic noise of the earth's surface. *Bull. Seism. Soc. Am.* **49**, 349-353.

Flinn,E.A.(1965), Signal analysis using rectilinearity and direction of particle motion. *Proc. I.E.E.E.* **53**, 1874-1876.

Fowler,R.A., Kotick,B.J. and Elliott,R.D.(1967), Polarization analysis of natural and artifically induced geomagnetic micropulsations. *J. Geophys. Res.* **72**, 2871-2883.

Griffin,J.N.(1966a), Application and development of polarization (REMODE) filters. *Seismic Data Laboratory Report 141*, Teledyne Inc., Alexandria, Va. (AD-630-515).

Griffin,J.N.(1966b), REMODE signal/noise tests in polarized noise. *Seismic Data Laboratory Report 162*, Teledyne, Inc., Alexandria, Va. (AD-800-039).

Ioannides,G.A.(1975), Application of multivariate autoregressive spectrum estimation to ULF waves. *Radio Science* **10**, 1043-1054.

Landau,L.D. and Lifshitz,E.M.(1962), The classical theory of fields. Pergamon Press, Oxford, Chapter 6, 404.

Lewis,B.T.R. and Meyer,R.P.(1968), A seismic investigation of the upper mantle to the west of Lake Superior. *Bull. Seism. Soc. Am.* **58**, 565-596.

Mal'cev,A.I.(1963), Foundations of Linear Algebra. H. W. Preeman and Co., San Francisco.

Means,J.D.(1972), The use of the three dimensional covariance matrix in analyzing the polarization preperties of plane waves. *J. Geophysical Research* **77**, 5551-5559.

Mims,C.H. and Sax,R.L.(1965), Rectilinear motion direction (REMODE). *Seismic Data Laboratory Report 118,* Teledyne, Inc., Alexandria, Va. (AD-460-631).

Montalbetti,J.F. and Kanasewich,E.R.(1970), Enhancement of teleseismic body phases with a polarization filter. *Geophys. J. R. Astr. Soc.* **21**, 119-129.

Olson,J.V. and Samson,J.C.(1979), On the detection of the polarization states of Pc micropulsations. *Geophysical Research Letters* **6**, 413-416.

Olson,J.V. and Samson,J.C.(1980), Generalized power spectra and the Stokes vector representation of ULF micropulsation states. *Canadian Journal of Physics in Press.*

Paulson,K.V., Egeland,A. and Eleman,F.(1965), A statistical method for quantitative analyses of geomagnetic giant pulsations. *Journal of Atmosphere and Terrestrial Physics* **27**, 943-967.

Pease III,M.C.(1965), Methods of matrix algebra. **4**, Academic Press, New York. p. 406.

Samson,J.C.(1973), Description of the polarization states of vector processes: Application to ULF magnetic fields. *Geophysical Journal* **34**, 403-419.

Samson,J.C.(1977), Matrix and Stokes vector representation of detectors for polarized waveforms. Theory and some application to Teleseismic waves. *Geophysical Journal* **51**, 583-603.

Samson,J.C. and Olson,J.V.(1979), Generalized Stokes vectors and generalized power spectra for second-order stationary vector precesses. *Siam Journal of Applied Mathematics, in Press.*

Samson,J.C. and Olson,J.V.(1980), Some comments on the description of the polarization states of waves. *Geophysical Journal* **61**, 115-129.

Sax,R.L. and Mims,C.H.(1965), Rectilinear motion detection (REMODE). Seismic Data Laboratory, P.O. Box 334. Teledyne Industries, Alexandria Virginia. Advanced Reasarch Project Agency report under Project Vela Uniform. p. 1-14.

Shimshoni,M. and Smith,S.W.(1964), Seismic signal enhancement with three-component detectors. *Geophysics* **29**, 664-671.

Simons,R.S.(1968), A surface wave particle motion discrimination process. *Bull. Seism. Soc. Am.* **58**, 629-637.

Stokes,G.G.(1852), On the composition and resolution of streams of polarized light from different sources. *Transactions Cambridge Philosophical Society* **9**, 399.

Wolf,E.(1959), Coherence properties of partially polarized electromagnetic radiation. *Il Nuovo Cimento* **13**, Series 10, 1165-1181.

20 Homomorphic deconvolution

20.1 Cepstrum and quefrency analysis for echoes

Echoes are often present in seismic recordings because of the presence of efficient reflecting planes such as the solid rock-fluid interface at the surface of the earth. If the echoes can be identified then they may be used to obtain the depth of earthquakes or in inverse convolution operators to simplify the seismograms. An autocovariance of the data may help in determining the time delay between the primary wave and its echo but this is seldom unambiguous because of the presence of many other peaks.

In 1959, Bogert found that spectograms of seismic signals displayed a periodicity. The periodic ripples, as illustrated in figure 20.1, were due to the data being made up of a signal and its echo. Tukey made the suggestion that the frequency spacing could be obtained by taking the logarithm of the spectrum and then make a power spectral analysis of this new frequency series. Tukey coined the word *cepstrum* for the spectrum of the log of the periodogram. The frequency of the cepstrum was called the *quefrency*. The work was published by Bogert, Healy and Tukey (1963) but they were not successful in obtaining the focal depth of earthquakes from the time difference between the direct wave and its echo from the surface of the earth.

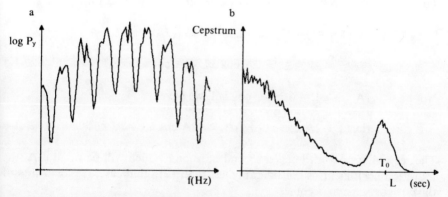

Figure 20.1 (a) The log of the periodogram showing a periodicity. (b) The cepstrum of (a) showing a separation of low and high quefrency components. The peak at T_0 is due to a reverberation or an echo in the original signal.

Consider a primary signal, x(t), which is followed by an echo with a smaller amplitude, A, at a time L seconds later. The recorded signal is

$$y(t) = x(t) + A\, x(t - L) \qquad 20.1\text{-}1$$

If $X(\omega)$ is the Fourier transform of x(t) then the transform of the signal y(t) is

$$Y(\omega) = X(\omega) + \int_{-\infty}^{\infty} A\, x(t - L)\, e^{-i\omega t}\, dt$$

$$= X(\omega)\,(1 + A\, e^{-i\omega L}) \qquad 20.1\text{-}2$$

The power spectrum is given by the squared modulus, YY*

$$P_y(\omega) = P_x(\omega)\,(1 + A\, e^{-i\omega L})\,(1 + A\, e^{i\omega L})$$

$$= P_x(\omega)\,(1 + [2A(e^{i\omega L} + e^{-i\omega L})/2] + A^2)$$

$$= P_x(\omega)\,(1 + 2A\cos\omega L + A^2) \qquad 20.1\text{-}3$$

where $P_x(\omega)$ is the power spectrum of the primary signal. Take the logarithm of both sides of 20.1-3.

$$\log P_y(\omega) = \log P_x(\omega) + \log(1 + 2A\cos\omega L + A^2) \qquad 20.1\text{-}4$$

The second term on the right side can be expanded as a series for log(1 + s).

$$\log(1 + 2A\cos\omega L + A^2) = (2A\cos\omega L + A^2) - [(2A\cos\omega L + A^2)^2]/2+..$$

$$= A^2 + 2A\cos\omega L - 2A^2\cos^2\omega L +.. \qquad 20.1\text{-}5$$

Noting that $2A^2\cos^2\omega L = A^2 + A^2\cos 2\omega L$ we have

$$\log(1 + 2A\cos\omega L + A^2) \cong 2A\cos\omega L - A^2\cos 2\omega L \qquad 20.1\text{-}6$$

The second term contributes a second harmonic which will be small if A, the reflection coefficient, is considerably less than 1. In that case, the logarithm of the power spectrum becomes

$$\log P_y(\omega) \cong \log P_x(\omega) + 2A\cos\omega L \qquad 20.1\text{-}7$$

Cepstrum and quefrency analysis for echoes

The effect of the echo on the log power spectrum is to add a cosinusoidal ripple. Since power is a function of frequency, $f = \omega/2\pi$, the *frequency* of the ripple, called *quefrency*, will be L in units of cycles per cycle per second or seconds. When operating on data in the frequency domain, Bogert, Healy and Tukey call the *amplitude*, 2A, the *gamnitude*. Its *phase*, which is 0 if A is positive and π if it is negative, is called the *saphe*.

The log power spectrum may be considered a time series and its autocovariance is called the *cepstrum* of the original time series. The cepstrum, which is plotted against quefrency in seconds, will contain a peak at the ripple frequency, which is the delay time between the primary and the echo. Before taking the autocovariance the log power spectrum may be passed through a high-pass filter to prewhiten the time series.

20.2 Homomorphic filtering

Homomorphic convolution is a generalization by Oppenheim (1965) of cepstrum analysis. It involves filtering by a non-linear system in which the input data undergoes a series of algebraically linear transformations to produce an output signal. Basic to it is the idea that some non-linear systems satisfy a principle of superposition under some operation other than addition as in linear systems. Non-linear systems which obey superposition are called *homomorphic* because they can be represented mathematically by an algebraically linear mapping between vector spaces. Problems solved by homomorphic filtering involve signals that have been combined multiplicatively and by convolution.

In homomorphic deconvolution a complex algebraic procedure is used to recover approximately an input signal. The technique, as discussed in section 20.1, is most valuable in physical systems where echoes dominate. Specifically it has been used to advantage in the seismic reflection method of exploration, in earthquake analysis of the source function, and in image enhancement and speech analysis. A large body of literature has been generated on the subject in the last decade and the interested reader is referred to the references at the end of the chapter.

A most important addition to homomorphic deconvolution was the definition of a *complex cepstrum* by Oppenheim (1967). The definition can be made in terms of a z transform or the angular frequency, ω. Let the input data be sampled and of finite length.

$$X(w) = \sum_{t=0}^{N-1} a^t x(t) \cdot \exp[(-i2\pi wt)/N] \qquad 20.2\text{-}1$$

$$\omega = 2\pi w / N \qquad 20.2\text{-}2$$

$$X(z) = \sum_{t=0}^{N-1} a^t x(t) z^t \qquad 0 < a < 1 \qquad 20.2\text{-}3$$

$$z = e^{-\sigma - i\omega} \qquad 20.2\text{-}4$$

The time series has been multiplied by a to stabilize it if it is not a minimum phase function, following the procedure of Schafer (1969). A value of a between 0.94 and 0.98 is often used according to Stoffa et al. (1974). The complex natural logarithm of the z transform or the frequency function is a multivalued quantity.

$$\log_e X(z) = \log_e |X(z)| + i \arg[X(z)] \pm i 2\pi j \qquad 20.2\text{-}5$$

Since $\exp(i2\pi j) = 1$ for any integer value of j, the argument of $X(z)$ can have infinitely many values with $j = 0, \pm 1, \pm 2, \ldots$. The value of j must be chosen so that the phase begins at zero at zero frequency and is continuous function for a z transform evaluated on the circular contour around $z = 0$. The unwrapping of the phase curve is critical and is the subject of much study. Reference should be made to Schafer (1969), Stoffa et al. (1974) and Tribolet (1977). Finally the complex cepstrum is defined as follows:

$$C(T) = (2\pi i) \oint_c [\log_e X(z)] z^{1-T} dz \qquad T = 0, \pm 1, \pm 2, \ldots \qquad 20.2\text{-}6$$

The actual computation is with an inverse discrete Fourier transform.

$$C(T) = N^{-1} \sum_{w=0}^{N-1} [\log_e Z(w)] \exp(i2\pi T w / N) \qquad 20.2\text{-}7$$

The complex cepstrum, $C(T)$, as computed using the discrete Fourier transform, evaluates the contour, c, on the unit circle, $|z| = 1$. Because the original time function is weight by a^t, $0 < a < 1$, the integration contour is actually inside the unit circle with radius e^A where $A = -\sigma$.

$$R = e^{-\sigma} \qquad \sigma = -\log_e a \qquad 20.2\text{-}8$$

All the poles and zeros of the original z transform, $X(z)$, should be outside the circle with this radius.

The complex cepstrum has a number of important properties. First of all, it is a real valued sequence with an even and an odd valued component. The

Homomorphic filtering

cepstrum defined by Bogert et al. (1963) is equal to the even part. The use of the word complex in the name stems from computation using a complex logarithm. The complex cepstrum of a convolution of two signals is equal to the addition of their individual cepstra. If the spectrum of the original signal is smooth then its complex cepstrum is zero for negative time. Time aliasing of the complex cepstrum is reduced if zeros are added to the data before making a discrete Fourier transform.

To recover the original function in the time domain from the complex cepstrum the inverse of the operations in equations 20.2-1, 20.2-5 and 20.2-6 must be taken.

$$\log_e X(z) = \sum_{T=-\infty}^{\infty} C(T) \, z^T \qquad 20.2\text{-}9$$

$$X(z) = \exp[\log_e X(z)] \qquad 20.2\text{-}10$$

$$x(t) = (2\pi i)^{-1} \, a^{-t} \oint_c X(z) \, z^{1-t} \, dz \qquad 20.2\text{-}11$$

The application of homomorphic deconvolution is discussed in detail by Ulrych (1971), Ulrych, Jensen, Ellis and Sommerville (1972), Buhl, Stoffa and Bryan (1974) and by Clayton and Wiggins (1976). Figure 20.2 from Ulrych (1971) illustrates the complex cepstrum of a signal containing a simple echo. The first peak in 20.2 (c) is due to the signal while the subsequent series is a result of the echo. This separation of peaks allows one to suppress the echoes in a deconvolution which yields the original input signal.

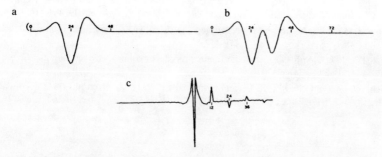

Figure 20.2 (a) A mixed-phase seismic wavelet. (b) A wavelet in (a) with an echo at 12 units of time. (c) A complex cepstrum of figure (b). (From Ulrych, 1971, courtesy of the Society of Exploration Geophysicists.)

References

Bhanu,B. and McClellan,J.H.(1980), On the computation of the complex cepstrum. *IEEE Transactions on Aucoustics,Speech and Signal Processing* **AASP 28**, 583-585.

Bogert,B.P., Healy,M.J.R. and Tukey,J.W.(1963), The quefrency analysis of time series for echoes: Cepstrum, Pseudo-autocovarinace, cross cepstrum and saphe cracking. *Ch. 15 in Time Series Analysis*. Ed.: M. Rosenblatt, 209-243, John Wiley and Sons, New York.

Bogert,B.P. and Ossanna,J.F.(1966), The heuristics of cepstrum analysis of a stationary complex echoed Gaussian signal in stationary Gaussian noise. *IEE Transactions on Information Theory* **IT-12**, 373-380.

Buhl,P., Stoffa,P.L. and Bryan,G.M.(1974), The application of homomorphic deconvolution to shallow water marine seismology. Part 2, Real data. *Geophysics* **39**, 417-426.

Clayton,R.W. and Wiggins,R.A.(1976), Source shape estimation and deconvolution of teleseismic bodywaves. *Geophysical Journal of the Royal Astronomical Society* **47**, 151-177.

Kemerait,R.C. and Childers,D.G.(1972), Signal extraction by cepstrum techniques. *IEEE Transactions on Information Theory* **IT-18**, 745-759.

Noll,A.M.(1966), Cepstrum pitch determination. *The Journal of the Acoustical Society of America* **41**, 293-309.

Oppenheim,A.V.(1965), Superposition in a class of nonlinear systems. Technical Report 432, Research Laboratory of Electronics, M.I.T., Cambridge, Mass. 1-62.

Oppenheim,A.V.(1967), Generalized linear filtering. *Information and Control* **11**, 528-536.

Oppenheim,A.V. and Schafer,R.W.(1975), Digital signal processing. **10**, 480-532. Prentice-Hall, Inc., Englewood Cliffs, New Jersey.

Otis,R.M. and Smith,R.B.(1977),Homomorphic deconvolution by log spectral averaging. *Geophysics* **42**, 1146-1157.

Schafer,R.W.(1969), Echo removal by discrete generalized linear filtering. Technical Report 466, Research Lab of Electronics, M.I.T., Cambridge, Mass. Also PhD Thesis, Dept. of Electrical Engineering, M.I.T., 1968.

Stoffa,P.L., Buhl,P. and Bryan,G.M.(1974), The application of homomorphic deconvolution to shallow water marine seismology. Part 1, Models. *Geophysics* **39**, 401-416.

Tribolet,J.M.(1977), A new phase unwrapping algorithm. *IEEE Transactions on Acoustics, Speech and Signal Processing* **77**, 170-177.

Ulrych,T.J.(1971), Applications of homomorphic deconvolution to seismology. *Geophysics* **36**, 650-660.

Ulrych,T.J.,Jensen,O.G.,Ellis,R.M. and Somerville,P.G.(1972), Homomorphic deconvolution of some teleseismic events. *Bulletin Seismological Society of America* **62**, 1269-1282.

21 The Hilbert transform

21.1 Minimum phase and delay

The research of Wiener (chapter 14) and Kolmogorov (1941) on linear operators implied that it was necessary to have all singularities outside the unit circle in the z plane as a condition for stability. This concept was introduced by Bode (1945) as a condition on the phase spectrum in the frequency domain. Bode defined a *minimum phase* operator as one that produces the minimum possible phase shift for its gain. Minimum phase is a description of the properties of the transfer function and Robinson (1962) has used the phrase *minimum delay* to describe these properties in terms of the impulse response function. Since the transfer function and impulse response are Fourier transforms of each other the definitions are synonymous. A minimum delay wavelet is formed by a convolution of n minimum delay dipoles and the resultant phase spectrum, $\phi(\omega)$, or phase lag, $-\phi(\omega)$, will be smaller than for the remaining 2^{n-1} family of maximum or mixed delay filters in the set. All the zeros of a minimum delay wavelet or minimum phase filter will lie outside the unit circle in the z plane.

From the definition of gain, equation 13.1-7, we find that the derivative of the transfer function with respect to angular frequency is

$$dY/d\omega = [d|Y|/d\omega - i|Y| \, d\theta/d\omega] \, e^{-i\theta(\omega)} \qquad 21.1\text{-}1$$

Robinson (1967) calls $d\theta/d\omega$ the group delay. The derivative in 21.1-1 can also be obtained from the Fourier transfer of the impulse response, $W(t)$ (see equation 5.3-3).

$$dY/d\omega = \int_0^\infty [-itW(t)] \, e^{-i\omega t} \, dt \qquad 21.1\text{-}2$$

Since $dY/d\omega$ and $-itW(t)$ are Fourier transforms, as seen in 21.1-2, it is possible to apply Parseval's equality to them.

$$\int_0^\infty t^2 |W(t)|^2 \, dt = \int_{-\infty}^\infty |dY/d\omega|^2 \, d\omega \qquad 21.1\text{-}3$$

or

$$\int_0^\infty t^2|W(t)|^2 \, dt = \int_{-\infty}^\infty [(d|Y|/d\omega)^2 + |Y|^2 \, (d\theta/d\omega)^2] \, d\omega \qquad 21.1\text{-}4$$

The quantity on the left is the second moment of $|Y(\omega)|^2$. Consider the group of impulse responses which all have the *same* amplitude spectra or gain, $|Y(\omega)|$. Because of the weighting by t^2, the second moment will be a minimum for the impulse response which has most of its energy close to the beginning. For such a wavelet, it is clear from 21.1-4 that $d\theta/d\omega$ must also be a minimum. Therefore this impulse response is called a minimum delay wavelet or it is said to have a minimum phase.

Solodovnikov (1952) has shown how it is possible to calculate the phase lag of a minimum phase operator if we can specify the amplitude spectra (or vice versa). This is of obvious advantage in designing stable filter operators. It also allows one to compute the complete response of a seismometer-galvanometer system when, as often happens, only the gain is known as a function of period. The process involved a Hilbert transform (Titchmarsh, 1937).

21.2 Hilbert transform equations

From Appendix 1 the Fourier transform of f(t) may be written as (see equation 1.33):

$$F(\omega) = \int_{-\infty}^\infty f(t) \, [\cos \omega t - i \sin \omega t] \, dt \qquad 21.2\text{-}1$$

or as

$$F(\omega) = F_R(\omega) - i \, F_I(\omega)$$

where

$$F_R(\omega) = \int_{-\infty}^\infty f(t) \, \cos \omega t \, dt \qquad 21.2\text{-}2$$

and

$$F_I(\omega) = \int_{-\infty}^\infty f(t) \, \sin \omega t \, dt$$

Hilbert transform equations

It also follows that

$$F(-\omega) = F_R(\omega) + i\, F_I(\omega) \qquad 21.2\text{-}3$$

and therefore

$$F_R(\omega) = [F(\omega) + F(-\omega)]/2$$
$$F_I(\omega) = i[F(\omega) - F(-\omega)]/2 \qquad 21.2\text{-}4$$

The inverse Fourier transform may be divided into regions of positive and negative frequency

$$f(t) = (2\pi)^{-1} \int_{-\infty}^{0} F(\omega)\, e^{i\omega t}\, d\omega + (2\pi)^{-1} \int_{0}^{\infty} F(\omega)\, e^{i\omega t}\, d\omega$$

$$= (2\pi)^{-1} \int_{0}^{\infty} F(-\omega)\, e^{-i\omega t}\, d\omega + (2\pi)^{-1} \int_{0}^{\infty} F(\omega)\, e^{i\omega t}\, d\omega \qquad 21.2\text{-}5$$

Upon substituting 21.2-4 we have equations with positive frequency only.

$$f(t) = (\pi)^{-1} \int_{0}^{\infty} F_R \cos \omega t\, d\omega + (\pi)^{-1} \int_{0}^{\infty} F_I \sin \omega t\, d\omega \qquad 21.2\text{-}6$$

This may be written in terms of a complex time $t_c = t + i\sigma$.

$$\lim_{\sigma \to 0} f(t) = \lim_{\sigma \to 0} (\pi)^{-1} \int_{0}^{\infty} [F_R \cos \omega t + F_I \sin \omega t]\, e^{-\sigma \omega}\, d\omega \qquad 21.2\text{-}7$$

Next define a complex function

$$f_c = f(t,\sigma) - i\, f_H(t,\sigma) \qquad 21.2\text{-}8$$

This complex functon is made up of our original real part, f(t), and a part called the *quadrature function*, $f_H(t)$, which will be seen to be the Hilbert transform of f(t). The quadrature is defined so that it introduces a 90 degree phase shift because we will want to use it to obtain the envelope of the real time function.

Therefore sines are converted to cosines and cosines are converted to negative sines in equation 21.2-6.

$$f_H(t,0) = (\pi)^{-1} \int_0^\infty [F_I \cos \omega t - F_R \sin \omega t] \, d\omega \qquad 21.2\text{-}9$$

and

$$f_c(t,\sigma) = (\pi)^{-1} \int_0^\infty [F_R(\omega) - i\, F_I(\omega)] \, e^{i\omega t} \, e^{-\sigma\omega} \, d\omega \qquad 21.2\text{-}10$$

Substituting F_R and F_I from 21.2-2 into 21.2-9 will relate f_H and f more clearly.

$$f_H(t) = (\pi)^{-1} \int_0^\infty \int_{-\infty}^\infty [f(T) \sin \omega T \cos \omega t -$$

$$- f(T) \cos \omega T \sin \omega t] \, dT \, d\omega$$

$$= \lim_{\omega' \to \infty} (\pi)^{-1} \int_0^{\omega'} \int_{-\infty}^\infty f(T) \sin \omega(T-t) \, dT \, d\omega \qquad 21.2\text{-}11$$

Integrating first with respect to $d\omega$ we obtain

$$f_H(t) = \lim_{\omega' \to \infty} (\pi)^{-1} \, \mathbf{P} \int_{-\infty}^\infty f(T) \, [[1 - \cos \omega'(T-t)]/(T-t)] \, dT \qquad 21.2\text{-}12$$

where **P** indicates the Cauchy principal value. It can be shown that the

$$\lim_{\omega' \to \infty} (\pi)^{-1} \, \mathbf{P} \int_{-\infty}^\infty f(T) \, [\cos \omega'(T-t)]/(T-t) \, dT = 0 \qquad 21.2\text{-}13$$

using the Riemann-Lebesque Lemma and therefore we have that

$$f_H(t) = -(\pi)^{-1} \, \mathbf{P} \int_{-\infty}^\infty f(T)/(t-T) \, dT \qquad 21.2\text{-}14$$

This equation is known as the Hilbert transform.

Hilbert transform equations

Substituting F_R and F_I from 21.2-4 into 21.2-9 gives

$$f_H(t) = \frac{i}{2\pi} \int_0^\infty [F(\omega) - F(-\omega)] \cos \omega t \, d\omega -$$

$$- (2\pi)^{-1} \int_0^\infty [F(\omega) + F(-\omega)] \sin \omega t \, d\omega$$

$$= \frac{i}{2\pi} \int_0^\infty [F(\omega) e^{i\omega t} - F(-\omega) e^{-i\omega t}] \, d\omega \qquad 21.2\text{-}15$$

$$= (2\pi)^{-1} \int_0^\infty i \, F(\omega) \, e^{i\omega t} \, d\omega - (2\pi)^{-1} \int_{-\infty}^0 i \, F(\omega) \, e^{i\omega t} \, d\omega \qquad 21.2\text{-}16$$

If we let

$$F_H(\omega) = i \, F(\omega) \, \text{sgn} \, \omega \qquad 21.2\text{-}17$$

where sgn is a signum function

$$\text{sgn} \, \omega = \begin{array}{ll} 1 & \omega > 0 \\ 0 & \omega = 0 \\ -1 & \omega < 0 \end{array} \qquad 21.2\text{-}18$$

then 21.2-16 shows that f_H and F_H are inverse Fourier transforms.

$$f_H(t) = \int_{-\infty}^\infty F_H(\omega) \, e^{i\omega t} \, d\omega \qquad 21.2\text{-}19$$

From 21.2-17 and 21.2-7 it is also possible to obtain the inverse Hilbert transform.

$$f(t) = (\pi)^{-1} \, \mathbf{P} \int_{-\infty}^\infty [f_H(T) / (t-T)] \, dT \qquad 21.2\text{-}20$$

In equation 21.2-17 we can define a new Fourier transform, $Q(\omega)$.

The Hilbert transform

$$Q(\omega) = i \text{ sgn } \omega \qquad 21.2\text{-}21$$

The Fourier series of $Q(\omega)$ is q_n.

$$q_n(n\Delta t) = (2\pi)^{-1} \int_{-\pi}^{\pi} Q(\omega) \exp(-i\omega n\Delta t) \, d\omega$$

$$= i(2\pi)^{-1} \int_{-\pi}^{0} \exp(-i\omega n\Delta t) - i(2\pi)^{-1} \int_{0}^{\pi} \exp(-i\omega n\Delta t) \, d\omega$$

$$= (2\pi n)^{-1} (-1 + \exp(in\pi \Delta t) + \exp(-in\pi \Delta t) - 1)$$

$$\begin{aligned} q_n(n\Delta t) &= 0 & \text{n even} \\ &= -1 / (\pi n \Delta t) & \text{n odd} \end{aligned} \qquad 21.2\text{-}22$$

As $\Delta t \to 0$ the Fourier series becomes a Fourier integral and $n\Delta t$ becomes the continuous time variable t.

$$q(t) = -1 / (\pi t) \qquad 21.2\text{-}23$$

The function $q(t)$ and its Fourier transform, $Q(\omega)$, are plotted in figure 21.1.

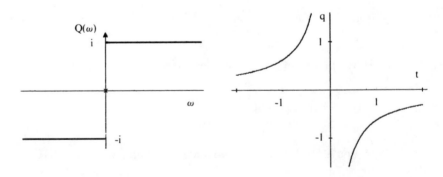

Figure 21.1 The function $i \text{ sgn } \omega$ and its inverse transform, $q(t)$.

By equation 21.2-17 the Hilbert transform in the frequency domain is given by the product of $F(\omega)$ and $Q(\omega)$.

Hilbert transform equations

$$F_H(\omega) = F(\omega) \, Q(\omega) \qquad 21.2\text{-}24$$

By the convolution theorem it is possible to obtain the Hilbert transform in the time domain by convolving q(t) with f(t).

$$f_H(t) = q(t) * f(t) = (-\pi t)^{-1} * f(t) \qquad 21.2\text{-}25$$

Writing out the convolution indicated by the star in 21.2-25 gives equation 21.2-14 again. It is seen also that if two successive Hilbert transforms are applied we should reverse all the phases in the Fourier harmonics and obtain the original function back.

$$f(t) = -(-\pi t)^{-1} * f_H(t) \qquad 21.2\text{-}26$$

This is seen to be identical to equation 21.2-20. Some Hilbert transforms are listed in Table 1.

Table 1 - Hilbert Transforms

f(t)	$F_H(t)$
$\delta(t)$	$-1/(\pi t)$
f(t) = 1 $-1/2 \leqslant t \leqslant 1/2$	
= 0 otherwise	$\pi^{-1} \log_e |(t-1/2) / (t+1/2)|$
(Box-car function)	
sin t	cos t
cos t	-sin t
(sin t)/t	(cos t-1)/t

21.3 Envelope of a function

The envelope, E, of any function, f(t), can be obtained by taking the modulus of equation 21.2-8.

$$E(t) = [f^2(t) + f_H^2]^{1/2} = [f_c \, f_c^*]^{1/2} \qquad 21.3\text{-}1$$

The instantaneous phase, ϕ, can be given also as a function of time (Bracewell, 1965).

$$\phi(t) = \tan^{-1} [f_H(t) / f(t)] \qquad 21.3\text{-}2$$

The instantaneous frequency, $\omega(t)$, is obtained from the rate of change of phase. The complex function f_c may be written as follows:

$$f_c = E(t)\, e^{-i\phi(t)} \qquad 21.3\text{-}3$$

Solving for the phase we have

$$\phi = i \log_e f_c / E \qquad 21.3\text{-}4$$

Letting
$$F_c = f_c / E \qquad 21.3\text{-}5$$

and differentiating, yields the instantaneous frequency.

$$\omega(t) = d\phi / dt = i(dF_c / dt) / F_c \qquad 21.3\text{-}6$$

Following Claerbout (1976) this may be approximated by the imaginary part, I_m, of the difference equation.

$$\omega(t) \approx I_m \frac{2[F_c(t) - F_c(t-1)]}{\Delta t [F_c(t) + F_c(t-1)]} \qquad 21.3\text{-}7$$

The instantaneous frequency may be a useful function to plot for data that shows dispersion. It is used in reflection seismology to detect zones of interference indicating rapid changes in the elastic impedance in the earth. Examples and details on the computation are given by Cizek (1970), Farnback (1975), Rabiner and Gold (1975), Taner et al. (1979) and McDonald et al. (1981).

21.4 The phase and modulus of causal filters
A linear system is said to be *physically realizable* and its impulse response is said to be *causal* if it vanishes for negative time.

$$f(t) = 0 \quad t < 0 \qquad 21.4\text{-}1$$

As noted in Appendix 1, equation 1.13, any function can be expressed as the sum of an even and odd function.

The phase and modulus of causal filters

$$f(t) = f_e(t) + f_o(t) \qquad 21.4\text{-}2$$

$$= [f(t) + f(t)]/2 + [f(+t) - f(-t)]/2 \qquad 21.4\text{-}3$$

But from 21.4-1

$$f_e(t) = -f_o(t) \qquad t < 0$$
$$f_e(t) = f_o(t) \qquad t > 0 \qquad 21.4\text{-}4$$

Therefore for causal signals we can introduce the signum function.

$$f_e(t) = f_o(t)\, \text{sgn}\, t$$
$$f_o(t) = f_e(t)\, \text{sgn}\, t \qquad 21.4\text{-}5$$

The real part of the Fourier transform of $F(\omega)$ is even while the imaginary part is odd.

$$F_e(\omega) = \frac{-i}{\pi}\, \mathbf{P} \int_{-\infty}^{\infty} \frac{F_o(w)}{\omega - w}\, dw = F_R(\omega)$$

$$F_o(\omega) = \frac{-i}{\pi}\, \mathbf{P} \int_{-\infty}^{\infty} \frac{F_e(w)}{\omega - w}\, dw = iF_I(\omega) \qquad 21.4\text{-}6$$

Thus as a result of 21.4-5 and the similarity with $F_H(\omega)$ in 21.2-17, the real and imaginary parts of a causal function are Hilbert transforms.

$$F_R(\omega) = (\pi)^{-1}\, \mathbf{P} \int_{-\infty}^{\infty} \frac{F_I(w)}{\omega - w}\, dw \qquad 21.4\text{-}7$$

$$F_I(\omega) = (\pi)^{-1}\, \mathbf{P} \int_{-\infty}^{\infty} \frac{F_R(w)}{\omega - w}\, dw \qquad 21.4\text{-}8$$

As previously stated the transfer function may be written as

$$Y(\omega) = |Y| \, e^{i\phi(\omega)} \qquad 13.1\text{-}7$$

Taking the log of both sides yields

$$\log_e Y(\omega) = \log_e |Y(\omega)| + i\phi(\omega) \qquad 21.4\text{-}9$$

Assuming that the impulse response is a causal function it follows that we can compare log Y to $F(\omega)$ in 21.2-2. Thus $\log |Y|$ and ϕ are Hilbert transforms just as $F_R(\omega)$ and F_I in 21.4-7 and 21.4-8.

$$\log |Y(\omega)| = -(\pi)^{-1} \, \mathbf{P} \int_{-\infty}^{\infty} \frac{\phi(w)}{w - \omega} \, dw \qquad 21.4\text{-}10$$

$$\phi(\omega) = (\pi)^{-1} \, \mathbf{P} \int_{-\infty}^{\infty} \frac{\log |Y(w)|}{w - \omega} \, dw \qquad 21.4\text{-}11$$

The log of the transfer function, $Y(\omega)$, must be analytic and bounded in the lower half plane. That is, the lower half plane must have no poles or zeros so that it will be a minimum phase system. The necessary and sufficient condition that the gain, $|Y(\omega)|$, is physically realizable is known as the Paley-Wiener criterion (Paley and Wiener, 1934). This states that

$$\int_{-\omega}^{\infty} \frac{|\ln |Y(\omega)||}{1 + \omega^2} \, d\omega$$

must be finite. Solodovnikov (1952) discusses practical considerations to be made in computing the phase spectrum of a minimum delay function.

The practical aspects of using Hilbert transforms in digital filter systems in terms of a z transform is discussed by Gold et al. (1969). Other aspects in one- and two-dimensional cases are treated by Cizek (1970) and by Read and Treitel (1973), Oppenheim and Schafer (1975).

References

Bode, H.W. (1945), Network analysis and feedback amplifier design. Princeton, Van Nostrand.

Bracewell, R.N. (1965), The Fourier transform and its applications. McGraw-Hill Book Co., New York. p. 268-271.

Cizek, V. (1970), Discrete Hilbert transform. *IEEE Transactions on Audio and Electroacoustics* **AU-18**, 340-343.

Claerbout, J. (1976), Fundamentals of Geophysical data processing. p. 20-23. McGraw-Hill, New York.

Farnback, J.S. (1975), The complex envelope in seismic signal analysis. *Bulletin seismological society of America* **65**, 951-962.

Gabor, D. (1946), Theory of communication, Part I. *Journal of Institute of Electrical Engineers* **93**, partIII, 429-441.

Gold, B., Oppenheim, A.V. and Radar, C.M. (1969), Theory and implementation of the discrete Hilbert transform. Symposium on Computer Processing in Communications, **19**, Editor J. Fox, Polytechnic Institute of Brooklyn.

Kolmogorov, A.N. (1941), Stationary sequences in Hilbert space (Russia). *Bull. Moscow State Univ. Math.* **2**, 40.

McDonald, J.A., Gardner, G.H.F. and Kotcher, J.S. (1981), Areal seismic methods for determining the extent of acoustic discontinuities. *Geophysics* **46**, 2-16.

Oppenheim, A.V. and Schafer, R.W. (1975), Digital signal processing, Chapter 7, 337-375. Prentice-Hall, Inc., Englewood Cliffs, New Jersey.

Paley, R.E.A.C. and Wiener, N. (1934), Fourier transforms in the complex domain. *Am. Math. Soc. Colloq. Pub. Vol.* **19**, Chapter 1.

Papoulis, A. (1962), The Fourier Integral and its application. McGraw-Hill, New York.

Rabiner, L.R. and Gold, B. (1975), Theory and application of digital signal processing. Prentice-Hall, Englewood Cliffs, New Jersey. p. 70-72.

Read, R.R. and Treitel, S. (1973), The stabilization of two dimensional recursive filters via the discrete Hilbert transform. *IEEE Transactions on Geoscience Electronics* **GE-11**, 153-160, Addendum: 205-207.

Robinson, E.A. (1962), Random wavelets and cybernetic systems. Hafner Publishing Co., New York.

Robinson, E.A. (1967), Statistical communication and detection. Appendix 3, Hafner Publishing Co., New York.

Solodovnikov, V.V. (1952), Introduction to the statistical dynamics of automatic control systems (Chapter 1). Translation edited by J.B. Thomas and L.A. Zadeh, Dover Publications, Inc., New York, 1960.

Taner, M.T., Koehler, F. and Sheriff, E.E. (1979), Complex seismic trace analysis. *Geophysics* **44**, 104-1063.

Titchmarsh, E.C. (1937, 1948), Introduction to the theory of Fourier integrals. Chapter 5, University Press, Oxford.

22 Acquisition of digital data

22.1 Components of a digital system

We are concerned with the measurement of physical signals in nature. Almost all of these will be continuous functions of time and space and they will be called analog data. The spatial sampling is carried out with a limited number of physical sensors. Often only one is used but occasionally as many as 1,000 sensors are distributed over a portion of the earth. The shortest wavelength that can be identified is related to twice the spatial sampling interval. Analog recording devices such as paper chart recorders and amplitude or frequency modulated magnetic tape recorders have a very limited dynamic range. Typically 64 to 1,024 different levels can be distinguished, giving a dynamic range of 36 to 60 decibels (db). One need only recall the difficulty of making measurements to 1/64 of an inch on a sinusoidal wave one inch high to verify the problems associated with achieving a higher level of resolution on this type of medium. Analog circuits and recorders produce signals with low noise immunity and they have problems with drift. Once recorded, the data is subject to introduction of additional noise and distortion. It is virtually impossible to duplicate the same signal exactly in two successive recordings let alone on two different play-back systems.

To obviate the problems associated with analog recording the signals may be sampled in time by the use of an analog to digital converter (A-D converter) and stored as discrete numbers on magnetic tape, a magnetic disc or some solid state memory device. The digital data will have a high immunity to noise, no drift and can be reproduced exactly with the help of a wide range of digital computers and plotting devices. The dynamic range of most digital recording systems ranges from 66 to 90 db. To preserve fidelity over a wide range of amplitudes and frequencies the sampling should be carried out in the field at the site of the sensor.

Typically the analog to digital device will produce a binary digit since the recording and reproduction of a quantity with only two possible states can be carried out with the least ambiguity. A binary digit is often called a *bit*. The term *word* is used to describe the complete set of bits making up one sensor's amplitude measurement at one time. In sampling the data it is necessary to make a choice of the sampling interval. This introduces a Nyquist or upper frequency limit on the information that is preserved. Analog recorders also have limitations on the high frequency content so this is not a disadvantage that is unique to digital data. The major disadvantage with digital recording is the high level of technical expertise required in designing, building and operating

Components of a digital system

the systems. It is evident that this problem will be overcome within a decade and that complete low power digital recording and play-back systems with microprocessor controls and a large memory capacity will be available. The cost and reliability may be similar to those of currently available pocket electronic computers and digital wrist-watches.

A typical system for digital data acquisition is shown in figure 22.1. In geophysics, typical sensors are seismometers, accelerometers, strain gauges, tiltmeters, gravimeters, magnetometers, potentiometers, thermocouples, thermistors, radiometers, photometers, spectrometers, etc. Band-pass filtering is usually required at an early stage to insure that noise outside the desired frequency range does not saturate any of the systems or degrade the resolution of the signals. At this initial stage as little filtering should be done as possible so as not to restrict the use of the data when carrying out the digital

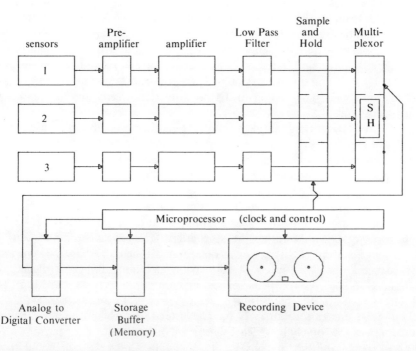

Figure 22.1 A block diagram of a multichannel system used to convert analog signals to digital form and store them in a recording device.

computation at a later stage. The low pass filtering should always be included to insure that unexpected high frequency noise is not aliased into the lower frequencies during the sampling. The low pass corner frequency is typically set at half the Nyquist frequency, $=1/($twice the sampling interval$)$. The power spectrum of the signal and noise should be down at least 48 to 60 db at the Nyquist frequency. In many geophysical measurements the earth acts as a low-pass filter; however this should not be relied upon without rigorous noise tests.

The amplifiers will be required to scale the signals up to ±10 volts for a full scale input into the analog to digital converter. High quality systems will allow the microprocessor to expand or compress the gain of the signal in a controlled manner so that weak inputs are measured with maximum resolution and strong inputs do not saturate any of the electronic devices.

The multiplexer is a programmed solid state switch which connects different input sensors to the sample and hold device which precedes the A-D converter. The sample and hold device is used to avoid inaccuracies made by the input signal changing during the period of a particular conversion time hence producing an average instead of an instantaneous value. Multiple sample and hold devices are used to capture simultaneous values on all input channels and hold them during the conversion time of all channels. For almost all geophysical measurements it is no longer necessary because converters are very fast at the present time and the delay in sampling different channels is negligible. Conversion rates of 50,000 channels per second are available at reasonable cost and the conversion of any signal only takes a fixed time of 20 to 50 microseconds. The analog to digital converter is only able to handle one signal at a time.

A-D converters sample an analog signal and measure its amplitude as a binary number by a successive approximation converter. This involves balancing an input voltage against some internally produced reference voltage. The comparator produces a bit on the basis of the polarity of the inequality. Typically a *zero*, *off* or *false* bit is produced if the reference voltage is larger. A *one*, *on* or *yes* bit is produced if the input signal is larger. This produces the most significant bit (MSB). The remaining signal is balanced against half the reference voltage and another bit is produced. The process continues until the least significant bit (LSB) has been determined. The cost of an A-D converter will depend upon the speed of conversion and the stability of the internal reference voltage. An inexpensive (≃$125) commonly used standard is a 12-bit A-D converter in which the supply must be stable to one part in 8,192. For about $800 one can obtain a 14-bit converter in which the reference voltage must be stable to one half the least significant bit or better than 1 part in 32,768. This is good standard for high quality field data. Converters to 15 and 16 bits are available but cost about $3,000. Converters to 15 or 16 bits based on a voltage to frequency conversion are inexpensive but their conversion time is in the 10's of milliseconds rather than microseconds. The least significant bit may not be reliable except in a closely controlled laboratory environment. Table 1 summarizes the resolution obtainable with A-D converters producing 1 to 16 bits.

Components of a digital system

Table 1   Resolution of A-D Converters

Bits	Resolution (Grey Scale)	Dynamic Range (db)	Amplitude Range	Voltage of LSB
1	$2=2^1$	6	±(sign bit)	10 v
2	$4=2^2$	12	±1	5
3	8	18	±3	2.5
4	16	24	±7	1.25
5	32	30	±15	0.625
6	64	36	±31	0.312
7	128	42	±63	0.156 v
8	256	48	±127	78.1 mv
9	512	54	±255	39.1 mv
10	1024	60	±511	19.5
11	2048	66	±1023	9.7
12	4096	72	±2047	4.88
13	8192	78	±4095	2.44 mv
14	16384	84	±8191	1.22
15	32768	90	±16383	0.610
16	65536	96	±32767	0.305 mv

A one bit converter is sometimes called a sign-bit converter because it simply gives the sign of the analog signal. That is, as shown in figure 22.2, the signal is characterized by a *one* or *yes* if the data has an amplitude between -10 and 0 volts and by a *zero* or *no* if it is between 0 and 10 volts. In special circumstances a great deal of useful information may be obtained by just recording the sign. A very large number of sensors may be recorded with a small sampling interval. As an example, a VIBROSEIS source can be used in 3 dimensional arrays. The earth is vibrated with a known amplitude over a different series of frequency ranges (20 to 50, 22 to 52, 24 to 54 Hz, etc.) and the reflected sign-bit output is correlated in the usual way (see section 6.3). When all the different source outputs are stacked it is possible to recover the relative amplitude at each sensor with a certain precision. Many other circumstances exist where sign-bit information is of value in the interpretation of physical data (Gimlin and Smith, 1980).

A two-bit converter is illustrated in figure 22.2 in which a ±10 volt signal is divided into 4 parts in 5 volt segments. The generalization to high resolution is obvious. A commonly used low resolution system involves a 6-bit A-D converter. Black and white or colour photographic detail is resolved very well by 64 shades of grey or a superposition of 64 shades of red, green and blue.

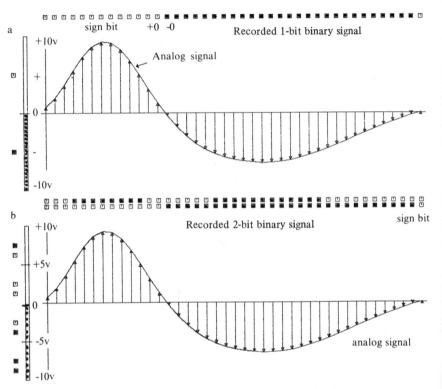

Figure 22.2 Binary signal from (a) a one-bit A-D converter and (b) a two-bit converter.

Examples are provided by satellite multispectral scanners for the ERTS and LANDSAT program.

High quality seismic field systems use a 15-bit A-D converter giving a dynamic range of 90 db. These will also have a binary gain switching system which expands the overall range to 132 db or more. The instantaneous signal can vary over a range of 4 million units of the least significant bit and be recorded as a 15-bit signal. This wide a range is necessary since reflection seismometers can have average outputs that range from 1 volt to less than a microvolt. Such large dynamic ranges are common in geophysics. As an example, earthquake magnitudes range from -2 to +9, a change of 10^{11} or 100 billion. However, most sensors are linear over a much smaller range and it is very difficult to calibrate and maintain a system so that it produces more than 4-figure accuracy. Therefore it is seldom possible to justify the use of an A-D converter with more than 14 bits. In a binary gain switching system the gains change suddenly by a factor of 2 (6db) when the signal reaches a certain level. Usually 15 such gain changes are allowed and 4 extra binary bits are recorded with the digital signal to keep track of the instantaneous gain.

Components of a digital system

The digital data is finally stored on some form of magnetic tape recorder at the present time. This can be a synchronous reel to reel tape deck, a cassette or cartridge drive or a magnetic disk. These units are characterized by high power consumption and large maintenance costs because of the prevalence of moving parts. Synchronous tape decks have a large storage capacity but require expert technicians to maintain them. Cassette systems have a small storage capacity but are operated easily by non-experts. In the near future various solid state memory devices should make it possible to store a large amount of digital data with low power requirements and little or no maintenance costs. Magnetic bubble memory devices are beginning to appear as a reliable storage medium for remote data acquisition terminals. It is anticipated that the present capacity of 10^6 bits per device will soon be expanded to exceed the formatted capacity of the cassette.

Table 2 Number Systems in Common Use

Base, b.	Name	Symbols (d_i)
2	binary	0,1
8	octal	0,1,2,3,4,5,6,7
10	decimal	0,1,2,3,4,5,6,7,8,9
12	duodecimal	0,1,2,3,4,5,6,7,8,9,A,B
16	hexadecimal	0,1,2,3,4,5,6,7,8,9,A,B,C,D,E,F

22.2 Conversion between number systems

In editing digital data it is often necessary to convert from one number system to another. This is particularly true if the editing is carried out on mini-computers in which the word length may be 8, 12, 16 or more bits. Digital recorders are based on the number 2 and we are most familiar with numbers having the base 10. In a digital computer a base 10 number such as 3,172 becomes 110 001 100 100. To facilitate the handling of such an awkward base 2 number it is easier to convert to a base 8 number in which system it becomes 6,144. It is useful to know how to convert from one base to another.

A number system consists of four elements:

(1) a base or radix, b.
(2) a set of unique symbols, d_i, from 0 to b-1.
(3) a period to separate the integral from the fractional parts of the number.
(4) a plus or minus sign.

Table 2 lists some commonly used numbering systems. The value of the number is given by

$$(\text{SIGN } [\sum_{i=0}^{n} d_i b^i] \cdot [\sum_{j=1}^{m} d_{-j} b^{-j}])_{\text{BASE}} \qquad 22.2\text{-}1$$

The SIGN is either + or -. The first square bracket contains the integer portion. It is separated by a period or decimal from the fractional portion.

Example 1

$[13.375]_{10} = +[1 \times 10^1] + [3 \times 10^0](.) + [3 \times 10^{-1}] + [7 \times 10^{-2}] + [5 \times 10^{-3}]$

The same number to base 2, 8, 10, 12 is as follows:

$[1101.011]_2 = (1 \times 2^3) + (1 \times 2^2) + (0 \times 2^1) + (1 \times 2^0)(.) + (0/2) + (1/4) + (1/8)$
$[15.3]_8 = (1 \times 8^1) + (5 \times 8^0)(.) + (3/8)$
$[11.46]_{12} = (1 \times 12^1) + (1 \times 12^0)(.) + (4/12) + (6/144)$
$[D.6]_{16} = (13 \times 16^0)(.) + (6/16)$

Octal numbers are favored frequently because they are close in magnitude to base 10 numbers and are easy to convert to binary form. Base 12 and 16 are less favored because of the combination of numeric and alphabetic characters. Base 10 numbers are converted to another base by repeated division of the integer portion by the new base and by repeated multiplication of the fractional part by the base.

Example 2

Convert $(3172.645)_{10}$ to a base 8 number. Consider first the integral portion which is decomposed by repeated division.

$3172 \div 8 = 396 + [4]$ (1st Remainder)
$396 \div 8 = 49 + [4]$ (2nd Remainder)
$49 \div 8 = 6 + [1]$ (3rd Remainder)
$6 \div 8 = 0 + [6]$ (4th Remainder)
Thus $(3172)_{10} = (6144)_8$

Next the fractional part is decomposed by repeated multiplication.

$.645 \times 8 = [5] + .160$
$.160 \times 8 = [1] + .28$
$.28 \times 8 = [2] + .24$
$.24 \times 8 = [1] + .92$

Conversion between number systems

Thus $(.645)_{10} = (.5121...)_8$

Finally

$$(3172.645)_{10} = (6144.5121...)_8$$

Example 3

Convert $(6144.5121)_8$ to the equivalent binary number. Each octal symbol, d_i, is converted into a group of 3 binary bits.

$$(\quad 6 \quad 1 \quad 4 \quad 4 \quad \cdot \quad 5 \quad 1 \quad 2 \quad 1 \quad)_8 =$$
$$(110 \quad 001 \quad 100 \quad 100 \quad \cdot \quad 101 \quad 001 \quad 010 \quad 001)_2$$

Example 4

Convert the binary number in example 3 to a hexadecimal number. This is accomplished by converting each group of 4 binary bits into the equivalent base 16 number.

$$(1100 \quad 0110 \quad 0100 \quad \cdot \quad 1010 \quad 0101 \quad 0001)_2 =$$
$$(\ C \quad 6 \quad 4 \quad \cdot \quad A \quad 5 \quad 1 \quad)_{16}$$

It is seen that conversion from binary to a base 8 or base 16 number is carried out easily by dividing the binary numbers into groups of 3 or 4 bits as one moves away from the decimal point.

22.3 Bipolar and other binary codes

Binary arithmetic may be facilitated on some computers by using a modification of the pure binary number system. These are computing systems that cannot subtract. In these, subtraction is carried out by first complementing a number and then adding.

One's complement numbers are the same as binary numbers if they are positive. Negative one's complement numbers are formed by taking the equivalent positive binary number and changing all the 0's to 1's and the 1's to 0's. This is called *complementing*. Thus +3 in one's complement is 0011 and -3 is 1100.

Two's complement positive numbers are the same as binary numbers. A negative number is formed by taking the positive binary form, complementing it and adding a 1 to the least significant bit. For example +3 is 0011 in two's complement. -3 is $1100 + 0001 = 1101$.

Offset binary does not have the ambiguous set of two values for +0 and -0. It is formed from a binary word by complementing the most significant bit

(MSB). If the new MSB = 0 then the other bits are complements and a 1 is added to the least significant bit.

In the Gray code only one bit changes as the decimal number changes by one. Table 3 illustrates some of the binary codes used by various instrument makers.

Binary coded decimal (BCD) is used in digital systems that must have a decimal number displayed as in a digital voltmeter. Each decimal digit is represented by 4 binary-coded digits. The various BCD codes are shown in Table 4 together with the bit value in each of the 4 positions. As an example the number 591 to base 10 is the following sequences of bits in 8421 BCD

$$(591)_{10} = (0101 \quad 1001 \quad 0001)$$

Excess 3 is a code in which a binary 3 (=0011) is added to the BCD numbers.

Table 3   Binary Bipolar Codes

Decimal	Pure Binary (sign + magnitude)	Two's Complement	Ones's Complement	Offset Binary	Gray Code
+8	0 1000	0 1000	0 1000	1 1000	0 1000
+7	0111	0111	0111	1111	0100
+6	0110	0110	0110	1110	0101
+5	0101	0101	0101	1101	0111
+4	0100	0100	0100	1100	0110
+3	0011	0011	0011	1011	0010
+2	0010	0010	0010	1010	0011
+1	0001	0001	0001	1001	0001
+0	0000	0000	0000	0000	0000
-0	1000	0000	1111	1000	1000
-1	1001	1111	1110	0111	1001
-2	1010	1110	1101	0110	1011
-3	1011	1101	1100	0101	1010
-4	1100	1100	1011	0100	1110
-5	1101	1011	1010	0011	1111
-6	1110	1010	1001	0010	1101
-7	1111	1001	1000	0001	1100
-8	1 1000	1 1000	1 0111	0 1000	1 1100

Bipolar and other binary codes

Table 4   Binary Coded Decimal Codes

Decimal	Pure Binary 8421	2421	Binary 8421	Coded 4221	Decimal Excess 3	Complementary
0	0000	0000	0000	0000	0011	1111
1	0001	0001	0001	0001	0100	1110
2	0010	0010	0010	0010	0101	1101
3	0011	0011	0011	0011	0110	1100
4	0100	0100	0100	1000	0111	1011
5	0101	0101	0101	0111	1000	1010
6	0110	0110	0110	1100	1001	1001
7	0111	0111	0111	1101	1010	1000
8	1000	1110	1000	1110	1011	0111
9	1001	1111	1001	1111	1100	0110

22.4   Reconstruction of analog signals

Sampled data may be reconstructed into continuous or analog form by some plotting device. The original signal that was input from the analog to digital converter can be reproduced to any degree of accuracy by convolving each sample by a sin t/t function. From equation 11.6-3 for a Daniell window with $t-t_0 = \tau/m$ the operator is as follows.

$$W(t) = [\sin \pi (t - t_0)] / [\pi (t - t_0)] \qquad 22.4\text{-}1$$

At time $t = t_0$ the function has a value of unity and when multiplied by the sampled amplitude at time t_0 it yields the exact sampled value. At discrete integer times, $t = \pm 1, \pm 2,...$, the function has a value of zero. At non-integer values of t the weighted sin t/t functions are summed to produce interpolated values at arbitrarily small spacing intervals. Figure 22.3b illustrates the operation. This method of reconstruction is almost never used because it is expensive on a computer.

Other methods are available which are simpler computationally but produce discontinuities in the function or some of its derivatives. For closely spaced samples in dominantly low frequency signals linear interpolation from sample to sample is adequate (figure 22.3c). This is particularly true of seismic reflection exploration data which has peak power at 50 Hz but is sampled at 1 or 2 millisecond intervals. Alternatively a higher order polynomial and some cubic spline subroutine may be used to interpolate values between the sampled points. A digital to analog converter may be used prior to plotting on a pen or galvonometer recorder. Each sample is held as a direct current (D.C.) step voltage of the appropriate amplitude for one sample period (figure 22.3d). The step voltages are then passed through a low pass filter prior to plotting to smoothen out the staircase-shaped signal.

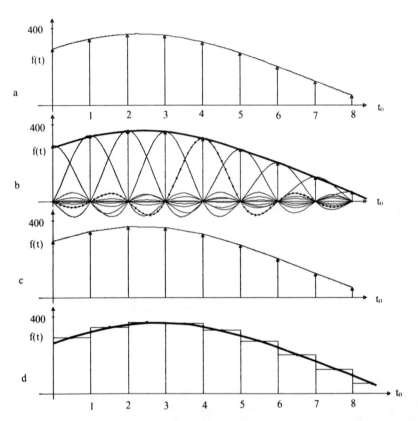

Figure 22.3 (a) The original analog signal and its sampled equivalent. (b) The exact reconstruction (dashed curve) of the digital signal with a sin t/t function (smooth curve). (c) A linear reconstruction of the analog signal with discontinuities in the first derivative. (d) A reconstruction of the sampled signal with an analog to digital converter (dashed line) followed by a smoothing with a low-pass filter (smooth curve).

References

Barry,K.M.,Carers,D.A. and Kneale,C.W.(1975), Recommended standards for digital tape formats. *Geophysics* **40**, 344-352.

Bell,L.W.(1968), Digital concepts. Tektronix Inc. Beaverton, Oregon.

Evenden,B.S., Stone, D.R. and Anstey,N.A.(1971), Seismic prospecting instruments. **2**, Gebruder Borntraeger, Berlin.

Gimlin,D.R. and Smith,J.W.(1980), A comparison of seismic trace summing techniques. *Geophysics* **45**, 1017-1041.

Lu,C.H. and Gupta,S.C.(1978), A multirate digital filtering approach to interpolation. *Geophysics* **43**, 877-885.

Northwood,E.J., Weisinger,R.C. and Bradley,J.J.(1967), Recommended standard for digital tape formats. *Geophysics* **32**, 1073-1084.

Peterson,R.A. and Dobrin,M.B.(1966), A pictoral digital atlas. United Geophysical Corp., Pasadena, Calif.

Savitt,C.H.(1966), A proposed standard format for nine-track digital tape. *Geophysics* **31**, 812-815.

Sheingold,D.H.(1972), Analog-digital conversion handbook. Analog devices Inc., Norwood Mass.

Texas Instruments (1966), Fundamentals of digitally recorded seismic data. Manual No. 174498-0001B., Dallas, Texas.

23 Walsh transforms and data compression

23.1 Introduction

The superposition of sines and cosines through a Fourier transform has been of great value in the analysis of many types of functions and of linear differential equations of second order. With the dominance of digital computers and solid state switching circuits it is possible to introduce complete systems of orthogonal functions based on only two states. These are in the form of square waves which take on only values of +1 and -1. Such transforms were introduced by Walsh (1923) and Paley (1932). They are a subset of the Hadamard (1893) matrices and the Reed-Muller (1954) codes which are themselves a part of a larger class of transformations called the Good (1958) transform of power 2. Walsh transforms are defined by a difference equation instead of by a differential equation. Another system known as the Rademacher (1922) transform is not a complete set.

The Walsh and Paley transforms have been shown by Wood (1974) to concentrate most of the energy of a seismic signal in a small part of the bandwidth. This allows one to consider methods of compressing sampled binary data without serious degradation of the waveform. In addition, filtering in the Walsh domain has proven to be of value in two-dimensional image enhancement. An advantage in the use of Walsh transforms is that they require only addition operations whereas Fourier transforms use complex arithmetic and multiplication. The fast Walsh transform is faster than the fast Fourier transform.

There is also a Haar (1910) function based on three values. These are +1, 0 and -1. The normalized Haar functions are multiplied by $2^{1/2}$. A Haar transform may be even faster than a Walsh transform but not much development has occurred with this class.

Data compression must be considered because of the vast amount of digital information generated at the present time. Satellites telemeter data to earth at rates of thousands of bits per second and may operate for years. In exploration seismology a single crew can generate over a 100 million bits per day and there are usually over a 100 such crews in operation, all sending tapes to half a dozen computing centers in the world. Data handling is time consuming and expensive so many methods of data compression are being considered. Apart from the use of special transforms a great reduction can be made by editing the data with microprocessors and minicomputers in the field.

Walsh function

23.2 Walsh function

The Walsh functions are illustrated in figure 23.1. They are subdivided into even and odd functions called Cal(j,t) and Sal(j,t) in much the same way as cosine and sine functions. The periodicity of the Walsh functions is specified by the number of zero crossings. Harmuth (1968) has coined the word *sequency* for half the number of sign changes per unit time. The unit for sequency is zps for zeros per second. The definition is similar to frequency for sine waves. Thus a 60 Hz sine wave has 120 zero crossings per second. The Walsh functions are

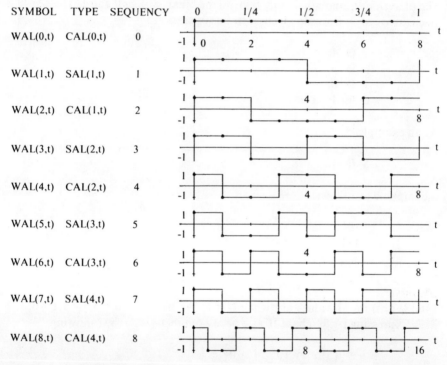

Figure 23.1 Walsh functions up to sequency 8 zps and their classification into Cal and Sal in the interval $0 < t < 1$ for n=3 and the first one for n=4.

given by the following equations in Boolean algebra.

$$W(j,t) = (-1)^{w(j,t)}$$

where

$$w(j,t) = \sum_{k=1}^{n} (j_{n-k+1} \oplus j_{n-k}) t_k \qquad 23.2\text{-}1$$

The time interval, t, is sampled N times.

$$N = 2^n$$ 23.2-2
$$t = 0, 1, \ldots, N-1$$
$$j = 0, 1, \ldots, N-1$$
$$k = 1, 2, \ldots, n$$

t_k is the n-k+1 binary bit of t
j_k is the n-k+1 binary bit of j

The quantities t_k and j_k can only take on the binary values 0 and 1. The sign $\oplus$ is an exclusive *or* in Boolean algebra. In this operation

$$1 \oplus 1 = 0$$
$$0 \oplus 0 = 0$$
$$1 \oplus 0 = 1$$
$$0 \oplus 1 = 1$$

As an example

$$5 \oplus 13 = 8$$

is computed as follows:

$$\begin{array}{r} 0101 \\ 1101 \\ \hline 1000 \end{array}$$

Example
Find W(5,t) for all values of t when N = 8. We have that n = 3, j = 5, t = 0, 1, 2, ... 7. Expressing j = 5 to base 10 as a base 2 number yields the following:

$$j = [5]_{10} = [0\ 1\ 0\ 1]_2$$
$$j_k = j_0 = 0, j_1 = 1, j_2 = 0, j_3 = 1$$
$$j_{n-k+1} = j_4 = 0, j_3 = 1, j_2 = 0, j_1 = 1$$

$$\sum_{k=1}^{3} [j_{n-k+1} \oplus j_{n-k}]\, t_k$$
$$= (j_3 \oplus j_2)\, t_1 + (j_2 \oplus j_1)\, t_2 + (j_1 \oplus j_0)\, t_3$$
$$= (1 \oplus 0)\, t_1 + (0 \oplus 1)\, t_2 + (1 \oplus 0)\, t_3$$
$$= 1\, t_1 + 1\, t_2 + 1\, t_3$$

Expressing t = 0 to base 10 as a base 2 number we can find t_1, t_2 and t_3.

Walsh function

$$[0]_{10} = [0 \; 0 \; 0]_2 \qquad t_1 = 0, t_2 = 0, t_3 = 0$$

$$w(5,0) = 1\cdot 0 + 1\cdot 0 + 1\cdot 0 = 0$$
$$w(5,0) = (-1)^0 = 1$$

Similarly for $t = [6]_{10} = [110]_2$, $t_1 = 1, t_2 = 1, t_3 = 0$.

$$w(5,6) = 1\cdot 1 + 1\cdot 1 + 1\cdot 0 = 2$$
$$W(5,6) = (-1)^2 = 1$$

In a similar manner $W(5,t)$ is evaluated for all t.

$$W(5,t) = [1, -1, -1, +1, -1, +1, +1, -1,]$$

This function may be viewed in figure 23.1. The Walsh function can also be obtained by a recursive scheme using the following difference equation in the interval $-1/2 < t < 1/2$:

$$W[2j + B, t] = [-1]^{B+(j/2)} [W[j, 2[t + 1/4]] \\ + [-1]^{j+B} W[j, 2[t - 1/4]] \qquad 23.2\text{-}3$$

The initial values for $W(0,t)$ are as follows:

$$W[0,t] = 1 \qquad -1/2 \leqslant t < 1/2$$
$$W[0,t] = 0 \qquad t < -1/2 \; ; \; t > 1/2$$
$$B = 0, 1 \quad ; \quad j = 0, 1, \ldots N-1$$

$(j/2)$ is the largest integer that is smaller or equal to $j/2$.

23.3 Hadamard matrix

The Hadamard matrix is constructed from a particular form of the Good (1958) transform. The Good transform is derived from a product of Kronecker matrices. These can always be decomposed into a product of matrices each with only two non-zero entries per row. That is, each 2^n by 2^n matrix is broken into n sparse matrices. For example, the Hadamard matrix, H_4, is composed as follows:

$$H_4 = \begin{bmatrix} +1 & +1 & +1 & +1 \\ +1 & -1 & +1 & -1 \\ +1 & +1 & -1 & -1 \\ +1 & -1 & -1 & +1 \end{bmatrix} = \begin{bmatrix} +1 & +1 & 0 & 0 \\ 0 & 0 & +1 & +1 \\ +1 & -1 & 0 & 0 \\ 0 & 0 & +1 & -1 \end{bmatrix} \begin{bmatrix} +1 & +1 & 0 & 0 \\ 0 & 0 & +1 & +1 \\ +1 & -1 & 0 & 0 \\ 0 & 0 & +1 & -1 \end{bmatrix} \begin{matrix} \text{sequency} \\ 0 \\ 3 \\ 1 \\ 2 \end{matrix}$$

An N by N Good matrix can be factored into a product of n ($N = 2^n$) matrices.

$$G_n = G_{n-1} \ G_{n-2} \ \ldots \ G_1 \ G_0 \qquad 23.3\text{-}1$$

where

$$G_0 = \begin{bmatrix} A_0 & B_0 \\ C_0 & D_0 \end{bmatrix}$$

$$G_r = \begin{bmatrix} A_r & B_r & 0 & 0 & \ldots & 0 & 0 \\ 0 & 0 & A_r & B_r & \ldots & 0 & 0 \\ \cdot & & & & & & \\ \cdot & & & & & & \\ 0 & 0 & 0 & 0 & \ldots & A_r & B_r \\ C_r & D_r & 0 & 0 & \ldots & 0 & 0 \\ 0 & 0 & C_r & D_r & \ldots & 0 & 0 \\ 0 & 0 & 0 & 0 & \ldots & C_r & D_r \end{bmatrix}$$

The core matrix for a Hadamard function with $N = 2$ is as follows:

$$H_2 = \begin{bmatrix} +1 & +1 \\ +1 & -1 \end{bmatrix} \qquad \begin{matrix} \text{sequency} \\ 0 \\ 1 \end{matrix} \qquad 23.3\text{-}2$$

For $N = 8$ the Hadamard function is given below.

$$H_8 = \begin{bmatrix} +1 & +1 & \cdot & +1 & +1 & \cdot & +1 & +1 & \cdot & +1 & +1 \\ +1 & -1 & \cdot & +1 & -1 & \cdot & +1 & -1 & \cdot & +1 & -1 \\ \hdashline +1 & +1 & \cdot & -1 & -1 & \cdot & +1 & +1 & \cdot & -1 & -1 \\ +1 & -1 & \cdot & -1 & +1 & \cdot & +1 & -1 & \cdot & -1 & +1 \\ \hdashline +1 & +1 & \cdot & +1 & +1 & \cdot & -1 & -1 & \cdot & -1 & -1 \\ +1 & -1 & \cdot & +1 & -1 & \cdot & -1 & +1 & \cdot & -1 & +1 \\ \hdashline +1 & +1 & \cdot & -1 & -1 & \cdot & -1 & -1 & \cdot & +1 & +1 \\ +1 & -1 & \cdot & -1 & +1 & \cdot & -1 & +1 & \cdot & +1 & -1 \end{bmatrix} \begin{matrix} \text{sequency} \\ 0 \\ 7 \\ 3 \\ 4 \\ 1 \\ 6 \\ 2 \\ 5 \end{matrix} \qquad 23.3\text{-}3$$

Note that

$$HH^* = NI \qquad 23.3\text{-}4$$

Hadamard matrix

where **I** is the identity matrix. The core matrices in H_8 are separated by dots for clarity. The Hadamard function is also given by the following equation in Boolean algebra:

$$H(j\ t) = (-1)^{h(j\ t)} \qquad 23.3\text{-}5$$

where

$$h(j\ t) = \sum_{k=1}^{n} t_k\ j_k \qquad 23.3\text{-}6$$

The Hadamard function is said to have a natural ordering because it can be built up from a matrix of order 2. Most importantly they have the same shape as the Walsh function but the sequency ordering is different. Equation 23.3-3 should be compared to figure 23.1.

23.4 Walsh and Hadamard transforms

If $f(x,y)$ is a two-dimensional set of intensities or a map of contours with N^2 points the Hadamard transform is obtained from the following matrix equation

$$\mathbf{F}_H(u,v) = \mathbf{H}_H(u,v)\ f(x,y)\ \mathbf{H}_H(u,v)\ /\ N^2 \qquad 23.4\text{-}1$$

The inverse Hadamard transform is similar except for a normalizing factor, N^2.

$$f(x,y) = \mathbf{H}(u,v)\ \mathbf{F}_H(u,v)\ \mathbf{H}(u,v) \qquad 23.4\text{-}2$$

A finite Walsh transform of any function, $f(t)$, $t = 0, 1, \ldots N-1$, can be obtained from the following equation:

$$F_w(j) = \sum_{t=0}^{N-1} W(j,t)\ f(t)\ /\ N \qquad 23.4\text{-}3$$

The inverse for the Walsh transform is

$$f(t) = \sum_{j=0}^{N-1} F_w(j\ t)\ W(j,t) \qquad 23.4\text{-}4$$

The computation may be carried out by matrix multiplication with a square N by N Walsh (or Hadamard) matrix, $\mathbf{W}_N$, with a column vector, **X**, of data.

$$\mathbf{F} = \mathbf{WX}/N \qquad 23.4\text{-}5$$

$$\mathbf{W} = \begin{bmatrix} W(0,0) & W(1,0) & W(2,0) & \ldots & W(N-1,0) \\ W(0,1) & W(1,1) & W(2,1) & \ldots & W(N-1,1) \\ W(0,2) & W(1,2) & W(2,2) & \ldots & W(N-1,2) \\ \vdots \\ W(0,N-1) & W(1,N-1) & W(2,N-1) & \ldots & W(N-1,N-1) \end{bmatrix} \qquad 23.4\text{-}6$$

Note that $W(j,t) = W(t,j)$ so the matrix **W** has a high degree of symmetry.

Example
Compute the Walsh transform of a sine wave with 1 period for $N = 8$. That is, $\mathbf{X} = [0, .707, 1., .707, 0, -.707, -1.0, -.707]^T$.

$$W(j,t) = \begin{bmatrix} 0 \\ 0.60355 \\ -0.25 \\ 0 \\ 0 \\ -0.25 \\ -0.10355 \\ 0 \end{bmatrix} = \begin{bmatrix} 1 & 1 & 1 & 1 & 1 & 1 & 1 & 1 \\ 1 & 1 & 1 & 1 & -1 & -1 & -1 & -1 \\ 1 & 1 & -1 & -1 & -1 & -1 & 1 & 1 \\ 1 & 1 & -1 & -1 & 1 & 1 & -1 & -1 \\ 1 & -1 & -1 & 1 & 1 & -1 & -1 & 1 \\ 1 & -1 & -1 & 1 & -1 & 1 & 1 & -1 \\ 1 & -1 & 1 & -1 & -1 & 1 & -1 & 1 \\ 1 & -1 & 1 & -1 & 1 & -1 & 1 & -1 \end{bmatrix} \begin{bmatrix} 0 \\ .707 \\ 1.0 \\ .707 \\ 0 \\ -1. \\ -1. \\ -.707 \end{bmatrix} \qquad 23.4\text{-}7$$

Figure 23.2 illustrates how closely the superposition of 4 non-zero Walsh functions approximate a sine wave. A computer program for Walsh transforming is at the end of the chapter.

23.5 Paley transform
The Paley functions are given in Boolean algebra by the following equation:

$$P(j,t) = (-1)^{p(j\,t)}$$

where

$$p(j\,t) = \sum_{k+1}^{n} j_{n-k+1} \cdot t_k$$

Although Paley (1932) does not acknowledge it, the Paley functions are the same as the Walsh functions except for the dyadic ordering instead of the sequency arrangement. The Paley functions are ordered according to g, the Gray code (see Chapter 22). In the Gray code the lowest possible binary digit changes without causing a repetition. The equivalent Paley and Walsh functions are shown in Table 1.

Paley transform

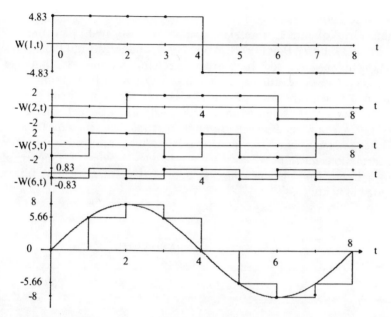

Figure 23.2 The superposition of weighted Walsh transforms W(1,t); W(2,t), W(5,t) and W(6,t) to approximate a sine wave.

Table 1 - Equivalent Paley and Walsh transform

Paley function	g	Gray Code (Binary)	Equivalent Decimal j	Equivalent Walsh function
P(0,t)	0	0	0	W(0,t)
P(1,t)	1	1	1	W(1,t)
P(2,t)	2	11	3	W(3,t)
P(3,t)	3	10	2	W(2,t)
P(4,t)	4	110	6	W(6,t)
P(5,t)	5	111	7	W(7,t)
P(6,t)	6	101	5	W(5,t)
P(7,t)	7	100	4	W(4,t)
P(8,t)	8	1100	12	W(12,t)
P(9,t)	9	1101	13	W(13,t)
P(10,t)	10	1111	15	W(15,t)

23.6 Data compression

Various forms of data compression have been investigated by Kortman (1967); Andrews et al.(1967); Bois (1972); Wood (1974) and Stigall and Panagas (1978). Some techniques employ first order predictor routines or eliminate low magnitude or low first difference samples. A system of book-keeping must be established to identify samples that are retained. Sequency band limiting has been shown to be of value with seismic exploration data. Figure 23.3 from Wood (1974) shows the Fourier amplitude spectra of some seismic traces before and after sequency band-limiting. It is seen that the introduction of Walsh transformation introduces spurious amplitudes beyond a quarter of the Nyquist limit. This may be eliminated by a stage of band-pass filtering as indicated in figure 23.4.

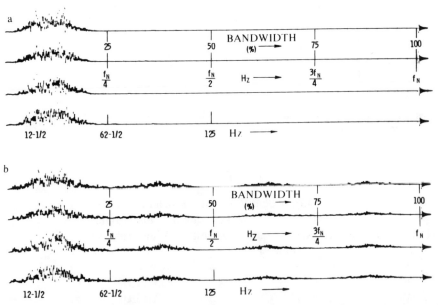

Figure 23.3 (a) Fourier amplitude spectra of some seismic traces. (b) Fourier amplitude spectra of the same seismic traces after sequency bandlimiting with a compression ratio of 7 to 1. Courtesy of the Society of Exploration Geophysicists; Wood (1974).

The partial energy curve of some seismic field data after various transformations may be seen in figure 23.5. It is seen that the Walsh and Paley transforms are suitable for data compression because over 95 percent of the energy is in the lower 50 percent of the sequency band-pass. About 14 percent of the bandwidth contains more than 75 percent of the energy so a compression ratio of 7 to 1 is feasible with this technique alone. It can be shown that an easy way of achieving some of the compression is to average adjacent samples in the

Data compression

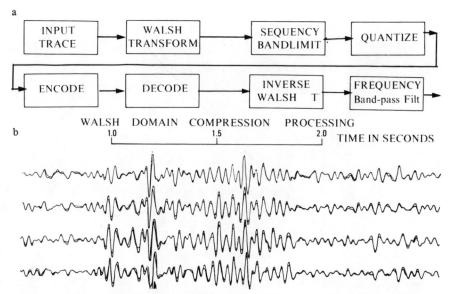

Figure 23.4 (a) Block diagram of a scheme for data compression, modified from Wood (1974). (b) An actual compressed seismic trace superimposed on the original recording. Courtesy of the Society of Exploration Geophysicists.

time domain before making a Walsh transformation. Thus if the adjacent 4 data points are averaged in the time domain, the input vector and the Walsh transform is reduced by a factor of 4. This will distort the signal and introduce high frequencies into the Fourier spectrum which must be removed by low-pass filtering. The output trace will also have some aliasing and it can be questioned if this extreme compression procedure is worthwhile.

The Hadamard transform distributes the energy equally over all sequences. This makes it valuable in transmitting complete data sets. Any noise or small loss of data will not significantly affect the image quality.

Further data compression can be carried out in the time domain using special codes which represent frequently occurring amplitudes by a small number of bits. One such coding scheme is the Shannon-Fanno-Huffman codes described by Rosenfeld (1969). The code consists of zeros and ones with the zero used as punctuation marks separating words. Wood (1974) uses such a scheme on the amplitude of 12-bit seismic data words to achieve a data compression of 4. Thus the actual recorded bit, including sign, has an average of 3 bits per sample. Stigall and Panagas (1978) achieve a similar compression rate using first differences instead of amplitudes. The sign (+ or -) bits are recorded separately at the beginning or end of each block of words. The complete binary bits for the first word in each block are also recorded. Table 23.2 summarizes the coding schemes, code-words recorded and probability of occurrence of each data range or difference.

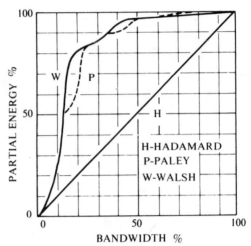

Figure 23.5 Partial energy curve of typical seismic field data. Courtesy of the Society of Exploration Geophysicists; Wood (1974).

Table 2 Compression Codes

Message No.	Probability	Message	Mean Value	Codeword
		First Difference		[Stigall and Panagas]
1	.525	0-8	4	0
2	.259	9-19	14	10
3	.107	20-39	30	110
4	.031	40-59	50	1110
5	.013	60-79	70	11110
6	.015	80-100	90	111110
7	.014	101-112	112	1111110
8	.015	123-153	138	11111110
9	.009	154-174	164	111111110
10	.005	174-211	192	111111111
		Amplitude [Wood]		
1	.5	0-50	25	0
2	.25	51-150	100	10
3	.125	151-325	238	110
4	.0625	326-550	438	1110
5	.03125	551-825	688	11110
6	.03125	826-2047	1436	11111

Data compression

Extremely high compression factors of up to 28 have been achieved by combining two or more methods (Wood, 1974). This is achieved at the expense of unusual or anomalous amplitudes or frequency changes. Since it is often the unusual event that is sought as an exploration target one must be careful in making use of extreme compression ratios lest one lose what is the original purpose for recording the data. Further sophisticated processing of seismic reflection results including deconvolution, wavelet decomposition and migration may be compromised if true amplitude recording over a wide dynamic range is not available.

23.7 Algorithm for Walsh transform

The following simple program may be used to obtain a Walsh transform on any data vector. It may be also used to give any Walsh function by using a data vector with one non-zero value. The examples illustrate the procedure.

```
C       PROGRAM FOR COMPUTING A FAST WALSH TRANSFORM
C       F(J) IS THE INPUT SIGNAL REPLACED BY ITS WALSH TRANSFORM
C       K IS RELATED TO THE NUMBER OF DATA POINTS, N, BY N=2**K
C       PROGRAM IS DIMENSIONED FOR K=1 TO 10
        DIMENSION F(1024),W(1024)
100     FORMAT(1X,I2)
101     FORMAT(8F8.0)
102     FORMAT(8(1X,F8.5))
103     FORMAT('1',' INPUT DATA '/)
104     FORMAT('0',' WALSH TRANSFORM IN SEQUENCY ORDER '/)
        READ(5,100)K
        N=2**K
        READ(5,101)(F(J),J=1,N)
        WRITE(6,103)
        WRITE(6,102)(F(KL),KL=1,N)
        DO 30 I=1,K
        NN=2**(I-1)
        IF(I-1)22,22,26
22      DO 24 J=1,N,2
        FF=F(J)
        F(J)=F(J)+F(J+1)
        F(J+1)=FF-F(J+1)
24      CONTINUE
        IF(N-2)32,32,25
25      GO TO 30
26      L=2*NN
        J=0
        DO 28 MA=1,N,L
        MJ=NN+MA-1
        DO 28 MW=MA,MJ,2
        J=J+1
        W(J)=F(MW)+F(NN+MW)
        J=J+1
        W(J)=F(MW)-F(NN+MW)
```

```
            J=J+1
            W(J)=F(MW+1)-F(NN+MW+1)
            J=J+1
            W(J)=F(MW+1)+F(NN+MW+1)
     28     CONTINUE
            DO 29 JJ=1,N
     29     F(JJ)=W(JJ)
     30     CONTINUE
     32     DO 33 JK=1,N
     33     F(JK)=F(JK)/N
     34     WRITE(6,104)
            WRITE(6,102)(F(K),K=1,N)
            STOP
            END
```

Example 1: Sine Wave
Input Data
0.0   0.70711   1.00000   0.70711   0.0   -.70711   -1.00000   -.70711
Walsh Transform in Sequency Order
0.0   0.60355   -0.25000   0.0   0.0   -0.25000   -0.10355   0.0

Example 2: The Walsh function, $W(0,t)$, may be obtained for 8 data points by placing an eight in the first position of the input vector. Walsh functions, $W(j,t)$, $j = 0, ..., 3$, $t = 0, 1, .., 7$, are also shown together with the input vector containing a single non-zero value in position $j + 1$.

Input Data
8.00000   0.0   0.0   0.0   0.0   0.0   0.0   0.0
Walsh Transform in Sequency Order
1.00000   1.00000   1.00000   1.00000   1.00000   1.00000   1.00000   1.00000
Input Data
0.0   8.00000   0.0   0.0   0.0   0.0   0.0   0.0
Walsh Transform in Sequency Order
1.00000   1.00000   1.00000   1.00000   -1.00000   -1.00000   -1.00000   -1.00000
Input Data
0.0   0.0   8.00000   0.0   0.0   0.0   0.0   0.0
Walsh Transform in Sequency Order
1.00000   1.00000   -1.00000   -1.00000   -1.00000   -1.00000   1.00000   1.00000
Input Data
0.0   0.0   0.0   8.00000   0.0   0.0   0.0   0.0
Walsh Transform in Sequency Order
1.00000   1.00000   -1.00000   -1.00000   1.00000   1.00000   -1.00000   -1.00000

Example 3 : In a similar manner, a 64-point Walsh function has been computed. The symbols p and n are used for +1 and -1.

Algorithm for Walsh transform

Cal(j,t)   Sequency
```
ppppppppppppppppppppppppppppppppppppppppppppppppppppppppppppppppp     0
ppppppppppppppppppppppppppppppnnnnnnnnnnnnnnnnnnnnnnnnnnnnnnnn        2
ppppppppppppppppppppnnnnnnnnnnnnnnnnppppppppppppppppppnnnnnnnnnnnnnnnpppppppp  4
ppppppppnnnnnnnnppppppppnnnnnnnnppppppppnnnnnnnnppppppppnnnnnnnnpppppppp      6
ppppnnnnnnnnppppppppnnnnnnnnppppppppnnnnnnnnppppppppnnnnnnnnppppppppnnnnnnnn    8
ppppnnnnnnnnppppppppnnnnnnnnppppppppnnnnnnnnppppppppnnnnnnnnpppp     10
ppppnnnnppppnnnnnnnnppppnnnnppppppppnnnnppppnnnnnnnnppppnnnnpppp     12
ppppnnnnppppnnnnnnnnppppnnnnppppppppnnnnppppnnnnnnnnppppnnnnpppp     14
ppnnnnppppnnnnppppnnnnppppnnnnppppnnnnppppnnnnppppnnnnppppnnnnpp     16
ppnnnnppppnnnnppnnppppnnnnppppnnnnppppnnnnppppnnnnppnnppppnnnnpp     18
ppnnnnppnnppppnnnnppnnppppnnnnppppnnnnppppnnnnppnnppppnnppnnnnpp     20
ppnnnnppnnppppnnnnppnnppppnnnnppppnnnnppppnnnnppnnppppnnppnnnnpp     22
ppnnppnnnnppnnppppnnppnnppppnnnnppnnppppnnppnnppppnnppnnnnppnnpp     24
ppnnppnnppnnppnnppppnnppnnppppnnnnppnnppppnnppnnppppnnppnnppnnpp     26
ppnnppnnppnnppnnppppnnppnnppppnnnnppnnppppnnppnnppppnnppnnppnnpp     28
ppnnppnnppnnppnnppnnppnnppnnppnnnnppnnppnnppnnppnnppnnppnnppnnpp     30
pnnppnnppnnppnnppnnppnnppnnppnnppnnppnnppnnppnnppnnppnnppnnppnn      32
pnnppnnppnnppnnppnnppnnppnnppnnppnnppnnppnnppnnppnnppnnppnnppnn      34
pnnppnnpnnppnnppnnppnnppnnppnnppnnppnnppnnppnnppnnppnnppnpppnnp      36
pnnppnnpnnppnnppnnppnnppnnppnnppnnppnnppnnppnnppnnppnnppnpppnnp      38
pnnpnnpnnppnnpnnppnnpnnppnnpnnppnnpnnppnnpnnppnnpnnppnnpnnppnnp      40
pnnpnnpnnppnnpnnppnnpnnppnnpnnppnnpnnppnnpnnppnnpnnppnnpnnppnnp      42
pnnpnnpnnpnnpnnpnnpnnpnnpnnpnnpnnpnnpnnpnnpnnpnnpnnpnnpnnpnnp       44
pnnpnnpnnpnnpnnpnnpnnpnnpnnpnnpnnpnnpnnpnnpnnpnnpnnpnnpnnpnnp       46
pnpnnpnpnnpnpnnpnpnnpnpnnpnpnnpnpnnpnpnnpnpnnpnpnnpnpnnpnpnnp       48
pnpnnpnpnnpnpnnpnpnnpnpnnpnpnnpnpnnpnpnnpnpnnpnpnnpnpnnpnpnnp       50
pnpnnpnpnnpnpnnpnpnnpnpnnpnpnnpnpnnpnpnnpnpnnpnpnnpnpnnpnpnnp       52
pnpnnpnpnnpnpnnpnpnnpnpnnpnpnnpnpnnpnpnnpnpnnpnpnnpnpnnpnpnnp       54
pnpnpnpnpnpnpnpnpnpnpnpnpnpnpnpnpnpnpnpnpnpnpnpnpnpnpnpnpnpnpn      56
pnpnpnpnpnpnpnpnpnpnpnpnpnpnpnpnpnpnpnpnpnpnpnpnpnpnpnpnpnpnpn      58
pnpnpnpnpnpnpnpnpnpnpnpnpnpnpnpnpnpnpnpnpnpnpnpnpnpnpnpnpnpnpn      60
pnpnpnpnpnpnpnpnpnpnpnpnpnpnpnpnpnpnpnpnpnpnpnpnpnpnpnpnpnpnpn      62
```

Sal(j,t)
```
ppppppppppppppppppppppppppppppppnnnnnnnnnnnnnnnnnnnnnnnnnnnnnnn      1
ppppppppppppppppppppnnnnnnnnnnnnppppppppppppppppppnnnnnnnnnnnnn      3
ppppppppppppppppnnnnnnnnnnnnnnnnppppppppppppppppnnnnnnnnnnnnnnn      5
ppppppppnnnnnnnnppppppppnnnnnnnnppppppppnnnnnnnnppppppppnnnnnnnn     7
ppppnnnnnnnnppppppppnnnnnnnnppppppppnnnnnnnnppppppppnnnnnnnnpppp     9
ppppnnnnnnnnppppppppnnnnnnnnppppppppnnnnnnnnppppppppnnnnnnnnpppp     11
ppppnnnnppppnnnnnnnnppppnnnnppppppppnnnnppppnnnnnnnnppppnnnnpppp     13
ppppnnnnppppnnnnnnnnppppnnnnppppppppnnnnppppnnnnnnnnppppnnnnpppp     15
ppnnnnppppnnnnppppnnnnppppnnnnppnnppnnnnppppnnnnppppnnnnppppnn       17
ppnnnnppppnnnnppnnppppnnnnppppnnppnnppnnnnppppnnnnppnnppppnnnnpp     19
ppnnnnppnnppppnnnnppnnppppnnnnppnnppnnppppnnnnppnnppppnnppnnnnpp     21
ppnnnnppnnppppnnnnppnnppppnnnnppnnppnnppppnnnnppnnppppnnppnnnnpp     23
ppnnppnnnnppnnppppnnppnnppnnnnppnnppnnppppnnppnnppppnnppnnppnnpp     25
ppnnppnnppnnppnnppppnnppnnppnnppnnppnnppnnppppnnppnnppnnppnnppnn     27
ppnnppnnppnnppnnppppnnppnnppnnppnnppnnppnnppppnnppnnppnnppnnppnn     29
ppnnppnnppnnppnnppnnppnnppnnppnnppnnppnnppnnppnnppnnppnnppnnppnn     31
pnnppnnppnnppnnppnnppnnppnnppnnpnnppnnppnnppnnppnnppnnppnnppnnp      33
pnnppnnppnnppnnppnnppnnppnnppnnpnnppnnppnnppnnppnnppnnppnnppnnp      35
pnnppnnppnnppnnppnnppnnppnnppnnpnnppnnppnnppnnppnnppnnppnnppnnp      37
pnnppnnpnnppnnppnnppnnppnnppnnppnnppnnppnnppnnppnnppnnppnpppnnp      39
pnnpnnpnnppnnpnnppnnpnnppnnpnnppnnpnnppnnpnnppnnpnnppnnpnnppnnp      41
pnnpnnpnnppnnpnnppnnpnnppnnpnnppnnpnnppnnpnnppnnpnnppnnpnnppnnp      43
pnnpnnpnnppnnpnnppnnpnnppnnpnnppnnpnnppnnpnnppnnpnnppnnpnnppnnp      45
pnnpnnpnnpnnpnnpnnpnnpnnpnnpnnpnnpnnpnnpnnpnnpnnpnnpnnpnnpnnp       47
pnpnnpnpnnpnpnnpnpnnpnpnnpnpnnpnpnnpnpnnpnpnnpnpnnpnpnnpnpnnp       49
pnpnnpnpnnpnpnnpnpnnpnpnnpnpnnpnpnnpnpnnpnpnnpnpnnpnpnnpnpnnp       51
pnpnnpnpnnpnpnnpnpnnpnpnnpnpnnpnpnnpnpnnpnpnnpnpnnpnpnnpnpnnp       53
pnpnpnpnpnpnpnpnpnpnpnpnpnpnpnpnpnpnpnpnpnpnpnpnpnpnpnpnpnpnpn      55
pnpnpnpnpnpnpnpnpnpnpnpnpnpnpnpnpnpnpnpnpnpnpnpnpnpnpnpnpnpnpn      57
pnpnpnpnpnpnpnpnpnpnpnpnpnpnpnpnpnpnpnpnpnpnpnpnpnpnpnpnpnpnpn      59
pnpnpnpnpnpnpnpnpnpnpnpnpnpnpnpnpnpnpnpnpnpnpnpnpnpnpnpnpnpnpn      61
pnpnpnpnpnpnpnpnpnpnpnpnpnpnpnpnpnpnpnpnpnpnpnpnpnpnpnpnpnpnpn      63
```

References

Andrews,C.A., Davis,J.M. and Schwarz,G.R.(1967), Adaptive data compression. *Proceedings IEEE* **55**, 267-277.
Andrews,H.C. and Pratt,W.K.(1969), Transform image coding. Symposium on Computer Processing in Communications. Editor, J. Fox. **19**, Polytechnic Institute of Brooklyn. 63-84.
Bois,P.(1972), Analyse sequentielle. *Geophysical Prospecting* **20**, 497-513.
Bowyer,D.E.(1970), Walsh functions, Hadamard matrices and data compression. *IEEE Transactions on Electromagnetic Compatibility* **EMC-12**, 33-37.
Brown,R.D.(1977), A recursive algorithm for sequency ordered fast Walsh transforms. *IEEE Transactions on Computers* **C26**, 819-822.
Good,I.J.(1958), The interaction algorithm and practical Fourier analysis. *Journal of the Royal Statistical Society, Series B* **20**, 361-372.
Haar,A.(1910), Zur Theorie der orthogonalen Functionensysteme. *Math. Ann.* **69**, 331-371. Also **71**, 38-53 (1912).
Hadamard,J.(1893), Resolution d'une question relative aux determinants. *Bull. Sci. Math. ser 2* **17**, 240-246.
Harmuth,H.F.(1968), A generalized concept of frequency and some applications. *IEEE Transactions on Information Theory* **IT-14**, 375-382.
Harmuth,H.F.(1969), Transmission of information by orthogonal functions. Springer Verlag, Berlin, New York.
Harmuth,H.F., Andrews,H.C. and Shibata,K.(1972), Two-dimensional sequency filters. *IEEE Transactions on Communications* **COM-20**, 321-330.
Kennett,L.N.(1970), A note on the finite Walsh Transform. *IEEE Transactions on Information Theory* **IT-16**, 489-491.
Kortman,C.M.(1967), Redundancy reduction. A practical method of data compression. *Proceedings IEEE* **55**, 253-263.
Paley,R.E.A.C.(1932), A remarkable series of orthogonal functions. *Proceedings of the London Mathematical Society* **34**, 241-279.
Pratt,W.K., Kane,J. and Andrews,H.C.(1969), Hadamard Transform Image Coding. *Proceedings of the IEEE* **57**, 58-68.
Rademacher,H.(1922), Einige Satze uber Reihen von allgemeinen orthogonalfunctionen. *Math. Annalen* **87**, 112-138.
Reed,I.S.(1954), A class of multiple-error correcting codes and the decoding scheme. *IRE Transactions on Information Theory* **IT-4**, 38-49.
Rosenfeld,A.(1969), Picture processing by computer. Academic Press, New York, 1-18.
Stigall,P.D. and Panagas,P.(1978), Data compression in microprocessor based data acquisition systems. *IEEE Transactions on Geoscience Electronics* **GE-16**, 323-332.
Walsh,J.L.(1923), A closed set of normal orthogonal functions. *American Journal of Mathematics* **45**, 5-24.
Wood,L.C.(1974), Seismic data compression methods. *Geophysics* **39**, 499-525.
Yuen,C.(1970-71), Walsh functions and Gray Code. *IEEE Transactions on Electromagnetic Compatibility* **EMC-12,13**, 68-73.

24 Generalized linear inverse method

24.1 Introduction
Geophysical observations are interpreted most often in the first instance by a *forward method* in which a simple model is advanced and the observed quantities are compared to the theoretical computations made using the model. The simple model may be perturbed many times to obtain a more satisfactory agreement. A major problem with the forward method of interpretation is that it is not known if there are other models which fit the problem and it is also difficult to estimate the errors in the solution or to optimize the resolution provided by the data. For instance, an earth model may be proposed with a great many layers and this may fit the observations quite well. However, it is possible that the data may be satisfied, within the constraints of the errors, by only a small number of layers. The generalized *linear inverse method* was developed by Backus and Gilbert (1967, 1968, 1970) to yield directly the collection of models which satisfy the observable quantities and quantify the degree of non-uniqueness inherent in the data set at hand. In addition they explored the resolving power attainable with any particular set of observations and the influence of the errors. The Backus-Gilbert papers are a rigorous and mathematically elegant presentation of the topic. It is recommended that the student read Parker's (1970, 1977) reviews as an introduction to their theory.

The Backus-Gilbert form of inverse theory treats the continuous case which is always underdetermined. It is also expensive computationally because the integral solutions require large amounts of core memory in a computer. A discrete formulation with a factoring of the problem into eigenvectors and eigenvalues has advantages which were pointed out by Smith and Franklin (1969) and Gilbert (1971). The most valuable matrix representations of the discrete form of linear inverse theory are the ones by Der and Landisman (1972), Wiggins (1972) and Jackson (1972). It is this form which will be discussed in the next sections because it is most readily programmed as an algorithm on a digital computer.

24.2 Discrete linear inverse problem
It is assumed that physical observations have been made which lead to a set of experimental data. These are related to the structure and properties of the body being observed by a set of known physical laws and mathematical equations. In geomagnetism one may want to find the conductivity of the earth from measurements of the electrical and magnetic field by the application of Maxwell's equations. In seismology the density and the velocity structure of the

earth may be obtained from measurements of the surface wave velocities, the periods of free oscillations, the mass and moment of inertia, together with an application of elastic wave theory. Many other applications of linear inverse theory, particularly in observations of potential fields, may be cited.

Let the m unknown parameters of the earth model be denoted by M_k. The n experimentally determined observations will be O_j. These data sets are arranged as column matrices **M** and **O**.

$$\mathbf{M}^T = (M_1, M_2, \ldots M_m)$$
$$\mathbf{O}^T = (O_1, O_2, \ldots O_n)$$
24.2-1

T indicates a transpose. It is assumed that there is a known functional relationship, A, between the model parameters and the observations.

$$O_j = A_j(M_1, M_2, \ldots M_m) \qquad j = 1, 2, \ldots n \qquad 24.2\text{-}2$$

An initial guess, M_k^a, is required for the model parameters.

$$\mathbf{M}^{Ta} = (M_1^a, M_2^a, \ldots M_m^a) \qquad 24.2\text{-}3$$

The function A_j will seldom be strictly linear but if it varies smoothly then a Taylor series expansion may be made about the initial guess.

$$O_j = A_j(M_1^a \ldots M_m^a) + \sum_{k=1}^{m} [\partial A_j / \partial M_k]_{M_k^a} (M_k - M_k^a) + \ldots \qquad 24.2\text{-}4$$

The higher order terms indicated at the end of 24.2-4 may be neglected in order to linearize the problem. It is very important to note that this linear inverse method will fail if there are discontinuities in the function $A_j(M_j)$. For instance, the inversion of seismic travel times must be approached cautiously because of the presence of cusps and discontinuities due to velocity gradients and low velocity layers in the earth (Nolet, 1978).

Having linearized the problem we can introduce notation which is more familiar in linear algebra. Let the difference between observations and model computations which are to be minimized be y.

$$y_j = O_j - A_j(M_1^a \ldots M_m^a) \qquad j = 1, \ldots n \qquad 24.2\text{-}5$$

The differences in the initial and the next approximation to the model parameters will be **x**.

Discrete linear inverse problem

$$x_k = M_k - M_k^a \qquad k = 1, \ldots m \qquad \text{24.2-6}$$

The first partial derivatives which are evaluated at the first guess, M_k^a, form an m by n matrix, **A**, with components A_{jk}.

$$\mathbf{A} = \begin{bmatrix} \partial A_1/\partial M_1 & \partial A_1/\partial M_2 & \ldots & \partial A_1/\partial M_m \\ \partial A_2/\partial M_1 & \partial A_2/\partial M_2 & \ldots & \partial A_2/\partial M_m \\ \partial A_n/\partial M_1 & \partial A_n/\partial M_2 & \ldots & \partial A_n/\partial M_m \end{bmatrix} \qquad \text{24.2-7}$$

$$A_{jk} = \partial A_j / \partial M_k \qquad \text{24.2-8}$$

The linearized approximation to 24.2-4 is the following set of n linear equations as written in matrix form:

$$\mathbf{y} = \mathbf{A}\,\mathbf{x} \qquad \text{24.2-9}$$

The solution of this set of equations will depend upon the number of model parameters as compared to the number of observations and also on the behavior of the matrix **A**. The overconstrained case (n>m) is a classic least squares problem and will be discussed in the next section. The underdetermined case (n<m) will be discussed in section 24.4. For the well-posed case (n=m) an exact solution may be possible. It is required that the matrix and its transpose be equal ($\mathbf{A}^T = \mathbf{A}$) and that it not be singular. That is, its determinant is not equal to zero.

$$det(\mathbf{A}) \neq 0 \qquad \text{24.2-10}$$

The solution is obtained from the inverse of the matrix.

$$\mathbf{x} = \mathbf{A}^{-1}\mathbf{y} \qquad \text{24.2-11}$$

A new approximation may be obtained at this stage if the initial guess was not close enough for the linear assumption to be valid.

$$M_k^b = x_k + M_k^a \qquad k = 1, \ldots m \qquad \text{24.2-12}$$

To obtain the variance it is useful to solve for the eigenvalues, λ_j, and the eigenvectors, u_j.

$$\mathbf{A}\mathbf{u}_k = \lambda_k \mathbf{u}_k \qquad \text{24.2-13}$$

Each eigenvector has n components.

$$\mathbf{u}_k = (u_{1k}, u_{2k}, \ldots u_{nk}) \qquad k = 1, \ldots m \qquad \text{24.2-14}$$
$$\mathbf{u}_k^T \mathbf{u}_i = \delta_{ki} \qquad \text{24.2-15}$$

If **A** is symmetric then the eigenvalues form an orthonormal system as indicated in 24.2-15 and they may be arranged as matrix **U**.

$$\mathbf{U} = \begin{bmatrix} u_{11} & u_{12} & \ldots & u_{1m} \\ u_{21} & u_{22} & \ldots & u_{2m} \\ \vdots & \vdots & & \vdots \\ u_{n1} & u_{n2} & \ldots & u_{nm} \end{bmatrix} \qquad \text{24.2-16}$$

The matrix is square (n=m) for the well-posed case and the eigenvector can be rearranged in order that the one with the largest eigenvalue is first and the smallest is last.

Let **L** be the diagonal matrix having eigenvalues, $\lambda_1, \lambda_2, \ldots \lambda_m$, on the main diagonal and zeros elsewhere.

$$\mathbf{AU} = \mathbf{UL} \qquad \text{24.2-17}$$

The product of U and its transform is equal to the identity matrix.

$$\mathbf{UU}^T = \mathbf{U}^T\mathbf{U} = \mathbf{I} \qquad \text{24.2-18}$$

It follows that we can write **A** and its inverse by multiplying 24.2-16 by $\mathbf{U}^T$.

$$\mathbf{A} = \mathbf{ULU}^T \qquad \text{24.2-19}$$
$$\mathbf{A}^{-1} = (\mathbf{ULU}^T)^{-1} = \mathbf{UL}^{-1}\mathbf{U}^T \qquad \text{24.2-20}$$

The inverse of **L** is a diagonal matrix with values $1/\lambda_1, \ldots 1/\lambda_m$ on the main diagonal.

$$\mathbf{L}^{-1} = \begin{bmatrix} 1/\lambda_1 & 0 & \ldots & 0 \\ 0 & 1/\lambda_2 & \ldots & 0 \\ \vdots & \vdots & & \vdots \\ 0 & 0 & \ldots & 1/\lambda_m \end{bmatrix} \qquad \text{24.2-21}$$

The variance of the model will depend upon how non-singular the matrix **A** is. If $1/\lambda_m$ is infinite the inverse of **L** does not exist. For statistically independent data the variance of the correction to the model parameters is as follows:

Discrete linear inverse problem

$$var(x_k) = \sum_{j=1}^{m} [\sum_{i=1}^{m} u_{ki} \lambda_i^{-1} u_{ji}]^2 \ var(y_j) \qquad 24.2\text{-}22$$

24.3 Inverse for an overconstrained system

When there are more observations than unknown model parameters (n>m) the solution, if it exists, should be determined by a least squares method. Let the error for each data set be **e**.

$$\mathbf{e} = \mathbf{Ax} - \mathbf{y} \qquad 24.3\text{-}1$$

The squared norm is obtained from a product of $\mathbf{e}^T$ and **e**.

$$e^2 = \mathbf{e}^T\mathbf{e} = (\mathbf{Ax} - \mathbf{y})^T (\mathbf{Ax} - \mathbf{y}) \qquad 24.3\text{-}2$$

$$e^2 = (\mathbf{x}^T\mathbf{A}^T - \mathbf{y}^T)(\mathbf{Ax} - \mathbf{y}) = \mathbf{x}^T\mathbf{A}^T\mathbf{Ax} - \mathbf{y}^T\mathbf{Ax} - \mathbf{x}^T\mathbf{A}^T\mathbf{y} + \mathbf{y}^T\mathbf{y} \quad 24.3\text{-}3$$

To minimize the squared norm we differentiate in a vector space with respect to **x** or $\mathbf{x}^T$ and set the result equal to zero.

$$\begin{aligned}\partial e^2 / \partial \mathbf{x} &= \mathbf{x}^T\mathbf{A}^T\mathbf{A} - \mathbf{y}^T\mathbf{A} = 0 \\ \partial e^2 / \partial \mathbf{x}^T &= \mathbf{A}^T\mathbf{Ax} - \mathbf{A}^T\mathbf{y} = 0 \end{aligned} \qquad 24.3\text{-}4$$

Both equations yield the same result so the solution for the estimate, $\mathbf{x}^b$, may be taken from the second one.

$$[\mathbf{A}^T\mathbf{A}] \mathbf{x}^b = \mathbf{A}^T\mathbf{y} \qquad 24.3\text{-}5$$
$$\mathbf{x}^b = [\mathbf{A}^T\mathbf{A}]^{-1} \mathbf{A}^T\mathbf{y} \qquad 24.3\text{-}6$$

It is necessary that the determinant be non-singular.

$$det\ (\mathbf{A}^T\mathbf{A}) \neq 0 \qquad 24.3\text{-}7$$

A computer algorithm similar to the one in the Levinson recursion for the deconvolution operator may be used to solve 24.3-5 for $\mathbf{x}^b$. The matrix $[\mathbf{A}^T\mathbf{A}]$ is obviously an m by m autocovariance matrix while $\mathbf{A}^T\mathbf{y}$ is a cross covariance.

To evaluate the solution a generalized eigenvector analysis is made following a method discovered by Moore (1920), rediscovered by Penrose (1955) and described by Lanczos (1961). The matrix **A** is factored into a product of orthonormal modal matrices **U** and **V** whose elements are the eigenvectors with non-zero eigenvalues.

$$\mathbf{A} = \mathbf{ULV}^T \qquad 24.3\text{-}8$$

This is called a spectral decomposition of the matrix **A**. The matrix **L** is a diagonal one with eigenvalues, λ_j, arranged in order of absolute value along the diagonal. The matrices **U** and **V** are made up from two sets of eigenvectors, $\mathbf{u}_j$ and $\mathbf{v}_k$.

$$\mathbf{Av}_k = \lambda_k \mathbf{u}_k \qquad k = 1, \ldots m \qquad 24.3\text{-}9$$
$$\mathbf{Au}_j = \lambda_j \mathbf{v}_j \qquad j = 1, \ldots n \qquad 24.3\text{-}10$$

The eigenvectors $\mathbf{u}_j$ are spectral components of the information content of the observational space, as will be shown in the derivation leading up to equation 24.3-29. Wiggins et al. (1976) point out that this spectral decomposition is similar to the transformation of an impulse response into a transfer function consisting of sinusoids (eigenvectors) having various amplitudes (eigenvalues). The matrix **A** is a filter that maps the input parameters into the output observations.

The following proof shows that $\lambda_k = \lambda_j$. Multiply 24.3-10 by $\mathbf{v}_k^T$ and note that $||v||^2 = ||u||^2 = 1$ since these are eigenvectors.

$$\mathbf{v}_k^T \mathbf{Au}_j = \lambda_j \mathbf{v}_k \mathbf{v}_j = \lambda_j ||v||^2 = \lambda_j \qquad 24.3\text{-}11$$

Take the transpose of 24.3-9 and multiply by $\mathbf{u}_j$ on the right.

$$[\mathbf{Av}_k]^T = \mathbf{v}_k^T \mathbf{A}^T = \lambda_k \mathbf{u}_k^T \qquad 24.3\text{-}9a$$
$$\mathbf{v}_k^T \mathbf{Au}_j = \lambda_k \mathbf{u}_k^T \mathbf{u}_j = \lambda_k ||u||^2 = \lambda_k \qquad 24.3\text{-}12$$

Comparing 24.3-11 and 24.3-12 we see that the left-hand sides are identical. Thus if $\lambda_j \neq 0$ or $\lambda_k \neq 0$ then $\lambda_j = \lambda_k$.

There are p non-zero eigenvalues and n-p eigenvectors with zero eigenvalues. If p = m the solution is unique and exact. If p < m there are many solutions to the problem and this underdetermined case will be discussed in the next section. **U** is an n by p matrix and **V** is an m by p matrix. By the orthonormality of the eigenvectors $\mathbf{u}_j$ and $\mathbf{v}_k$ we have the following relation in which $\mathbf{I}_p$ is an identity matrix with p one's along the diagonal.

$$\mathbf{U}^T \mathbf{U} = \mathbf{I}_p \qquad \mathbf{V}^T \mathbf{V} = \mathbf{I}_p \qquad 24.3\text{-}13$$

If we multiply 24.3-8 by **V** the following relations are found with the use of the identity matrix:

Inverse for an overconstrained system

$$AV = UL \qquad 24.3\text{-}14$$

It is also possible to find the following relation by taking the transpose of 24.3-8 and multiplying by **U**:

$$A^T U = VL \qquad 24.3\text{-}15$$

Multiplying 24.3-14 by A^T and substituting 24.3-15 for the righthand side yields the following equation:

$$A^T AV = VL^2 \qquad 24.3\text{-}16$$

The m by m matrix $A^T A$ is non-negative definite and symmetric. Equation 24.3-16 may be used to solve for the eigenvalues and eigenvectors if m < n. Similarly one can find the following relation and use it for the underconstrained case, n < m, in finding the eigenvalues and eigenvectors.

$$AA^T U = UL^2 \qquad 24.3\text{-}17$$

Let us substitute 24.3-8 for **A** into the solution for x^b in 24.3-6.

$$x^b = [(ULV^T)^T (ULV^T)]^{-1} [ULV^T]^T y$$

Using the properties of a matrix transpose, $(A^T)^T = A$; $(AB)^T = B^T A^T$, and the identity matrix in 24.3-13, this equation may be simplified.

$$x^b = [VLU^T ULV^T]^{-1} [VLU^T] y = [VL^2 V^T]^{-1} [VLU^T] y$$

The inverse of $VL^2 V^T$ is $VL^{-2} V^T$.

$$x^b = VL^{-2} V^T VLU^T y$$
$$x^b = VL^{-1} U^T y \qquad 24.3\text{-}18$$

The product $VL^{-1} U^T$ is given the symbol **H** and is called the Moore-Penrose operator or the Lanczos generalized inverse operator.

$$H = VL^{-1} U^T \qquad 24.3\text{-}19$$
$$x^b = Hy \qquad 24.3\text{-}20$$

Since by equation 24.2-9 $y = Ax$, the new generalized solution, x^b, is obtained by iteration and can be expressed in terms of the true solution, **x**.

$$\mathbf{x}^b = \mathbf{HAx} = \mathbf{Rx} \qquad 24.3\text{-}21$$

$$x_i^b = \sum_{j=1}^{n} r_{ij} x_j \qquad 24.3\text{-}22$$

The quantity **HA** was defined by Backus and Gilbert (1968) as the *resolution matrix*, **R**. If $\mathbf{x}^b$ is close to x then the components r_{ij} of **R** are close to the Dirac delta function. Since the Lanczos inverse is the best available the following quantity is minimized:

$$r_q = \sum_{k=1}^{m} (r_{qk} - \delta_{qk})^2 = \sum_k (H_{qi} A_{ik} - \delta_{qk})^2 \qquad 24.3\text{-}23$$

The resolution matrix may be obtained more directly by substituting 24.3-8 for **A** and 24.3-19 for **H**.

$$\mathbf{R} = \mathbf{HA} = (\mathbf{VL}^{-1}\mathbf{U}^T) \mathbf{ULV}^T \qquad 24.3\text{-}24$$

This reduces to $\mathbf{VV}^T$ since $\mathbf{U}^T\mathbf{U}$ is equal to the identity matrix.

$$\mathbf{R} = \mathbf{VV}^T \qquad 24.3\text{-}25$$

It is also possible to examine the data in terms of the information content. Let us rewrite equation 24.2-9 in terms of the actual data, $\mathbf{y}^b$, and substitute 24.3-20.

$$\mathbf{y}^b = \mathbf{Ax}^b \qquad 24.3\text{-}26$$

$$\mathbf{y}_b = \mathbf{AHy} = \mathbf{Sy} \qquad 24.3\text{-}27$$

Perfect or error-free data is represented by **y**. The quantity **AH** is given the symbol **S** and was defined by Wiggins (1972) as the *information density matrix* or the covariance matrix for the observations. The Lanczos inverse gives us the best information density matrix and minimizes the following quantity:

$$s_q = \sum_{k=1}^{n} (s_{qk} - \delta_{qk})^2 = \sum_k (A_{qi} H_{ik} - \Delta_{qk})^2 \qquad 24.3\text{-}28$$

The information density matrix is useful in letting us know if some of the data is redundant. From the relation for **A** in 24.3-8 and **H** in 24.3-19 the information density matrix is found to be equal to $\mathbf{UU}^T$.

Inverse for an overconstrained system

$$\mathbf{S} = \mathbf{AH} = (\mathbf{ULV}^T)(\mathbf{VL}^{-1}\mathbf{U}^T)$$
$$\mathbf{S} = \mathbf{UU}^T \qquad 24.3\text{-}29$$

For statistically independent data which has been normalized (var y_j = 1) the variance of one component of $\mathbf{x}^b$ is given the relation below.

$$var\ (x_q^b) = \sum_{i=1}^{n} H_{qi}^2\ var\ (y_i) \qquad 24.3\text{-}30$$

Making use of 24.3-19 for **H** we obtain the variance as follows for normalized data:

$$var\ (s_q^b) = \sum_{i=1}^{p} (v_{iq}/\lambda_i)^2 \qquad 24.3\text{-}31$$

Since the eigenvalues, λ_i, are in the denominator, the variance can become very large for small eigenvalues. It is possible to have very good resolution but a very large variance if the errors are large. The procedure recommended by Wiggins and Jackson is that the eigenvalues be arranged in order and the very small ones should be disregarded. This has the effect of decreasing the resolution. A trade-off curve of resolution versus variance is shown in figure 24.1. It is recommended that the number of eigenvalues be reduced from p to q to optimize the solution.

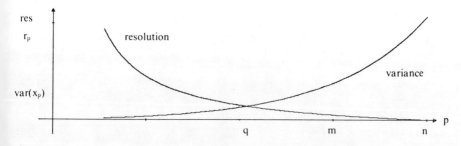

Figure 24.1 The trade-off curve between resolution and variance for the kth parameter of a hypothetical model. The generalized inverse procedure would choose p = m yielding an unstable solution. An optimal solution would retain q eigenvalues.

It is also possible to place additional constraints on the solution in 24.2-9.

$$y = Ax \qquad 24.2\text{-}9$$

For instance, there may be an auxilliary condition expressed by the following equation which must be satisfied exactly:

$$Cx - z = 0 \qquad 24.3\text{-}32$$

The new squared norm, replacing 24.3-2 becomes the following:

$$E^2 = e^T e + \gamma^T(Cx - z) + (Cx - z)^T \gamma \qquad 24.3\text{-}33$$

The quadratic form of the equation is summed to the auxilliary equation which is multiplied by a Lagrange multiplier γ. The result is minimized by taking a derivative with respect to x^T and setting it equal to zero.

$$A^T A x - A^T y + C^T \gamma = 0 \qquad 24.3\text{-}34$$

This set of simultaneous equations in 24.3-32 and 24.3-34 may be used to eliminate the Lagrange multiplier and solved for x as in the previous case.

24.4 Inverse of the underdetermined case

When $n < m$ there are not enough equations to solve for all the unknown parameters of the model. It is necessary to add some new conditions or constraints to allow us to place some limits on the solution. It is also usual to restrict the number of iterations so that the initial guess should be close to the final answer. The problem is treated in the same manner as the solution with the Lagrange multiplier in equations 24.2-9 and 24.3-34 of the last section. However the role of **C** and **A** are interchanged. The auxilliary condition to be satisfied exactly is equation 24.2-9.

$$y = Ax \qquad 24.2\text{-}9$$

The principal equation consists of the difference in the initial and the computed model parameters

$$Ix = 0 \qquad 24.4\text{-}1$$

in which **I** is the identity matrix. The least squares solution must minimize the norm of the differences in model parameters.

Inverse of the underdetermined case

$$\mathbf{e}^T\mathbf{e} = \sum_{k=1}^{m} x_k^2 \qquad 24.4\text{-}2$$

The new squared norm replacing equation 24.4-33 is written with α's being the Lagrange multipliers.

$$E^2 = (\mathbf{Ix})^T \mathbf{Ix} + \alpha^T(\mathbf{Ax} - \mathbf{y}) + (\mathbf{Ax} - \mathbf{y})^T \alpha \qquad 24.4\text{-}3$$
$$E^2 = \mathbf{x}^T\mathbf{x} + \alpha^T(\mathbf{Ax} - \mathbf{y}) + (\mathbf{x}^T\mathbf{A}^T - \mathbf{y}^T) \alpha \qquad 24.4\text{-}4$$

Differentiate with respect to $\mathbf{x}^T$ and set the result equal to zero to obtain a least squares solution.

$$\mathbf{x} + \mathbf{A}^T\alpha = 0 \qquad 24.4\text{-}5$$

The equation is multiplied by $\mathbf{A}$.

$$\mathbf{Ax} + \mathbf{AA}^T\alpha = 0 \qquad 24.4\text{-}6$$

Solve for the Lagrange multiplier, and replace $\mathbf{Ax}$ by $\mathbf{y}$ from 24.2-9.

$$\alpha = -(\mathbf{AA}^T)^{-1} \mathbf{Ax} = -(\mathbf{AA}^T)^{-1} \mathbf{y} \qquad 24.4\text{-}7$$

Substitute the result into 24.4-5 to obtain a solution we will call $\mathbf{x}^c$.

$$\mathbf{x}^c = \mathbf{A}^T (\mathbf{AA}^T)^{-1} \mathbf{y} \qquad 24.4\text{-}8$$

This equation for the solution of an underdetermined case was given by Smith and Franklin (1969). The matrix $\mathbf{AA}^T$ must be non-singular for a solution.

The generalized eigenvector analysis may also be carried out in this case. The result with $\mathbf{A} = \mathbf{ULV}^T$ in 24.3-8 is substituted in 24.4-8 and it is assumed that the inverse exists.

$$\mathbf{x}^c = (\mathbf{ULV}^T)^T [\mathbf{ULV}^T (\mathbf{ULV}^T)^T]^{-1} \mathbf{y} \qquad 24.4\text{-}9$$
$$\mathbf{x}_c = \mathbf{VLU}^T [\mathbf{ULV}^T\mathbf{VLU}^T]^{-1} \mathbf{y} = \mathbf{VLU}^T [\mathbf{UL}^2\mathbf{U}^T]^{-1} \mathbf{y}$$

The inverse of $\mathbf{UL}^2\mathbf{U}^T$ is $\mathbf{UL}^{-2}\mathbf{U}^T$.

$$\mathbf{x}^c = \mathbf{VLU}^T\mathbf{UL}^{-2}\mathbf{U}^T \mathbf{y}$$
$$\mathbf{x}^c = \mathbf{VL}^{-1}\mathbf{U}^T \mathbf{y} \qquad 24.4\text{-}10$$

Let the Lanczos generalized inverse operator be $\mathbf{H}$.

$$H = VL^{-1}U^T \qquad 24.4\text{-}11$$

The least squares solution for the underdetermined case with a minimum norm for **x** is then as follows:

$$x^c = Hy \qquad 24.4\text{-}12$$

24.5 Conclusions

The linear inverse method is a powerful technique for obtaining a solution to a physical problem and to evaluate the final model. Its greatest disadvantage is that a reasonable first guess is required to linearize the problem (Jackson, 1973; Sabatier, 1974; Anderssen, 1975; and Thornton, 1979). The iterative computations do not always converge. Much of the research effort to the present time has been devoted toward the formulation of a realistic function which relates observations and physical models and not to the subsequent method of iterating to an acceptable model. Indeed there are a number of geophysical problems in which the inversion solution can be obtained in a closed form without any iteration (Loewenthal et al., 1978). Given a limited number of measurements the inverse method of determining the subsurface properties of the earth is non-unique in seismological, electromagnetic wave and in all potential field problems.

Several non-linear methods of inversion are being investigated at the present time. The inverse Sturm Liouville theory of Gel'fand and Levitan (1955) using the scattering matrix Kay (1960) may be applied to electromagnetic problems. Other approaches involving integral equations have been attempted by Bailey (1970), Johnson and Smylie (1970), Weidelt (1972) and Kalaba and Zagustin (1974).

The linear inverse method has been applied to a number of interesting problems in geophysics. Some of these are listed in the references and, to conserve space, they will not be discussed here. It is probable that in the future most geophysical problems will be solved through a use of some form of a generalized linear inverse method. The interpretation of solutions made from a limited number of inaccurate observations requires us to quantify the degree of non-uniqueness in the data set and establish the resolving power along with the solution of some possible models. Of greater importance at the present time is the adequecy of the assumptions in the model. Simplifications such as the assumption of horizontal stratification have a profound effect on the final models in both seismology and the magnetotelluric method. The question of a sufficient condition for existence and the difficulty of proving the uniqueness is discussed at length by Parker (1977).

References

References

Anderssen,R.S.(1975), Inversion of global electromagnetic induction data. *Physics of the Earth and Planetary Interiors* **10**, 292-298.

Backus,G.E. and Gilbert,J.F.(1967), Numerical application of a formalism for geophysical inverse problems. *Geophysical Journal* **13**, 247-276.

Backus,G.E. and Gilbert,J.F.(1968), The resolving power of gross earth data. *Geophysical Journal* **16**, 169-205.

Backus,G.E. and Gilbert,J.F.(1970), Uniqueness in the inversion of inaccurate gross earth data. *Philosophical Transactions of the Royal Society of London* **266**, 123-192.

Bailey,R.C.(1970), Inversion of the geomagnetic induction problem. *Proceedings of the Royal Society of London* **A315**, 185-194.

Berryman,J.G. and Greene,R.R.(1980), Discrete inverse methods for elastic waves in layered media. *Geophysics* **45**, 213-233.

Braile,L.W.(1973), Inversion of crustal seismic refraction and reflection data. *Journal of Geophysical Research* **78**, 7738-7744.

Braile,L.W., Keller,G.R. and Peeples,W.J.(1974), Inversion of gravity data for two dimensional density distributions. *Journal of Geophysical Research* **79**, 2017-2021.

Cribb,J.(1976), Application of the generalized linear inverse to the inversion of static potential data. *Geophysics* **41**, 1365-1369.

Cuer,M. and Bayer,R.(1980), Fortran routines for linear inverse problems. *Geophysics* **45**, 1706-1719.

Der,Z.A., Masse,R. and Landisman,M.(1970), Resolution of surface waves at intermediate distances. *Journal of Geophysical Research* **75**, 3399-3409.

Der,Z.A. and Landisman,M.(1972), Theory of errors, resolution and separation of unknown variables in inverse problems. *Gelphysical Journal* **27**, 137-178.

Dorman,L.M.,Jacobson,R.S. and Shor,G.G.(1981),Linear inversion of body wave data.Part 1:Velocity structure from travel-times and ranges; Part 2:Attenuation. *Geophysics* **46**, 138-162.

Gel'fand,I.M. and Levitan,B.M.(1955), On the determination of a differential equation by its spectral function. *American Mathematical Society Translation Ser 2* **1**, 253-305; *Izvestia Akademia Nauk SSSR* **77**, 557-560, (1951).

Gilbert,J.F.(1971), Ranking and winnowing gross earth data for inversion and resolution. *Geophysical Journal* **23**, 125-128.

Glenn,W.E., Ward,S.H., Peeples,W.J., Ryu,J. and Phillips,R.J.(1973), The inversion of vertical magnetic dipole sounding data. *Geophysics* **38**, 1109--1129.

Green,W.R.(1975), Inversion of gravity profiles by use of a Backus-Gilbert approach. *Geophysics*, 763-772.

Jackson,D.D.(1972), Interpretation of inaccurate, insufficient and inconsistent data. *Geophysical Journal* **28**, 97-109.

Jackson,D.D.(1973), Marginal solutions to quasi-linear inverse problems in geophysics. Edgehog method. *Geophysical Journal* **35**, 121-136.

Johnson,I.M. and Smylie,D.E.(1970), An inverse theory of the electrical conductivity of the lower mantle. *Geophysical Journal* **22**, 41-53.

Jordan,T.H. and Franklin,J.N.(1971), Optimal solutions to a linear inverse problem in geophysics. *Proceedings of the National Academy of Sciences* **68**, 291-293.

Kalaba,R. and Zagustin,E.(1974), An initial value method for an inverse problem in wave propagation. *Journal of Mathematical Physics* **15**, 289-290.

Kay,I.(1960), The inverse scattering problem when the reflection coefficient is a rational function. *Communications on Pure and Applied Mathematics* **13**, 371.

Lanczos,C.(1961), *Linear differential operators*. D. Van Nostrand Co., London, 140.

Levy,S. and Clowes,R.M.(1980), Debubbling: A generalized linear inverse approach. *Geophysical Prospecting* **28**, 840-858

Loewenthal,D., Gutowski,P.R. and Treitel,S.(1978), Direct inversion of the transmission synthtic seismogram. *Geophysics* **43**, 886-898.

Moore,E.H.(1920), (Untitled). *Bulletin American Mathematical Society* **26**, 394-395.

Nolet,G.(1978), Simultaneous inversion of seismic data. *Geophysical Journal* **55**, 679-691.

Oldenburg,D.W.(1976), Calculations of Fourier transforms by the Backus-Gilbert method. *Geophysical Journal* **44**, 413-431.

Oldenburg,D.W.(1979), One-dimensional inversion of natural source magnetotelluric observations. *Geophysics* **44**, 1218-1244.

Parker,R.L.(1970), The inverse problem of electrical conductivity in the mantle. *Geophysical Journal* **22**, 121-138.

Parker,R.L.(1977), Understanding inverse theory. *Annual Reviews of Earth and Planetary Sciences* **5**, 35-64.

Penrose,R.(1955), A generalized inverse for matrices. *Proceedings of Cambridge Philosophical Society* **51**, 406-413, **52**, 17-19.

Sabatier,P.C.(1974), Remarks on approximate methods in geophysical inverse problems. *Proceedings of the Royal Society London* **A337**, 49-71.

Safon,C., Vasseur,G. and Cuer,M.(1977) Some applications of linear programming to the inverse gravity problem. *Geophysics* **42**, 1215-1229.

Smith,M.L. and Franklin,J.N.(1969), Geophysical application of generalized inverse theory. *Journal of Geophysical Research* **74**, 2783-2785.

Thornton,B.S.(1979), Inversion of the geophysical inverse problem for n layers with non-uniqueness reduced to n cases. *Geophysics* **44**, 801-819.

Voith,R.P., Vogt,W.G. and Mickle,MIH.(1972), On the computation of the generalized inverse by classical minimization. *Computing* **9**, 175-187.

Vozoff,K. and Jupp,D.L.B.(1975), Joint inversion of geophysical data. *Geophysical Journal* **42**, 977-991.

Weidelt,P.(1972), The inverse problem of geomagnetic induction. *Zeitschrift fur Geophysik* **38**, 257-289.

Wiggins,R.A.(1972), The general linear inverse problem. Implications of surface waves and free oscillations for earth structure. *Reviews of Geophysics and Space Physics* **10**, 251-285

Wiggins,R.A., Larner,K.L. and Wisecup,R.D.(1976), Residual statics analysis as a general linear inverse problem. *Geophysics* **41**, 922-938.

Appendix 1

1.1 Fourier series and integrals

It is possible to represent any function f(t) in terms of a set of functions, $y_n(t)$, which are orthogonal in a given interval (0,T) with respect to the weighting function, r(t), in equation 3.1-2 as

$$f(t) = \sum_{n=0}^{\infty} b_n y_n(t) \qquad 1.1$$

Let us assume such an expansion exists and represent f(t) by a sine function in which case the equation is known as the Fourier sine series. Multiply both sides by $y_k = \sin k\pi t/T$ where y_k is the k-th function in the set.

$$f(t) \sin \frac{k\pi t}{T} = \sum_{n=0}^{\infty} b_n \sin \frac{n\pi t}{T} \sin \frac{k\pi t}{T} \qquad 1.2$$

Integrating both sides over the interval (0,T), we can interchange the summation and integration if the series is uniformly convergent in the interval.

$$\int_0^T f(t) \sin \frac{k\pi t}{T} dt = \sum_{n=0}^{\infty} b_n \int_0^T \sin \frac{n\pi t}{T} \sin \frac{k\pi t}{T} dt \qquad 1.3$$

Because the functions are orthogonal, all the terms on the right are zero except when $n = k$ in which case the integral is equal to $T/2$. Therefore

$$b_n = \frac{2}{T} \int_0^T f(t) \sin \frac{n\pi t}{T} dt \qquad 1.4$$

If f(t) is an odd function of t, the series

$$f(t) = \sum_{n=0}^{\infty} b_n \sin \frac{n\pi t}{T} \qquad 1.5$$

turns out to represent the function in the larger interval (-T,T). For example, take the function $f(t) = t$. Substituting into equation (1.4), the Fourier coefficients are found to be

$$b_n = (-1)^{n+1} \cdot 2/n \qquad 1.6$$

The Fourier sine series is

$$f(t) = 2 \sin t - \sin 2t + \frac{2}{3} \sin 3t - \frac{1}{2} \sin 4t + \ldots \qquad 1.7$$

Figure A1.1(a) is a graph of the function above both inside the interval $\pm T$ equal to π and also outside of this interval.

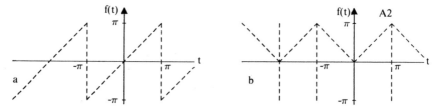

Figure A1.1 $f(t) = t, 0 \leqslant t \leqslant \pi$ as (a) an odd function and (b) an even function.

Similarly if f(t) is an even function it is possible to construct a Fourier cosine series which is valid in the interval (-T,T) (see figure A1.1b).

$$f(t) = a_0 + \sum_{n=1}^{\infty} a_n \cos \frac{n\pi t}{T} \qquad 1.8$$

$$a_n = 2T^{-1} \int_0^T f(t) \cos \frac{n\pi t}{T} dt \qquad n=1,2,\ldots \qquad 1.9$$

$$a_0 = T^{-1} \int_0^T f(t) \, dt \qquad 1.10$$

The coefficient a_0 is the average value or DC level of the function f(t). The coefficients a_n and b_n have the factor 2 in them, indicating that they are twice the average value of the product of f(t) and $\cos n\pi t/T$ in the interval 0 to

Fourier series and integrals

T. An alternate and perhaps better way of viewing this is to consider that the angular frequencies

$$\omega_n = 2n\pi/T \qquad 1.11$$

may take on both positive and negative values which are physically indistinguishable. Half the power is in the negative frequencies and half is in the positive ones. This interpretation will become apparent when we consider the complex Fourier series.

The cosine series of $f(t) = t$ is

$$f(t) = \frac{\pi}{2} - \frac{4}{\pi} \cos t - \frac{4}{9\pi} \cos 3t - \frac{4}{25\pi} \cos -5t - \ldots \qquad 1.12$$

Any function may be expressed as the sum of an even and odd function

$$f(t) = f_e(t) + f_o(t) \qquad 1.13$$

since $f(t)$ may be written

$$f(t) = [f(t) + f(-t)]/2 + [f(t) - f(-t)]/2 \qquad 1.14$$

For example, the unit step function may be formed from a sum of two other functions as shown in figure A1.2.

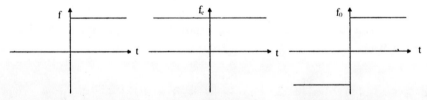

UNIT STEP FUNCTION = EVEN FUNCTION + ODD FUNCTION

Figure A1.2

For any specific frequency the sum of a sine and cosine term can always be reduced to a single cosine (or sine) with a particular amplitude and phase. This is illustrated in figure A1.3 and by the following trigonometric identity.

$$y = A \cos(t + \phi) = (A \cos \phi) \cos t - (A \sin \phi) \sin t \qquad 1.15$$

or
$$y = A\cos(t+\phi) = A'\cos t + B'\sin t \qquad 1.16$$

As a specific example, let $A = 1$.

$$\phi = 45°$$
$$y = \cos(t+\phi) = \frac{1}{(2)^{1/2}}\cos t - \frac{1}{(2)^{1/2}}\sin t \qquad 1.17$$

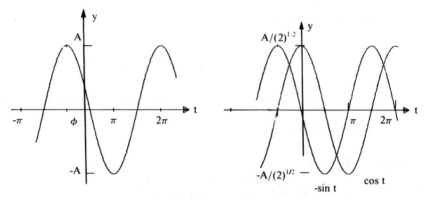

Figure A1.3 As long as the frequency is the same, the sum of any number of sines and cosines can be reduced to a single cosine with a particular amplitude and phase.

A proof is now given that the Fourier expansion in trigonometric functions is equivalent to one using exponentials. Equation 3.1-6 may be expanded as

$$f(t) = a_0 + \sum_{n=1}^{\infty}\left[a_n\left(\frac{e^{i\omega_n t} + e^{-i\omega_n t}}{2}\right) - ib_n\left(\frac{e^{i\omega_n t} - e^{-i\omega_n t}}{2}\right)\right]$$

or

$$f(t) = a_0 + 1/2\sum_{n=1}^{\infty}\left[(a_n - ib_n)e^{i\omega_n t} + (a_n + ib_n)e^{-i\omega_n t}\right] \qquad 1.18$$

Using 3.1-7 and 3.1-8

$$a_n - ib_n = 2T^{-1}\int_{-T/2}^{T/2} f(t)(\cos\omega_n t - i\sin\omega_n t)\,dt$$

Fourier series and integrals

$$a_n + ib_n = 2T^{-1} \int_{-T/2}^{T/2} f(t)(\cos \omega_n t + i \sin \omega_n t)\, dt$$

or

$$a_n - ib_n = 2T^{-1} \int_{-T/2}^{T/2} f(t)\, e^{-i\omega_n t}\, dt \qquad 1.19$$

$$a_n + ib_n = 2T^{-1} \int_{-T/2}^{T/2} f(t)\, e^{i\omega_n t}\, dt \qquad 1.20$$

Since $\omega = \pi n/T$, we see that the two equations above differ only by the sign of n. Therefore we may let

$$F_n = (a_n + ib_n) = (a_n + ib_n)_{n=-n} \qquad 1.21$$

and

$$F_n = (a_n^2 + b_n^2)^{1/2}\, e^{i \tan^{-1}(-b_n/a_n)} \qquad 1.22$$

If we allow n to be negative, equation (1.20) implies

$$F(\omega) = F^*(\omega)$$

where F^* is the complex conjugate of F. Then 1.19 and 1.20 can be expressed by 3.1-12.

$$F_n = T^{-1} \int_{-T/2}^{T/2} f(t)\, e^{-i\omega_n t}\, dt \qquad 1.23$$

Thus by artificially introducing negative frequencies it is possible to simplify the equations. The amplitude of the sinusoidal oscillations e^A (where $A = i\omega t$) must be complex if the signal f(t) is to be real. The signal, f(t), from 1.18 becomes

$$f(t) = \sum_{n=-\infty}^{n=+\infty} F_n e^{i\omega_n t} \qquad 1.24$$

which is identical to 3.1-11.

Suppose we generate an infinite series of square pulses of width L (see figure A1.4).

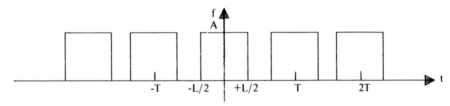

Figure A1.4   A function consisting of a series of narrow square waves.

In equation 3.1-12 we can let $T/2 = L/2$ since the function is zero elsewhere in the interval.

$$F_n = T^{-1} \int_{-L/2}^{L/2} A\, e^{-i\omega_n t}\, dt = \frac{1}{T} \frac{-A}{i\omega_n} \left[e^{-i\omega_n t} \right]_{-L/2}^{L/2}$$

$$= \frac{1}{T} \frac{2A}{\omega_n} \frac{e^{i\omega_n L/2} - e^{-i\omega_n L/2}}{2i} = \frac{1}{T} \frac{2A}{\omega_n} \sin \frac{\omega_n L}{2}$$

$$F_n = \frac{AL}{T} \left[\frac{\sin(\omega_n L/2)}{(\omega_n L)/2} \right]$$
$$\phi_n = 0 \qquad\qquad 1.25$$

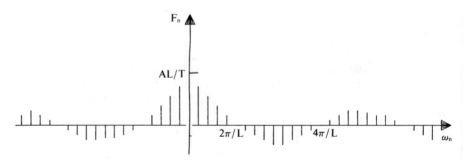

Figure A1.5   The Fourier amplitude spectrum of a series of square waves.

Fourier series and integrals

The spacing between the spectral lines is

$$\Delta\omega = 2\pi/T \qquad 1.26$$

Taking a specific case where $L = \pi$, $T = 2\pi$ and $A = 1$, as in figure A1.6

$$F_n = \frac{1}{\pi n} \sin \frac{n\pi}{2} \qquad n \neq 0$$

$$F_0 = 1/2$$

Substituting in 3.1-11

$$f(t) = \sum_{n=-\infty}^{n=+\infty} F_n e^{i\omega_n t} = \sum_{n=-\infty}^{n=+\infty} F_n [\cos nt + i \sin nt]$$

We find that the term i sin nt drops out when plus and minus n are considered. This is what is expected since the function f(t) is an *even* function and so it can be expressed in terms of cosines only. In this case

$$f(t) = \frac{1}{2} - \frac{2}{\pi} \sum_{1}^{\infty} (-1)^n \frac{\cos(2n-1)t}{(2n-1)} \qquad 1.27$$

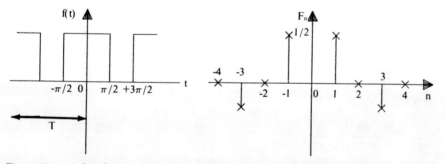

Figure A1.6 A function consisting of a series of wide square waves and their Fourier amplitude spectrum.

The Fourier cosine series is

$$f(t) = \frac{1}{2} + \frac{2}{\pi} \cos t - \frac{2}{3\pi} \cos 3t + \frac{2}{5\pi} \cos 5t - \ldots \qquad 1.28$$

The frequencies present are

$$\omega_n = (2n - 1) \qquad n=1,2,... \qquad 1.29$$

The 2 in $2/\pi \cos t$, etc., comes from power in positive and negative frequencies. Comparing with equation 3.1-3 to 3.1-10

$$\begin{aligned} F_n &= a_n \\ b_n &= 0 \\ \phi_n &= 0 \end{aligned} \qquad 1.30$$

Thus there are only cosine terms and the phase shift is zero for even functions in general. Note that both the function and its spectrum are even and real.

The effect of phase shift is illustrated in figure A1.7 if the rectangular pulses are displaced by $t = \pi/2$.

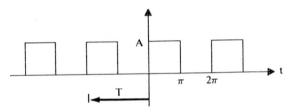

Figure A1.7 Square waves with a phase shift of $\pi/2$.

$$F_n = \frac{A}{n\pi} \cdot \sin \frac{n\pi}{2} \, e^{in\pi/2} \qquad 1.31$$

If $A = 1$

$$f(t) = \frac{1}{2} + \frac{2}{\pi} \cos(t - \pi/2) - \frac{2}{3\pi} \cos(3t - 3\pi/2)$$

$$\cdot + \frac{2}{5\pi} \cos(5t - 5\pi/2) - ....$$

$$= \frac{1}{2} + \frac{2}{\pi} \sin t + \frac{2}{3\pi} \sin 3t + \frac{2}{5\pi} \sin 5t + ... \qquad 1.32$$

which can be compared with 1.28. Notice that the magnitude of the frequencies is not affected by the translation of the function. All that happens is that a phase lag is introduced that is proportional to frequency. For the example given above the spectrum is shown in figure A1.8.

Fourier series and integrals

$$F_0 = 1/2$$
$$F_n = 0 \quad n=\pm 2, \pm 4, \ldots$$
$$F_n = 1/in\pi \quad n=\pm 1, \pm 3, \ldots$$

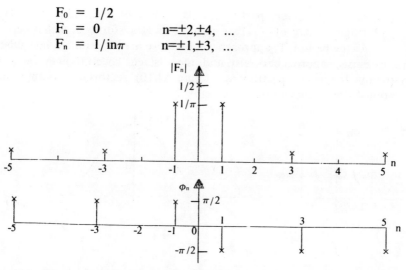

Figure A1.8 Phase and amplitude spectrum of the odd function in figure A1.7.

$$a_n = 0$$
$$b_n = 1/n\pi$$
$$\phi_n = \tan^{-1}(-b_n/a_n)$$
$$\phi_n = -\pi/2 \quad n = +1, +3, \ldots$$
$$\phi_n = +\pi/2 \quad n = -1, -3, \ldots$$

If we use the cosine series only and restrict ourselves to positive frequencies the amplitude spectrum is shown in figure A1.9.

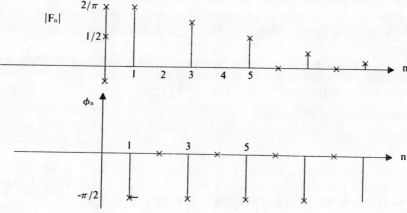

Figure A1.9

If $F(\omega_n)$ had an *imaginary* part to it, it has to be *odd*. The odd part of f(t) and the odd imaginary part of $F(\omega_n)$ are negative sine transforms of each other.

As the period, T, approaches infinity we are left with a single pulse. In this case $\Delta\omega_n$ approaches zero and the discrete spectral lines become a continuous frequency spectrum (see figure A1.10). As this limit is approached equation 4.10 becomes a Fourier integral.

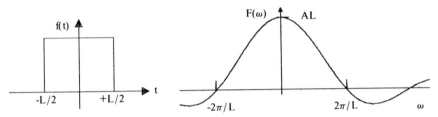

Figure A1.10

The Fourier transform or spectrum of any signal which satisfies 2.1-1 is given by the formula

$$F(\omega) = \int_{-\infty}^{\infty} f(t)\, e^{-i\omega t}\, dt \qquad 1.33$$

The inverse Fourier transform is

$$f(t) = \frac{1}{2\pi} \int_{-\infty}^{\infty} F(\omega)\, e^{i\omega t}\, d\omega \qquad 1.34$$

The transformation above is usually done with complex variables and integration by the method of residues. This is illustrated for the pulse beginning at time t = 0 as shown in figure A1.11.

Example

The pulse will be similar to one produced by closing a switch for a time L.

$$f(t) = A \qquad 0 < t < L$$

$$f(t) = 0 \qquad t < 0 \qquad\qquad t > L$$

The function above is neither odd nor even. However, it can be considered the sum of an even and odd function as seen in figure A1.12. The spectrum will have real and imaginary terms in it. From 1.33

Fourier series and integrals

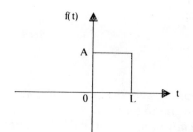

Figure A1.11

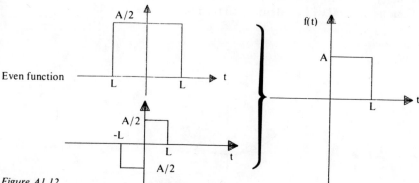

Even function

Figure A1.12

$$F(\omega) = \frac{1}{i\omega} \int_0^L A e^{-i\omega t} dt(-i\omega) = -\frac{A}{i\omega} [e^{-i\omega t}]_0^L$$

$$F(\omega) = \frac{A}{i\omega} (1 - e^{-i\omega t}) \qquad 1.35$$

The inverse transform over the limits $\pm\infty$ is given by 1.34.

$$f(t) = \frac{1}{2\pi} \int \frac{1 - e^{-i\omega t}}{i\omega} e^{i\omega t} d\omega$$

$$= (2\pi i)^{-1} \int [e^{i\omega t} - e^{+i\omega(t-L)}] d\omega/\omega$$

$$(2\pi i)^{-1} \int e^{i\omega t} d\omega/\omega - (2\pi i)^{-1} \int e^{i\omega(t-L)} d\omega/\omega \qquad 1.36$$

This integral is difficult to solve directly so we shall proceed in an indirect fashion by solving another integral in which complex variables are present. Let

Appendix 1

us evaluate the integral of

$$\frac{1}{2\pi i} \frac{e^{izt}}{z}$$

around the closed contour, c, in the lower half of the z plane. We shall see that restricting the line integral to the *lower* half of the z plane allows us to evaluate the transform for *negative* time. Later in the derivation we will let $z = \omega + iy$ and note that $y = 0$ in the interval -R to R. As R approaches infinity this part of our continuous integral will be identical to the final integral in 1.36 above.

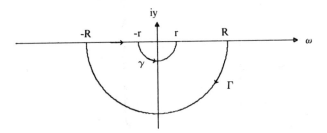

Figure A1.13

$$(2\pi i)^{-1} \int_c \frac{e^{izt}}{z} dz = 0 \qquad 1.37$$

Each pole of any function of a complex variable has a number called a residue associated with it. The number is a particular coefficient in the series expansion of the function near the pole. Thus according to the calculus of residues the integral is zero because there are no singularities of the integrand inside the contour.

The line integral may be broken down into its component sections around the contour c.

$$(2\pi i)^{-1} \Big[\int_{-R}^{-r} \exp(i\omega t) \, d\omega/\omega + \int_\gamma \exp(izt) \, dz/z$$

$$+ \int_r^R \exp(i\omega t) \, d\omega/\omega + \int_\Gamma e^{izt} \, dz/z \Big] = 0 \qquad 1.38$$

Fourier series and integrals

or

$$I_1 + I_2 + I_3 + I_4 = 0$$

On the semi-circle, γ, we let $z = re^A$ (where $A = i\Theta$) as shown in figure A1.14.

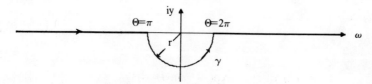

Figure A1.14

$$z = \omega + iy = r \cdot \exp(i\Theta)$$
$$dz = ir \cdot \exp(i\Theta) \cdot d\Theta$$

$$\exp(izt) \cdot dz/z = \exp[irt \cdot \exp(i\Theta)] \cdot d\Theta / [r \exp(i\Theta)]$$

$$I_2 = (2\pi i)^{-1} \int_\gamma \exp(izt) \, dz/z = (2\pi)^{-1} \int_\pi^{2\pi} \exp[irt \cdot \exp(i\Theta)] \cdot i d\Theta$$

As r approaches zero $\exp[irt \cdot \exp(i\Theta)]$ approaches one.

$$\lim_{r \to 0} (2\pi i)^{-1} \int_\gamma \exp(izt) \, dz/z = i(2\pi i)^{-1} \lim_{r \to 0} \int_\pi^{2\pi} d\Theta = 1/2 \qquad 1.39$$

Next we evaluate I_4 which is the integral along the large semi-circle, Γ, as R approaches infinity.

$$I_4 = (2\pi i)^{-1} \int_\Gamma \exp(izt) \cdot dz/z = (2\pi)^{-1} \int_0^\pi \exp[iRt \cdot \exp(i\Theta] \cdot d\Theta$$

$$I_4 = (2\pi)^{-1} \int_0^\pi \exp[iRt(\cos\Theta + i \sin\Theta)] \, d\Theta$$

As $R \to \infty$ the portion e^A (where $A = iRt \cos\Theta$) oscillates very rapidly and the positive and negative contributions cancel.

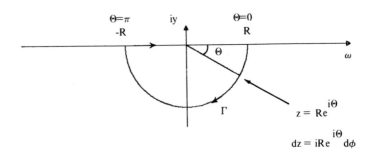

Figure A1.15

$$I_4 = (2\pi)^{-1} \int_0^\pi \exp(-Rt \sin\Theta) \, d\Theta$$

Sin Θ is negative between 0 and $-\pi$ so this integral is finite as R increases without limit if the time is always negative. This is the reason for the restriction on the time in the lower z plane. We can integrate from 0 to $\pi/2$ and multiply by 2 if we use the absolute value of t.

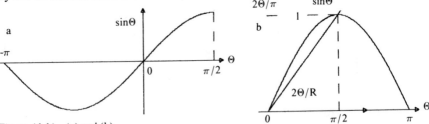

Figure A1.16 (a) and (b)

The maximum value of $\sin\Theta$ is 1 so $\sin\Theta$ can be replaced by $2\Theta/\pi$ which has this maximum when $\Theta = \pi/2$. In fact, for the range $0 < \Theta < \pi/2$, $\sin\Theta > 2\Theta/\pi$. Therefore

$$I_4 < \pi^{-1} \int_0^{\pi/2} \exp[-2R|t|\Theta/\pi] \cdot d\Theta$$

$$I_4 < [\pi \cdot [-2R|t|/\pi]^{-1} \, [\exp-2R|t|\Theta/\pi]_\Theta^{\pi/2}$$

$$I_4 < [1 - \exp(-R|t|)]/(2R|t|) \to 0 \quad \text{as } R \to \infty$$
$$t < 0 \qquad 1.40$$

This conclusion is known as Jordan's Lemma (Hildebrand, 1948, p. 529).

Fourier series and integrals

Substituting 1.39 and 1.40 into 1.38

$$\lim_{\substack{r \to 0 \\ R \to \infty}} (2\pi i)^{-1} \left[\int_{-R}^{-r} \exp(i\omega t) \, d\omega/\omega + \int_{r}^{R} \exp(i\omega t) \, d\omega/\omega \right] = -1/2$$

Letting **P** stand for the principal part of the integral

$$(2\pi i)^{-1} \, \mathbf{P} \int_{-\infty}^{\infty} \exp(i\omega t) \, d\omega/\omega = -1/2 \quad \text{for } t < 0 \qquad 1.41$$

For positive time the contour is drawn principally in the upper half of the z plane.

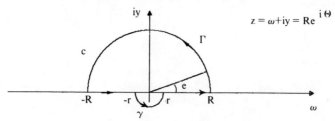

Figure A1.17

The integrand has a pole of order 1 at $z = 0$. Its residue is

$$\text{Res}(0) = [z \cdot \exp(izt)/z]_{z=0} = 1$$

Therefore the value of the contour integral is

$$I = (2\pi i)^{-1} \int_c \exp(izt) \, dz/z = 1$$

The individual line integrals are

$$I = \left[\int_\Gamma \exp(izt) \, dz/z + \int_{-R}^{-r} \exp(i\omega t) \, d\omega/\omega \right.$$

$$\left. + \int_\gamma \exp(izt) \, dz/z + \int_r^R \exp(i\omega t) \, d\omega/\omega \right] (2\pi i)^{-1} \qquad 1.42$$

or

$$I = I_1 + I_2 + I_3 \, I_4$$

Appendix 1

The integral around the large semi-circle vanishes as before when $R \to \infty$ and the time is positive. Similarly on the small semi-circle

$$I_3 = \int_\gamma \exp(izt) \, dz/z = 1/2 \quad \text{(see 1.39)}$$

Substituting in 1.42

$$\lim_{\substack{r \to 0 \\ R \to \infty}} (2\pi i)^{-1} \cdot \left[\int_{-R}^{-r} \exp(i\omega t) \, d\omega/\omega + \int_{r}^{R} \exp(i\omega t) \, d\omega/\omega \right] = 1 - 1/2$$

The principal value is

$$(2\pi i)^{-1} \, P \int_{-\infty}^{\infty} \exp(i\omega(t-L)) \, d\omega/\omega = \begin{array}{l} +1/2 \quad t < L \\ \pm 1/2 \quad t > L \end{array} \qquad 1.43$$

Figure A1.18 illustrates the addition of the two integrals in 1.36.
$$I = (2\pi i)^{-1} \int \exp(i\omega t) \, d\omega/\omega$$

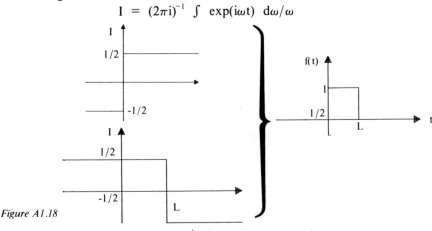

$$I = -(2\pi i)^{-1} \int \exp(i\omega(t-L)) \, d\omega/\omega$$

Figure A1.18

Therefore

$$(2\pi i)^{-1} \, P \int_{-\infty}^{\infty} [\exp(i\omega t) - \exp(i\omega(t-L))] \, d\omega/\omega = \begin{array}{ll} 0 & t < 0 \\ 1 & 0 < t < L \\ 0 & t > L \end{array} \qquad 1.44$$

The integrand has no singularities in the range of integration so the principal value of the integral is equal to the ordinary Riemann integral.

References

Copson,E.T.(1935), *An introduction to the theory of functions of complex variable*. Oxford University Press, London.

Hildebrand,F.B.(1948), *Advanced calculus for engineers*. Prentice Hall, New York.

Appendix 2

2.1 The Dirac delta function

A useful *function* or measure is the Dirac delta function which is defined as

$$\delta(t) = 0 \quad t \neq 0$$

$$\int_{-\infty}^{\infty} \delta(t) \, dt = 1$$

2.1

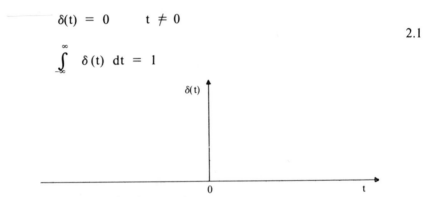

Figure A2.1 A delta Dirac function at $t = 0$.

An alternative definition is one in the form of a convolution integral, where $f(t)$ is any arbitrary function.

$$f(t) = \int_{-\infty}^{\infty} f(L) \, \delta(t - L) \, dL \qquad 2.2$$

The *function* $\delta(t - L)$ is zero everywhere except at one point. It is an improper function because it is not continuous or differentiable at $t = L$. The delta function was introduced as the derivative of a unit step $[d/dt \, U(t - L)]$ by Heaviside (1925) and reintroduced by Dirac (1935) in its presently used form.

The shape of the delta function before reducing its width to zero in the limit is arbitrary. For instance, it is often thought of a Gaussian-shaped function as $\sigma \to 0$.

$$G(t) = \lim_{\sigma \to 0} (2\pi\sigma^2)^{-1/2} \exp(-t^2/2\sigma^2) \qquad 2.3$$

The Dirac delta function

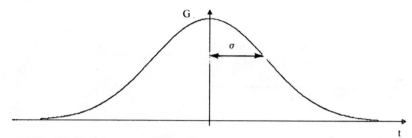

Figure A2.2   The Gaussian or normal function.

The Fourier transform of a delta function $\delta(t_0 - t)$ is given directly, letting $L = t$ in the Fourier transform below and comparing to equation 2.2.

$$F(\omega) = \int_{-\infty}^{\infty} \delta(t_0 - t)\, e^{-i\omega t}\, dt$$

$$= \int_{-\infty}^{\infty} e^{-i\omega L}\, \delta(t_0 - L)\, dL = e^{-i\omega t_0} \qquad 2.4$$

or

$$F(\omega) = \cos \omega t_0 - i \sin \omega t_0 \qquad 2.5$$

The amplitude spectrum is

$$|F(\omega)| = (\cos^2 \omega t_0 + \sin^2 \omega t_0)^{1/2} = 1 \qquad 2.6$$

The phase spectrum is

$$\phi(\omega) = \tan^{-1} \frac{\sin \omega t_0}{\cos \omega t_0} = \tan^{-1}(\tan \omega t_0)$$

$$\phi(\omega) = \omega t_0 \qquad 2.7$$

It is seen that the Dirac function may be viewed as a superposition of sine waves with all frequencies represented and with an amplitude of one. The waves are in phase at time t_0 only. At any other time the waves interfere destructively to give zero output. Note that if $t_0 = 0$

$$F(\omega) = 1 \qquad 2.8$$

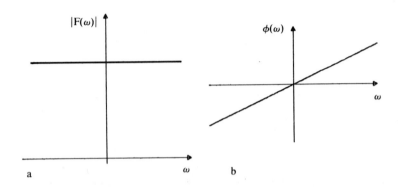

Figure A2.3 (a) Amplitude spectrum of a Dirac delta function. (b) Phase spectrum of a Dirac delta function. The slope of the line is equal to t_0.

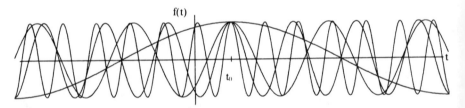

Figure A2.4 Superposition of sine waves to make a Dirac delta function $\delta(t - t_0)$.

Another definition of the Dirac delta function is given by the inverse Fourier transform (equation 1.34, Appendix 1).

$$\delta(t_0 - t) = (2\pi)^{-1} \int_{-\infty}^{\infty} e^{i\omega(t_0-t)} \, d\omega \qquad 2.9$$

References

Dirac,P.A.M.(1935), *The Principles of Quantum Mechanics*, 2nd ed., p. 71, Oxford University Press, New York; 3rd ed., p. 58, 1947.

Heaviside,O.(1925), *Electrical Papers*, Copley Publishers, Boston.

Appendix 3

3.1 The Wiener-Hopf equation

It is often necessary to find a function which is a maximum or a minimum for a given set of boundary conditions. An example is the brachistochrone (shortest time) problem posed by John Bernoulli in 1696. The quest is to determine the shape of the slide so that a body under gravitational attraction but no friction takes the least time to travel from $A(x_1,y_1)$ to $B(x_2,y_2)$ (figure A3.1).

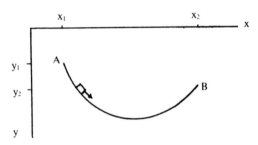

Figure A3.1 A brachistochrone problem which requires that we find the minimum time for a body to travel from $A(x_1,y_1)$ to $B(x_2,y_2)$.

The problem can be stated for a general case by requiring that we minimize the functional

$$J = \int_{x_1}^{x_2} L(x,y,\partial y/\partial x) \, dx \qquad 3.1$$

subject to the boundary conditions at y_1 and y_2.

$$y(x_1) = y_1$$
$$y(x_2) = y_2 \qquad 3.2$$

Consider a change in the shape of the curve $y(x)$ so that J is a maximum or a minimum as desired. The curve then becomes

The Wiener-Hopf equation

$$y_{p\eta} = y(x) + p\eta(x) \qquad 3.3$$

p   is a real parameter independent of x;
η is a differentiable function with $\eta(x_1) = \eta(x_2) = 0$.

By varying p (figure A3.2) it is possible to form a whole family of curves about y(x).

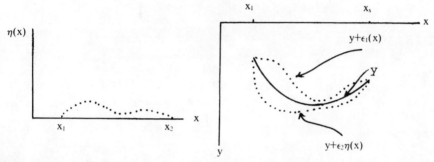

Figure A3.2 A possible $\eta(x)$ and the variation on the curve y(x) to generate the family of curves designated by y(p).

Let the variation of y be called δy.

$$\delta y = p\eta(x) \qquad 3.4$$

When y is changed by δy, the variation in L, the Lagrangian density, is

$$J + \delta J = \int_{x_1}^{x_2} L(x, y + p\eta, y' + p\eta') \, dx \qquad 3.5$$

where the prime indicates differentiation with respect to x(n' = dn/dx). It is necessary to find a value p such that $J + \delta J$ is a minimum. That is, we demand that

$$\frac{d}{dp}(J + \delta J) = 0 \qquad 3.6$$

Let us suppose that y(x) is the actual curve that gives an extremum. Then

$$\delta y = p\eta(x) = 0$$

regardless of the choice of $\eta(x)$. Then p = 0 for equation 3.6

$$\frac{d}{dp}(J + \delta J) = 0 \quad (p=0) \qquad 3.7$$

for all η. This is the *necessary* condition.

Let
$$\left.\begin{array}{l} L_p = L(x, y + p\eta, y' + p\eta') \\ y_p = y + p\eta \\ y_p' = y' + p\eta' \end{array}\right\} \qquad 3.8$$

Then 3.5 becomes

$$J + \delta J = \int_{x_1}^{x_2} L_p(x, y_p, y_p') \, dx$$

Differentiating

$$\frac{d}{dp}(J + \delta J) = \int_{x_1}^{x_2} \left[\frac{\partial L_p}{\partial x} \frac{dx}{dp} + \frac{\partial L_p}{\partial y_p} \frac{\partial y_p}{\partial p} + \frac{\partial L_p}{\partial y_p'} \frac{dy_p'}{dp}\right] dx$$

$$= \int_{x_1}^{x_2} \left[\frac{\partial L_p}{\partial y_p} \eta + \frac{\partial L_p}{\partial y_p} \eta'\right] dx \qquad 3.9$$

When $\epsilon = 0$, equation 3.9 becomes

$$\frac{d}{dp}(J + \delta J)_{p=0} = \int_{x_1}^{x_2} \left[\frac{\partial L}{\partial y} \eta + \frac{\partial L}{\partial y'} \eta'\right] dx \qquad 3.10$$

The last integral can be simplified by differentiating by parts.

$$\int_{x_1}^{x_2} \frac{\partial L}{\partial y'} \eta' dx = \left[\frac{\partial L}{\partial y'} \cdot \eta\right]_a^b - \int_{x_1}^{x_2} \eta \frac{d}{dx}\left(\frac{\partial L}{\partial y'}\right) dx \qquad 3.11$$

The first part is zero by the boundary conditions on η.

The Wiener-Hopf equation

$$\eta(x_1) = \eta(x_2) = 0$$

Using 3.7, equation 3.10 becomes

$$\int_{x_1}^{x_2} [\frac{\partial L}{\partial y} - \frac{d}{dx}(\frac{\partial L}{\partial y'})] \eta \ dx = 0 \qquad 3.12$$

for all possible η. Clearly the coefficient of η in 3.12 must be zero. Therefore

$$\boxed{\frac{\partial L}{\partial y} - \frac{d}{dx}(\frac{\partial L}{\partial y'}) = 0} \qquad 3.13$$

This is Euler's equation or sometimes it is called the Euler-Lagrange equation for an extremum.

In Bernoulli's sliding problem a particle with mass m has its kinetic energy equal to its potential energy by the law of conservation of energy.

$$mv^2/2 = mg\ x$$

The velocity is v and the acceleration of gravity is g. From the definition of velocity and the geometry we can write down velocity squared as

$$v^2 = (dx^2 + dy^2)/dt^2$$

Therefore

$$dt = [(dx^2 + dy^2)/2gx]^{1/2}$$

The left-hand side can be integrated from 0 to time t and the right-hand side can be multiplied and divided by dx and an integral set up between 0 and x_2. This is the functional, J, that is to be minimized and it is found that the integrand L is

$$L = [(1 + (y')^2)/x]^{1/2}$$

The student may verify that substitution into the Euler-Lagrange equation leads, after two integrations, to the following equation for the path:

$$y = (2c_1)^{-1} \cos^{-1}(1 - 2c_1 x) - (x/c_1 - x^2)^{1/2} + c_2$$

The constant $c_2 = 0$ if x_1 and y_1 are at the origin. The constant c_1 is chosen so that $y = y_2$ when $x = x_1$ at point B. The path followed by the particle is a mathematical curve called an inverted cycloid. A similar derivation will be used to determine the Wiener-Hopf equation for an optimum filter.

Let

x be an input signal consisting of a desired wavelet and some noise
d = desired output
y = actual output (see figure A3.3)
W^{-1} = impulse response of an optimum filter.

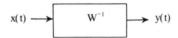

Figure A3.3   Deconvolution with an optimum filter.

The difference between the actual output and the desired output is called the instantaneous error.

$$e(t) = y(t) - d(t) \qquad 3.14$$

Wiener required that the mean square error be as small as possible. The mean square error is defined as

$$E^2 = \lim_{T \to \infty} (2T)^{-1} \int_{-T}^{T} [y - d]^2 \, dt \qquad 3.15$$

The output signal is given by a convolution equation.

$$y(t) = \int_{-\infty}^{\infty} W^{-1}(L) \, x(t-L) \, dL \qquad 3.16$$

Substituting 3.16 into 3.15 gives

$$E^2 = \lim_{T \to \infty} (2T)^{-1} \int_{-T}^{T} [\int_{-\infty}^{\infty} W^{-1}(L) x(t-L) dL - d]^2 \, dt \qquad 3.17$$

Expanding the square and changing the order of integration yields

$$E^2 = \int_{-\infty}^{\infty} W^{-1}(L) dL \int_{-\infty}^{\infty} W^{-1}(\sigma) \, a_x(L-\sigma) d\sigma$$

The Wiener-Hopf equation

$$-2 \int_{-\infty}^{\infty} W^{-1}(L)\, c_{xd}(L)\, dL + a_d(0) \qquad 3.18$$

where the a(L) denotes an autocovariance function with lag L

$$a_x(L-\sigma) = \lim_{T \to \infty} (2T)^{-1} \int_{-T}^{T} x(t-L)x(t-\sigma)\, dt \qquad 3.19$$

$$a_d(0) = \lim_{T \to \infty} (2T)^{-1} \int_{-T}^{T} d^2(t)\, dt \qquad 3.20$$

and c(L) is the cross covariance function.

$$c_{xd}(L) = \lim_{T \to \infty} (2T)^{-1} \int_{-T}^{T} d(t)x(t-L)\, dt \qquad 3.21$$

The optimum filter, W^{-1}, is determined by making the functional, E^2, in 3.18 as small as possible. Comparing with 3.1 we have that W^{-1} plays the same role as y. As in 3.3 let
p be a real parameter
$\eta(t)$ be a differentiable function satisfying the condition

$$\eta(t) = 0 \qquad t < 0 \qquad 3.22$$

This condition is necessary to obtain a causal or physically realizable operator, W^{-1}. That is, W^{-1} must be zero for negative times.

Introduce a variation in $W^{-1}(L)$.

$$\delta W^{-1} = p\eta(t) \qquad 3.23$$

This produces a change in the functional equal to δE^2.

$$E^2 + \delta E^2 = \int_{-\infty}^{\infty} [W^{-1}(L) + p\eta(L)\, dL \int_{-\infty}^{\infty} [W^{-1}(\sigma) + p\eta(\sigma)]\, d\sigma \cdot a_x(L-\sigma)$$

$$-2 \int_{-\infty}^{\infty} [W^{-1}(L) + p\eta(L)]\, dL\, c_{xd}(L) + a_d(0) \qquad 3.24$$

Expand the equation.

$$E^2 + \delta E^2 = \int W^{-1}(L)dL \int W^{-1}(\sigma)d\sigma a_x(L-\sigma) + p \int W^{-1}(L)dL \int \eta(\sigma)d\sigma a_x(L-\sigma)$$
$$+ p \int \eta(L)dL \int W^{-1}(L)d\sigma a_x(L-\sigma) + p^2 \int \eta(L)dL \int \eta(\sigma)d\sigma a_x(L-\sigma)$$
$$- 2 \int W^{-1}(L)dL c_{xd}(L) - 2p \int \eta(L)dL c_{xd}(L) + a_d(0) \qquad 3.25$$

Rearrange term containing p so as to get δE^2 alone and change the order of integration.

$$\delta E^2 = p \underbrace{\int \eta(\sigma)d\sigma \int W^{-1}(L)a_x(L-\sigma)dL}_{P} + p \underbrace{\int \eta(L)dL \int W^{-1}(\sigma)a_x(L-\sigma)d\sigma}_{P}$$
$$+ p^2 \underbrace{\int \eta(L)dL \int \eta(\sigma)a_x(L-\sigma)d\sigma}_{Q} - 2p \underbrace{\int \eta(L)c_{xd}(L)\, dL}_{R} \qquad 3.26$$

The first two terms are equal since a_x is an even function. Substituting letters for the integral alone, equation 3.26 becomes

$$\delta E^2 = 2pP + p^2 Q - 2pR \qquad 3.27$$

For an extremum of the function, E^2, we require that

$$(d/dp)\,[\delta E^2]_{p=0} = 0 \qquad 3.28$$

for all possible η.

$$(d/dp)\,[\delta E^2] = [2P + 2pQ - 2R]_{p=0} = 0 \qquad 3.29$$

or

$$[P - R] = 0 \qquad 3.30$$

for all η. That is,

$$\int \eta(L)\,[\int W^{-1}(\sigma)a_x(L-\sigma)\,d\sigma - c_{xd}(L)]\,dL = 0 \qquad 3.31$$

The Wiener-Hopf equation

For a minimum

$$\int_{-\infty}^{\infty} W^{-1}(\sigma) a_x(L-\sigma) d\sigma - c_{xd}(L) = 0 \quad \text{for} \quad L \geq 0 \qquad 3.32$$

This will not be zero for the case, $L < 0$. This equation is the Wiener-Hopf equation for an optimum linear system.

$$\boxed{c_{xd}(L) = \int_0^{\infty} W^{-1}(\sigma) a_x(L-\sigma) \, d\sigma} \quad L \geq 0 \qquad 3.33$$

Note that the equation only involves the autocovariance function and the cross covariance function between the actual signal and the desired signal. The lower limit is zero because W^{-1} is zero for negative times.

The solution of the integral equation in 3.33 is difficult and we shall follow Bode and Shannon (1950) by proceeding in the frequency domain. Multiply both sides of the Wiener-Hopf equation by $\exp(-i\omega L)$ and integrate from $L = 0$ to ∞.

$$\int_0^{\infty} c_{xd}(L) e^{-i\omega L} dL = \int_0^{\infty} e^{-i\omega L} dL \int_0^{\infty} W^{-1}(\sigma) a_x(L-\sigma) \, d\sigma \qquad 3.34$$

Let $T = L - \sigma$ on the right-hand side and interchange order of integration.

$$\int_0^{\infty} c_{xd}(L) e^{-i\omega L} dL = \int_0^{\infty} W^{-1}(\sigma) e^{-i\omega \sigma} d\sigma \int_{-\sigma}^{\infty} a_x(T) e^{-i\omega T} dT \qquad 3.35$$

Since the lower limit is $-\sigma$ the right-hand integral is not an autopower spectral density. This is why there is difficulty in also solving this in the frequency domain. To overcome the problem two new autocovariance functions, A^+ and A^-, are convolved to generate $a_x(L)$

$$a_x(L) = \int_{-\infty}^{\infty} A^+(L-t) \, A^-(t) \, dt \qquad 3.36$$

where the functions are only defined for positive or negative lags.

$$A^+(L) = 0 \quad L \leq 0$$
$$A^-(L) = 0 \quad L > 0 \quad \quad 3.37$$

Since a_x is obtained from a convolution, it will be made up of a product in the frequency domain

$$P_x(\omega) = P^+(\omega) \; P^-(\omega) \quad \quad 3.38$$

where P^+ and P^- are autopower spectral density functions. All of the singularities of P^+ are in the upper complex frequency domain while P^- has all of its singularities in the lower domain. It is found that $P_x(\omega)$ can always be factored into a product of polynomials. The poles and zeros are arranged symmetrically about both axes (figure A3.4). This symmetry requirement arises because the spectral estimate, $P_x(\omega)$, must be a real, positive and even function of ω.

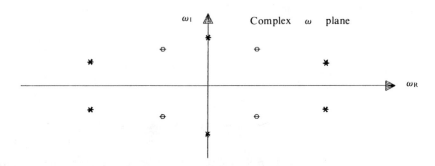

Figure A3.4 Poles and zeros of a power spectral density function.

Let us define a filter operator, B, which, when it is convolved with A^-, will yield the cross covariance function between the input signal and the desired signal.

$$c_{xd}(L_1) = \int_{-\infty}^{0} B(L_1 - t) \; A^-(t) \; dt \quad -\infty < L < \infty \quad \quad 3.39$$

Taking the Fourier transform of both sides gives

$$\int c_{xd}(L_1) \, e^{-i\omega L_1} \, dL_1 = \int_{-\infty}^{\infty} e^{-i\omega L_1} \, dL_1 \int_{-\infty}^{0} B(L_1-t) A^-(t) \, dt \quad \quad 3.40$$

The Wiener-Hopf equation

or since the left-hand side is the cross-power spectral density, $C_{xd}(\omega)$

$$C_{xd}(\omega) = \int_{-\infty}^{0} A^-(t)\, dt \int_{-\infty}^{\infty} B(L_1-t)\, e^{-i\omega L_1}\, dL_1 \qquad 3.41$$

The limits on the integral with A^- can be expanded from $-\infty$ to $+\infty$ since A^{-1} is zero for positive time. Then let $L = L_1 - t$

$$C_{xd}(\omega) = \int_{-\infty}^{\infty} A^-(t)\, e^{-i\omega L}\, dt \int_{-\infty}^{\infty} B(L)\, e^{-i\omega t}\, dL \qquad 3.42$$

In terms of auto- and cross-power spectral density,

$$C_{xd}(\omega) = P^-(\omega) \int_{-\infty}^{\infty} B(L)\, e^{-i\omega L}\, dL \qquad 3.43$$

Taking the inverse Fourier transform we can solve for $B(L)$

$$B(L) = (2\pi)^{-1} \int_{-\infty}^{\infty} ([C_{xd}(\omega_1)\exp(i\omega_1 t)] / P^-(\omega_1))\, d\omega_1 \qquad 3.44$$

Substituting 3.39 and 3.36 into the Wiener-Hopf equation, 3.33

$$\int_{-\infty}^{0} B(L_1-t)A^-(t)dt = \int_{0}^{\infty} W^{-1}(\sigma)d\sigma \int_{-\infty}^{\infty} A^+(L_1-\sigma-t)A^{-1}(t)dt \qquad L_1 \geq 0$$

$$= \int_{-\infty}^{\infty} A^{-1}(t)dt \int_{0}^{\infty} A^+(L_1-\sigma-t)W^{-1}(\sigma)d\sigma \qquad L_1 \geq 0$$

$$3.45$$

The term may be rearranged as follows:

$$\int_{-\infty}^{\infty} A^-(t)[B(L_1-t) - \int_{0}^{\infty} A^+(L_1-\sigma-t)W^{-1}(\sigma)d\sigma]dt = 0 \qquad L_1 \geq 0$$

$$3.46$$

For this to hold, it is sufficient that the terms in square brackets be zero for all negative time, t.

$$B(L_1-t) - \int_0^\infty A^+(L_1-\sigma-t)W^{-1}(\sigma)d\sigma = 0 \qquad \begin{matrix} L_1 > 0 \\ t < 0 \end{matrix} \qquad 3.47$$

We can let $L = L_1 - t$ and the integral is still valid because of the non-zero ranges of A^+ and W^{-1}.

$$B(L) = \int_0^\infty A^+(L-\sigma) \, W^{-1}(\sigma) \, d\sigma \qquad L \geqslant 0 \qquad 3.48$$

Taking the Fourier transform of both sides, (with $t = L - \sigma$), gives

$$\int_0^\infty B(L) \, e^{-i\omega L} \, dL = \underbrace{\int_0^\infty W^{-1}(\sigma) \, e^{-i\omega\sigma} \, d\sigma}_{Y_{opt}(\omega)} \underbrace{\int_0^\infty A^+(t) \, e^{i\omega t} dt}_{P^+(\omega)} \qquad 3.49$$

or the solution of the transfer function, Y_{opt} (which is the Fourier transform of the impulse response, W^{-1}) is

$$Y_{opt}(\omega) = \left[\int_0^\infty B(t) \cdot \exp(-i\omega t) \cdot dt\right] / P^+(\omega) \qquad 3.50$$

Substituting $B(L)$ from 3.44 gives the final result.

$$\boxed{Y_{opt}(\omega) = [2\pi P^+(\omega)]^{-1} \int_0^\infty \exp(-i\omega t) \left[\int_{-\infty}^\infty C_{xd}(\omega_1) \cdot \exp(i\omega_1 t) \cdot [P^-(\omega_1)]^{-1} \cdot d\omega_1\right] dt} \qquad 3.51$$

Integration with respect to ω_1 is in the complex plane ($\omega_1 = \omega_{1R} + i\omega_{1I}$) (see figure A3.5).

The Wiener-Hopf equation

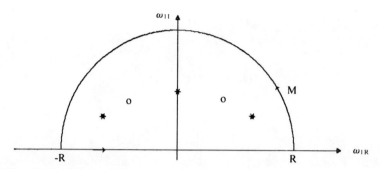

Figure A3.5 Integration with respect to ω_1 is made by a contour integration around the singularities.

$$\oint_C = \int_M + \lim_{R \to \infty} \int_{-R}^{R} = 2\pi i \Sigma \quad \text{(Residues)}$$

Note that P^+ and P^- are complex conjugates of each other since the autocovariance function and the power spectral density P_x have even symmetry.

$$P^+(\omega) = P^{-*}(\omega) \qquad 3.52$$

Equation 3.51 may be written to produce a prediction filter where t_0 is a time delay that is positive if a prediction filter is desired and negative or zero if a smoothing filter is required.

Let the desired signal at time $t + t_0$ in the future be related to the input signal by

$$d = x(t + t_0) \qquad 3.53$$

The cross covariance between the input and the desired signal is

$$c_{xd}(L) = \lim_{T \to \infty} (2T)^{-1} \int_{-T}^{T} x(t + t_0 + L) x(t) dt \qquad 3.54$$

The cross-power density is related to the auto power of the input signal by a delay function.

$$P_{xd}(\omega) = \int_{-\infty}^{\infty} c_{xd}(L)\ e^{-i\omega L}\ dL$$

$$= \int_{-\infty}^{\infty} a_x(L + t_0)\ e^{-i\omega L}\ dL$$

$$= e^{i\omega t_0}\ P_x(\omega)$$

or

$$P_{xd}(\omega) = e^{i\omega t_0}\ P^+(\omega)\ P^-(\omega) \qquad 3.55$$

This is the case if there is no noise in the input signal. If noise is present, equation 3.51 for a prediction optimum transfer function is

$$Y_{opt}(\omega) = [2\pi P^+(\omega)]^{-1} \int_0^{\infty} \exp(-i\omega t) \Big[\int_{-\infty}^{\infty} P_{xd}(\omega_1)$$

$$\cdot \exp[i\omega(t+t_0)] \cdot [P^-(\omega)]^{-1} \cdot d\omega_1 \Big]$$

References are given in Chapter 14.

Appendix 4

4.1 Stationary time series and white noise

A series of data that is independent of the starting time is called a stationary time series. In other words it must have properties which are independent of translation of the origin of time. The definition must be developed in greater detail because we are dealing with a function that is random in time and whose future cannot be described exactly. It must be described in terms of the whole set of possible values using probability distributions.

The random variable may be voltage fluctuations due to thermal noise generated in resistors or transistors of an amplifier, the microseismic activity recorded on a seismograph or the variations of amplitude of ocean waves. If the fluctuations are studied by a similar group of instruments at a large number of places some statistical similarities will be noted and the probability of any particular amplitude can be given. For simplicity, suppose the quantity measured is transformed into a voltage and is quantized so it can only take on discrete values $x_{-m}, \ldots x_{-1}, x_0, x_1, \ldots x_n$.

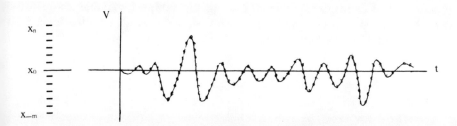

Figure A4.1 Graph of a random variable, v, as a function of time.

The probability of each voltage amplitude is

$$P_v(x_{-m}) \quad \ldots \quad P_v(x_0), \quad \ldots \quad P_v(x_n)$$

The function $P_v(x_i)$ is a *probability distribution* and satisfies the relation that the total probability of all events occurring is one.

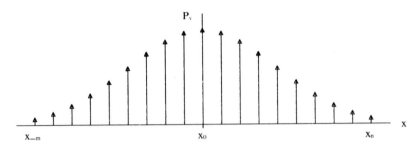

Figure A4.2 The probability that the voltage $v = x_i$ is given by the probability distribution P_v.

$$\sum_{i=-m}^{n} P_v(x_i) = 1 \qquad 4.1$$

A random variable that can be described only by a probability distribution is called a *stochastic variable* or a stochastic process (see Parzen, 1962). This is in contrast to a *deterministic process* (such as a sequence of square waves or sine waves) which contains no features of randomness. Stochastic processes were discussed by Einstein (1905) and Smoluchowski (1906), particularly with respect to Brownian motion, and Wiener (1923) proved that $x(t)$ is a continuous function. The concepts used in discussing the sampled variable above are readily extended to a continuous random variable in which case the function $P_v(x)$ is called the *probability density*.

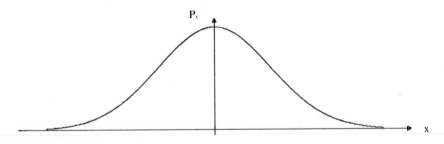

Figure A4.3 Probability density for a continuous voltage $v = x$.

The area under the curve must also be equal to 1.

$$\int_{-\infty}^{\infty} P_v(x)\, dx = 1 \qquad 4.2$$

Stationary time series and white noise

The class of all possible messages or noise which may be generated by a single source or a group of similar sources forms an *infinite aggregate of random functions*. This aggregate need not be sampled for an infinite time but may consist of signal and noise recorded during repeated experiments as in geophysical seismic or electromagnetic exploration.

If the data is recorded by a large number of similar receivers the collection of time series which is obtained is called an *ensemble*. In actual practice one can only obtain a sample from the infinite aggregate of time series forming the ensemble. If the probability for a particular amplitude of an ensemble is the same regardless of the time at which the observations are made, the ensemble is said to be *stationary*. The ensemble of signal and noise which is received may be called a *stationary random process*. A second type of random process is called a *Markov process*. This is a stochastic process in which the future development depends only upon the most recent state. An autoregressive process was discussed in Chapter 12.

The random variations of ground displacement called microseismic activity is due to the superposition of a large number of independent events. These may be due to storms in distant oceans where energy is transmitted into the solid earth and propagated as an elastic wave between interfaces in the crust, or they may be due to waves striking the shores around the continent, winds causing trees and other obstructions to vibrate and cultural noise due to traffic and machines in factories. Random voltages across a resistor are due to fluctuations in the motion of ions and electrons and is called Johnson noise.

The *central limit theorem* (Cramer, 1946) states that the sum of a number of independent random variables $x = v_1+v_2+v_3....v_n$ has a distribution which approaches the *Gaussian function* as n approaches infinity. A *Gaussian* or *normal* density distribution is defined as

$$P_v(x) = (2\pi)^{-1/2}\sigma^{-1} \exp[-(x-x_0)^2/(2\sigma^2)] \qquad 4.3$$

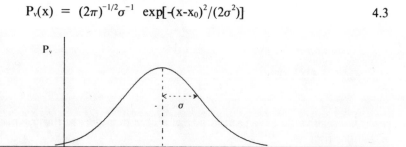

Figure A4.4 The Gaussian or normal distribution.

The function was found by DeMoivre in 1733 and used in probability theory later by Gauss. The random variable, v, takes on any value in the range $-\infty < x < \infty$. The standard deviation is σ, the variance is σ^2. Experimentally it is found that most noise, such as thermal noise, obeys a normal density function with a mean value, x_0, of zero.

It is possible to show that a linear combination of random variables which are distributed as a Gaussian function is also distributed as a Gaussian function. Most signals of interest are not normally distributed.

The *ergodic theory* stems from Gibbs' theories in statistical mechanics dealing with six-dimensional phase space. The temperature and pressure of a gas are time averages of the motion of molecules in the system. To obtain this time average Gibbs assumed that in a closed system where the total energy is constant, a time average over the motion of the particles could be obtained by an integration over a surface in phase space called the ergodic surface. *Ergodic* is a Greek word meaning *work path*. The hypothesis, in modified form, proposes that *the system*, if left to itself in its actual state of motion, will, sooner or later, pass infinitely close to every phase ($dp_x \cdot dp_y \cdot dp_z \cdot dx \cdot dy \cdot dz$), which is consistent with the equation of motion.

As applied to ensembles of signals and noise, the *ergodic theory* is that in the stationary ensemble of random functions produced by identical sources and having a continuous range of possible values, the observed amplitudes of an ensemble member come infinitely close, sooner or later, to every point of the continuous range of possible values.

We are concerned with an ensemble of functions, of the type f(t), each generated by a random stationary Gaussian process. Ergodic theory insures that the average of the ensemble is equivalent to the time average of a single function.

$$\int_{-\infty}^{\infty} x \; P_v(x) \; dx = \lim_{T \to \infty} (T)^{-1} \int_{-T}^{0} f(t) \; dt \qquad 4.4$$

The expression

$$\int_{-\infty}^{\infty} x \; P_v(x) \; dx$$

is the average value of the amplitude of all the members of the ensemble at one particular time.

The expression

$$\lim_{T \to \infty} (T)^{-1} \int_{-T}^{0} f(T) \; dt$$

is the time average amplitude of an ensemble member on the basis of its infinite

Stationary time series and white noise

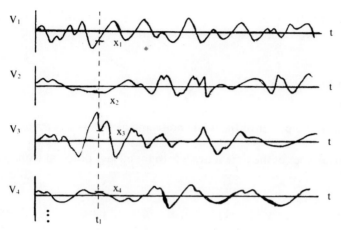

Figure A4.5 Ensemble of random functions.

past history. The subject is pursued at greater length in Lee (1960) and Thomas (1969). A signal which has a flat power density spectrum over the frequency response of the recording instrument is called *white noise* or *Gaussian white noise*.

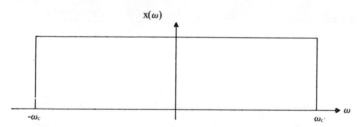

Figure A4.6 Fourier transform of band-limited white noise. The cut-off frequency for the instrument is ω_c.

If the signal is sampled then the time series is defined as

$$\mathbf{x} = (\ \ldots\ x_{-2},\ x_{-1}\ x_0,\ x_1,\ x_2\ \ldots\ )$$

This signal is white noise if it satisfies the following three criteria:
(1) The time average of the series is zero.

$$E[x_t] \equiv \lim_{N \to \infty} (2N+1)^{-1} \sum_{i=-N}^{N} x_i = 0 \qquad 4.5$$

The E indicates that this is the *expectation* value or first moment.
(2) The mean squared value or autocovariance with zero lag is finite.

$$a_x(0) = \lim_{N \to \infty} (2N+1)^{-1} \sum_{i=-N}^{N} x_i^2 \equiv \sigma^2 < \infty \qquad 4.6$$

The quantity σ^2 is the power of the white noise and also the *variance* or second moment.
(3) The autocovariance of the time series is zero for all lags different from zero.

$$a_x(j) = \lim_{T \to \infty} (2N+1)^{-1} \sum_{i=-N}^{N} x_{i+j} \; x_i = 0 \qquad j \neq 0 \qquad 4.7$$

The random noise which was discussed above is an example of white noise and, in general, all white noise will have Gaussian distribution. The power density of white noise is a constant over all frequencies.

$$P(\omega) = \sigma^2 \qquad 4.8$$

Taking the transform of 4.8, the autocovariance for white noise is

$$a(L) = (2\pi)^{-1} \int_{-\infty}^{\infty} \sigma^2 \, e^{i\omega L} \, d\omega \qquad 4.9$$

But from Appendix 2, equation 2.9 we have that the Dirac delta function at time zero is

$$\delta(t) = (2\pi)^{-1} \int_{-\infty}^{\infty} e^{i\omega t} \, d\omega \qquad A2.9$$

Therefore by 4.9

$$a(L) = \sigma^2 \, \delta(L) \qquad 4.10$$

The autocovariance of white noise is an impulse function at $L = 0$. This establishes criteria (2) and (3) given above.
    Examples of band-limited white noise are found in many physical processes. Examples include:

Stationary time series and white noise

(a) Johnson noise (1928) which is due to random electronic motion in a conductor.
(b) Brownian motion of pollen or dust when bombarded at random by molecules.
(c) Shot noise due to the random arrival of electrons at the anode of an electronic vacuum tube.
(d) Microseismic activity due to energy transfer between ocean waves and the shore.
(e) Amplitude of waves in an ocean basin
(f) Electromagnetic micropulsation waves due to the interaction of the solar wind with the earth's outer atmosphere of charged particles.

Examples in probability theory include the tosses of a coin (Heads = 1; Tails = -1) and the throws of a dice (1 = 1; 2 = 2; 3 = 3; 4 = -1; 5 = -2; 6 = -3). The geophysical examples given in (d), (e) and (f) are imperfect examples of white noise since these are only approximately stationary.

White noise may be convolved with a basic wavelet to represent the actual recorded signal after modification by natural processes (Wold, 1938; Robinson 1962, 1964). This gives us a signal called a stationary time series.

$$x = W * n \qquad 4.11$$

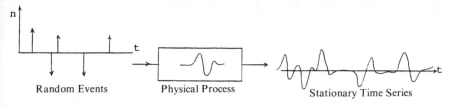

Figure A4.7 Wold decomposition. White noise produced by a convolution of random pulses and a basic wavelet.

$$x_t = \sum_{i=0}^{\infty} W_i \, n_{t-i} \qquad 4.11a$$

Let us find the autocovariance of a stationary time series. Replace t by $t+k$ to obtain a lagged time series.

$$x_{t+k} = \sum_{j=0}^{\infty} W_j \, n_{t+k-j} \qquad 4.12$$

Appendix 4

The autocovariance is then the mean value of $(x_t \cdot x_{t+k})$

$$a_k = E[\sum_{j=0}^{\infty} \sum_{i=0}^{\infty} W_i W_j n_{t+k-j} n_{t-i}] \qquad 4.13$$

$$\lim_{N \to \infty} (2N+1)^{-1} \sum_{t=-N}^{N} \sum_{j=0}^{\infty} \sum_{i=0}^{\infty} W_i W_j n_{t+k-j} n_{t-i}$$

$$= \sum_{j=0}^{\infty} \sum_{i=0}^{\infty} W_i W_j [\lim_{N \to \infty} (2N+1)^{-1} \sum_{t=-N}^{N} n_{t+k-j+i} n_t] \qquad 4.14$$

The quantity in brackets is the autocovariance of the white noise by equation 4.6.

$$\begin{matrix} a_n(k-j+i) = \sigma^2 & k-j+i = 0 \\ 0 & k-j+1 \neq 0 \end{matrix} \qquad 4.15$$

The power of the white noise is σ^2 and the equation for a_k has a value only when $j = i + k$. Hence the autocovariance is

$$a_k = \sigma^2 \sum_{i=0}^{\infty} W_{i+k} W_i \qquad 4.16$$

If we let a_k^w be the autocovariance of the wavelet W, then

$$a_k = \sigma^2 a_k^w \qquad 4.17$$

The white noise is completely uncorrelated so the autocovariance is only due to the wavelet as weighted by the power of the white noise.

References

Cramer,H(1946), *Mathematical methods of statistics*, Princeton University Press, Princeton, N.J.

Einstein,A.(1905), Investigations on the theory of the Brownian movement, *Ann. d. Physik* **17**, ; also **19**, 371-381, (1906). Translated in Dover Publications 1956, New York.

Helstrom,C.W.(1960), *Statistical theory of signal detection*, Pergamon Press, New York.

Johnson,J.B.(1928), Thermal agitation of electricity in conductor, *Phys. Res.* **32**, 97-109.

Lee,Y.W.(1960), *Statistical theory of communication*, John Wiley & Sons, Inc. New York.

Parzen,E.(1962), *Stochastic Processes*, chapter 1, Holden-Day Inc., San Francisco.

Rice,S.O.(1944-45), Mathematical analysis of random noise, *Bell System Technical Journal* **23** and **24** . See also Selected Papers on Noise and Stochastic Processes, Ed. ;N. Wax, 1954, 133-295, Dover Publ., New York.

Robinson,E.A.(1962), *Random wavelets and cybernetic systems*. Griffin Statistical Monographs and Course, 6, Charles Griffin, London.

Robinson,E.A.(1964), Wavelet composition of time-series, 37-106 , *Econometric Model Building*, Editor, A.O.A. Wold, North-HOlland Publishing Co., Amsterdam.

Smoluchowski,M.V.(1906), Zur kinetischen Theorie der Brownschen Molekularbewegegung und der Suspensionen. *Ann. d. Physik* **21**, 756-780.

Thomas,J.B. *An introduction to statistical communication theory*. John Wiley & Sons, Inc., New York, 1969.

Wiener ,N.(1923), Differential space, *J. Math. Phys., Mass. Inst. Tech.* **2**, 131-174.

Wold,H.(1938), A study in the analysis of stationary time series, Thesis, University of Stockholm, Almgvist and Wiksells, Uppsala, 2nd Ed. (1954).

Appendix 5

5.1 The Hanning and Hamming windows

Blackman and Tukey (1958) have recommended that either one of two special but empirical weighting functions called the Hanning and Hamming windows be used in computing the power spectrum with equation 7.2-10. The truncated autocovariance function, a(L), is multiplied by the window to produce a modified apparent autocovariance function, $a_m(L)$.

$$a_m(L) = W(L)\ a(L) \qquad 5.1$$

The autocovariance is *apparent* because only a finite length of data was used in obtaining it. It is *modified* by the window and also because only lags between 0 and L_m are used. Blackman and Tukey (1958) recommended that only 10 percent to 20 percent of the total available lags be calculated. Longer lags are unreliable since they depend critically upon the sample of record chosen initially. Computation with the Hanning window is seldom used now but another weighting function called the Parzen window is still used on occasion. These windows should be used only with the autocovariance function in the time (lag) domain to obtain the power spectrum. They are not very suitable for computation in the frequency domain. With the availability of the fast Fourier algorithm it is recommended that some form of Bartlett's or Daniell's spectral estimate (sections 9.4 and 9.6) be used. However, to appreciate the spectral estimators in the literature of the 1960s the various windows or weighting functions will be discussed in this appendix.

The Hanning window was named after Julius von Hann, an Austrian meteorologist, and consists of a truncated cosine wave with a DC shift as it approaches zero without discontinuites. Alternatively it is known as a cosine bell.

$$\begin{aligned} w(L) &= 0.5 + 0.5 \cos \pi L/L_m & |L| &< L_m \\ &= 0 & |L| &> L_m \end{aligned} \qquad 5.2$$

Its Fourier transform consists of three terms containing a sin x/x function.

$$W_{Hanning} = 0.5 W_0(\omega) + 0.25 W_0(\omega + \pi/L_m) + 0.25 W_0(\omega - \pi/L_m) \qquad 5.3$$

The Hanning and Hamming windows

where

$$W_0(\omega) = 2L_m \sin \omega L_m / (\omega L_m) \qquad 5.4$$

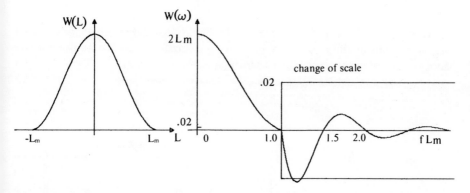

Figure A5.1   The Hanning window and its Fourier transform.

The Hanning window was used to a considerable extent when computers were slow and built with vacuum tubes, or with the first generation of transistors, because it is reasonably effective and very inexpensive in terms of computing time. The power estimate has objectionable side lobes (figure A5.1) although they are very small and this gives rise to some loss of resolution in the frequency domain. The asymptotic variance is given by equation 9.4-2.

$$\text{A.V.} = (T)^{-1} \int_{-L_m}^{L_m} W^2(L) \, dL = 0.75 \, L_m/T \qquad 5.5$$

If L_m is made large, the spectral estimate will have a high resolution but a low reliability according to 5.5. The length of data is specified by T.

A function which is similar to the Hanning window has been named after R.W. Hamming, an associate of J.W. Tukey.

$$W(L) = 0.54 + 0.46 \cos \pi L / L_m \qquad |L| < L_m$$
$$W(L) = 0 \qquad |L| > 0 \qquad 5.6$$

The Fourier transform is

$$W_{Hamming} = 0.54\ W_0(\omega) + 0.23\ W_0(\omega + \pi/L_m)$$
$$+ 0.23\ W_0(\omega - \pi/L_m) \qquad 5.7$$

where $W_0(\omega)$ is given by 5.4. The first side lobe of the Hamming window is less than 1 percent of the main peak but the lobes farther out become quite large since they must generate the discontinuity at $L = L_m$.

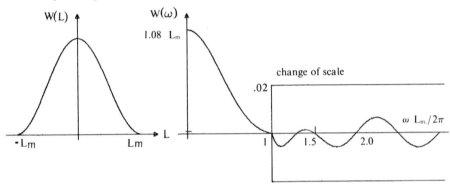

Figure A5.2 The Hamming window and its Fourier transform.

5.2 The Bartlett window

The Bartlett (1950) window is a triangular-shaped function in the time domain and uses all possible lags, 0 to T.

$$W(L) = 1 - |L|/T \qquad |L| \leq T$$
$$= 0 \qquad |L| > T \qquad 5.8$$

Its Fourier transform is the product of two $(\sin x)/x$ functions. This product is often described as the convolution of two box-cars (figure A5.3).

$$W(\omega) = T\ [(\sin \omega T/2) / (\omega T/2)]^2 \qquad 5.9$$

The Bartlett window is always used in shaping the autocovariance function although this is sometimes not realized. The autocovariance in digital analysis should be defined.

The Bartlett window

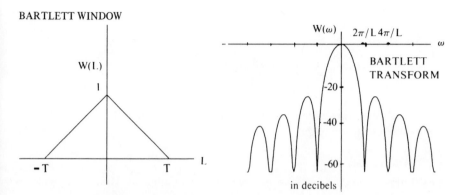

Figure A5.3 The Bartlett window and its Fourier transform.

$$a(L) = (N-L)^{-1} \sum_{t=1}^{N-L} f_{t+L} \, f_t \qquad 0 \leq L < N \qquad 5.10$$

Note that there are N-L products and these are summed and divided by (N-L) to obtain the autocovariance with lag L. In actual practice a(L) is calculated by equation 6.1-10.

$$a(L) = (N)^{-1} \sum_{t=1}^{N-L} f_{t+L} \, f_t \qquad 6.1\text{-}10$$

The extra weighting forces the autocovariance to decrease to zero as very large lags are calculated and so insures that the function does not fluctuate erratically. This is a desirable property since it is expected for a transient signal or for data which approximates Gaussian noise. Equation 6.1-10 may be rewritten as

$$a(L) = (1 - |L|/N) \, [(N-L)^{-1} \sum_{t=1}^{N-L} f_{t+L} \, f_t] \qquad 5.11$$

In this form the appearance of Bartlett's window is clearly evident. The effect of a window with this shape and length is very slight and any additional weighting function will completely predominate in defining the characteristics of the power spectrum. The asymptotic variance for Bartlett's window is

$$A.V. = \int_{-\infty}^{\infty} [W(L)]^2 \, dL = -\frac{2}{3} \frac{L_m}{T} \qquad 5.12$$

Note that the variance decreases as L_m is made smaller but unfortunately this causes the resolution to get worse.

5.3 The Parzen window

The Parzen window is favored over the Hanning or Hamming windows because it involves only a small amount for extra computation but gives non-negative estimates with extremely low side lobes (Parzen, 1961). The Parzen window is built from a truncated symmetric cubic equation.

$$W(L) = \begin{cases} 1 - 6(|L|/L_m)^2 + 6(|L|/L_m)^3 & |L| < 0.5L_m \\ 2(1 - |L|/L_m)^3 & 0.5L_m \leq |L| \leq L_m \\ 0 & |L| > L_m \end{cases} \qquad 5.13$$

The Fourier transform is a product of the transform of four box-car or $(\sin x)/x$ functions.

$$W(\omega) = \frac{3L_m}{4} \left[(\sin \omega L_m/4)/(\omega L_m/4) \right]^4 \qquad 5.14$$

The variance is found to be

$$A.V. = (T)^{-1} \int_{-\infty}^{\infty} W^2(\omega) \, d\omega = 151 L_m/(280T) \simeq 0.54 L_m/T \qquad 5.15$$

A different window with smaller variance has been found by Papoulis (1973).

References

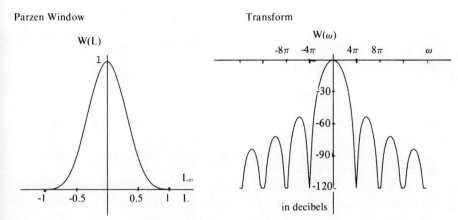

Figure A5.4 The Parzen window and its Fourier transform.

References

Bartlett,M.S.(1950), Periodogram analyses and continuous spectra, *Biometrika* **37**, 1-16.

Blackman,R.B. and Tukey,J.W.(1958), *The measurement of power spectra*, Dover Publication, New York.

Papoulis,A.(1973), Minimum-bias windows for high resolution. *IEEE Transactions on Information Theory* **19**, 9-12.

Parzen,E.(1961), Mathematical considerations in the estimations of spectra, *Technometrics* **3**, 167-189.

Appendix 6

6.1 Stability of filters

A stable filter will produce a bounded output as long as the input never becomes infinitely large. As stated in Chapter 2, the necessary and sufficient condition for the stability of a digital filter is that the sum of its coefficients converge absolutely. That is

$$\sum_{t=-\infty}^{\infty} |W_t| = S < \infty \qquad 2.1\text{-}4$$

The W_t are weighting coefficients obtained from the passage of a unit impulse through the filter. The output, y_L, from a filter is given by a convolution of the coefficients with the input x

$$y_L = \Delta t \sum_{t=-\infty}^{\infty} W_t \, x_{L-t} \qquad 2.1\text{-}6$$

where for convenience we can set the sampling interval, Δt, equal to unity. To establish the proof in a manner similar to Hurewicz (1947), let the maximum amplitude, $|W_t|$ ($t = \ldots$-2, -1, 0, 1, 2, ...), be M. Then the output has a maximum value given by

$$|y_L| < \sum_{t=-\infty}^{\infty} M \, |W_t| = MS \qquad 6.1$$

Since the output is bounded by the number MS, a sufficient condition for stability is that S be a finite positive number.

Since all computer-based digital filters must be realizable it will be assumed that the coefficients are one sided ($\mathbf{W} = W_0, W_1, W_2, \ldots$). The proof for the necessity of absolute convergence will proceed on this assumption although the more general case follows from a similar development. Let the sequence, $\mathbf{W}$, be bounded by a positive number, B.

$$|W_t| < B \qquad t = 0, 1, 2, \ldots \qquad 6.2$$

Stability of filters

Let us make the premise that the filter is *stable* but that the sum of absolute values *diverges*. Then there is a positive integer, T(1), which makes the sum exceed unity.

$$\sum_{t=0}^{T(1)} |W_t| > 1 \qquad 6.3$$

Another integer, T(2) - T(1), will make the sum exceed BT(1)+2. If T(1) is even then let T(2) be odd or vice-versa.

$$\sum_{t=0}^{T(2)-T(1)} |W_t| > BT(1) + 2 \qquad 6.4$$

In general one can find an integer T(i) - T(i-1) such that the sum exceeds BT(i-1)+i.

$$\sum_{t=0}^{T(i)-T(i-1)} |W_t| > BT(i-1) + i \qquad 6.5$$

The input signal will be chosen to be a simple well-behaved one such as a rectangular function with values of only 0 or ±1. In each interval, 0 to T(1), T(1) to T(2), etc., the value of x is determined by the sign of the filter coefficient.

$$\begin{aligned} x_t &= 0 & t &\leq 0 \\ x_t &= \operatorname{sgn} W_{T(1)-t} & 1 &\leq t < T(1) \\ x_t &= \operatorname{sgn} W_{T(1)-t} & T(i-1) &\leq t < T(i) \end{aligned} \qquad 6.6$$

where

$$\operatorname{sgn} W = \begin{cases} 1 & W > 0 \\ 0 & W = 0 \\ -1 & W < 0 \end{cases} \qquad 6.7$$

Using the convolution equation in 2.1-6, the partial sum for the output is

$$y_{T(i)} = \sum_{t=0}^{T(i)-1} W_t \, x_{T(i)-t} \qquad 6.8$$

From equation 6.5 and the form of the input signal which makes $W_t x_{T(i)-t}$ equal to $|W_t|$ one has the following partial sum:

$$\sum_{t=0}^{T(i)-T(i-1)} W_t x_{T(i)-t} > BT(i-1) + i \qquad 6.9$$

For the input series, x_t above, one also has that

$$\sum_{t=T(i)-T(i-1)+1}^{T(i)-1} W_t \, x_{T(i)-t} > -BT(i-1) \qquad 6.10$$

The sum of equation 6.9 and 6.10 gives the following inequality for $y_{T(i)}$.

$$y_{T(i)} > i \qquad 6.11$$

Although the input signal is bounded, we find that the output signal diverges which is contrary to the assumption that the filter is stable. Therefore, the necessary and sufficient condition for stability is that the sum of the coefficients converges absolutely as in equation 2.1-4.

References are listed in Chapter 2.

Appendix 7

7.1 Cooley-Tukey fast Fourier transform

For an understanding of the way in which a fast Fourier transform is carried out the matrix formulation given in section 3.4 should be followed. The actual equations which are incorporated in the computer algorithm in section 3.8 are derived in this appendix. Cooley and Tukey (1965) compute a form of the discrete Fourier transform equation using a base 2 or binary number system. If N is the number of data points then the product of sums is given by n.

$$n = LOG_2 N \qquad 3.4\text{-}12$$

The discrete Fourier transform to be computed is as follows:

$$F(W) = \sum_{T=0}^{N-1} f(T) \, e^{-2\pi i WT/N} \qquad 3.2\text{-}18$$

Let

$$\Omega = -2\pi i / N$$

as in equation 3.4-3.

$$F(W) = \sum_{T=0}^{N-1} f(T) \, e^{\Omega WT} \qquad 7.1$$

The binary equivalent of equation 7.1 is determined using an example in which $N = 4$. The indices W and $T = 0, 1, 2, 3$ in base 10 are rewritten as a binary system in which T_a, T_b, W_a and W_b can assume only values of 0 or 1.

$$T_{10} = (T_b, T_a)_2 \qquad W_{10} = (W_b, W_a)_2 \qquad 7.2$$

That is, for $N = 4$, $0_{10} = (0, 0)_2$, $1_{10} = (0, 1)_2$, $2_{10} = (1, 0)_2$, $3_{10} = (1, 1)_2$. The equation for the discrete Fourier transform may be written as a double summation as in equation 3.3-3.

$$F(W_b, W_a) = \sum_{T_a=0}^{1} \sum_{T_b=0}^{1} f(T_b, T_a) \cdot exp[\Omega(2W_b + W_a)(2T_b + T_a)] \qquad 7.3$$

The exponential is factored as follows:

$$e^{\Omega WT} = exp[\Omega(2W_b+W_a) \; 2T_b] \cdot exp[\Omega(2W_b+W_a)T_a]$$
$$= exp[\Omega 4W_bT_b] \cdot exp[\Omega 2W_aT_b] \cdot exp[\Omega(2W_b+W_a)T_a]$$

But $exp[4\Omega W_bT_b] \equiv 1$ so equation 7.3 can be written in the following way. All the summations in the following section are for T_a and T_b from 0 to 1.

$$F(W_b, W_a) = \sum_{T_a} [\sum_{T_b} f(T_b, T_a) \; exp[2\Omega W_aT_b] \cdot exp \; [\Omega(2W_b+W_a)T_a]] \quad 7.4$$

The equation in brackets is the T_b point Fourier transform, $P(W_a, T_a)$.

$$P(W_a, T_a) = \sum f(T_b, T_a) \; exp \; (2\Omega \; W_aT_b) \quad W_a=0,1, \; T_a=0,1 \quad 7.5$$

The outer summation in 7.4 can be written in a scrambled (bit reversed) order, $F_S(W_a, W_b)$.

$$F_S(W_a, W_b) = \sum_{T_a} P(W_a, T_a) \; exp[\Omega(2W_b+W_a)T_a]$$
$$W_a=0,1; \; W_b=0,1 \quad 7.6$$

The final unscrambled equation is obtained by reversing the binary bit order, an effect similar to that of the permutation matrix in 3.4-13.

$$F(W_b, W_a) = F_S(W_a, W_b) \quad 7.7$$

The process in equation 7.5 may be visualized more easily by writing the T_b point Fourier transform, $P(W_a, T_a)$, as the following four equations for $N=4$:

$$\begin{aligned}
P_0 &= P(0,0) = f(0,0) + f(1,0) \; e^0 \\
P_1 &= P(0,1) = f(0,1) + f(1,1) \; e^0 \\
P_2 &= P(1,0) = f(0,0) + f(1,0) \; e^{2\Omega} \\
P_3 &= P(1,1) = f(0,1) + f(1,1) \; e^{2\Omega}
\end{aligned} \quad 7.8$$

There is no multiplication in this P point transform because $e^0 = 1$ and $e^{2\Omega} = -1$. In matrix form the equations may be written as follows with $e^0 = 1$

$$\begin{bmatrix} P(0,0) \\ P(0,1) \\ P(1,0) \\ P(1,1) \end{bmatrix} = \begin{bmatrix} 1 & 0 & e^0 & 0 \\ 0 & 1 & 0 & e^0 \\ 1 & 0 & e^{2\Omega} & 0 \\ 0 & 1 & 0 & e^{2\Omega} \end{bmatrix} \begin{bmatrix} f(0,0) \\ f(0,1) \\ f(1,0) \\ f(1,1) \end{bmatrix} \quad 7.9$$

The outer sum in equation 7.6 can also be written in matrix form.

Cooley-Tukey fast Fourier transform

$$\begin{bmatrix} F_S(0,0) \\ F_S(0,1) \\ F_S(1,0) \\ F_S(1,1) \end{bmatrix} = \begin{bmatrix} 1 & e^0 & 0 & 0 \\ 1 & e^{2\Omega} & 0 & 0 \\ 0 & 0 & 1 & e^{\Omega} \\ 0 & 0 & 1 & e^{3\Omega} \end{bmatrix} \begin{bmatrix} P(0,0) \\ P(0,1) \\ P(1,0) \\ P(1,1) \end{bmatrix} \quad 7.10$$

There are Nn or 8 complex additions and complex multiplications contained in equations 7.8 and 7.10. However, some of the multiplications are by $e^0 = 1$ and $e^{2\Omega} = -1$. Note that $f(T_b, T_a)$ is not used after equation 7.9 so it can be replaced by $P(W_a, T_a)$ in the computer memory. The order in which $F_S(W_a, W_b)$ is computed is such that it replaces $P(W_a, T_a)$.

For the general case where $N = 2^n$ the integers W and T are written in binary form with T_j and W_j taking on values of only 0 or 1.

$$T_{10} = 2^{n-1}T_{n-1} + 2^{n-2}T_{n-2} + \ldots + T_0 = (T_{n-1}, T_{n-2}, \ldots, T_0)_2$$
$$W_{10} = 2^{n-1}W_{n-1} + 2^{n-2}W_{n-2} + \ldots + W_0 =$$
$$(W_{n-1}, W_{n-2}, \ldots W_0)_2 \quad 7.11$$

The discrete Fourier transform in 3.2-18 is written in binary form as follows with the summations from 0 to 1:

$$F(W_{n-1}, \ldots W_0) = \sum_{T_0} \sum_{T_1} \cdots \sum_{T_{n-1}} f(T_{n-1}, \ldots T_0) \, e^{\Omega WT} \quad 7.12$$

Using equation 7.11 the exponented term is expanded.

$$e^{\Omega WT} = exp\,[\Omega(2^{n-1}W_{n-1} + \ldots + W_0)\,(2^{n-1}T_{n-1})]$$
$$\cdot exp\,[\Omega(2^{n-1}W_{n-1} + \ldots + W_0)\,(2^{n-2}T_{n-2})] \cdots$$
$$\cdot exp\,[\,\Omega(2^{n-1}W_{n-1} + \ldots + W_0)\,T_0] \quad 7.13$$

The first exponential term is expanded and then simplified by noting that $exp(\Omega 2^n K) = exp(-2\pi i K N/N) \equiv 1$ since $K = W_j T_j$ is an integer.

$$exp\,[\Omega 2^n(2^{n-2}W_{n-1}T_{n-1})] \cdot exp\,[\Omega\, 2^n(2^{n-3}W_{n-2}T_{n-1})] \cdot$$
$$\cdots exp\,[\Omega 2^n W_1 T_{n-1}] \cdot exp\,[\Omega 2^{n-1} W_0 T_{n-1}]$$
$$= exp\,[\Omega 2^{n-1} W_0 T_{n-1}] \quad 7.14$$

The second exponential in 7.13 is expanded in a similar manner.

$$exp\,[\Omega 2^{n-1} W_{n-1} + 2^{n-2}W_{n-2} + \ldots + W_0)2^{n-2}T_{n-2}] =$$
$$exp\,[\Omega 2^n 2^{n-3}W_{n-1}T_{n-2}] \cdot exp\,[\Omega 2^n 2^{n-4}W_{n-2}T_{n-2}] \cdot$$
$$\cdots exp\,[\Omega 2^{n-1} W_1 T_{n-2}]$$
$$\cdot exp\,[\,\Omega 2^{n-2} W_0 T_{n-2}] = exp\,[\Omega(2W_1 + W_0)2^{n-2}T_{n-2}]$$

Appendix 7

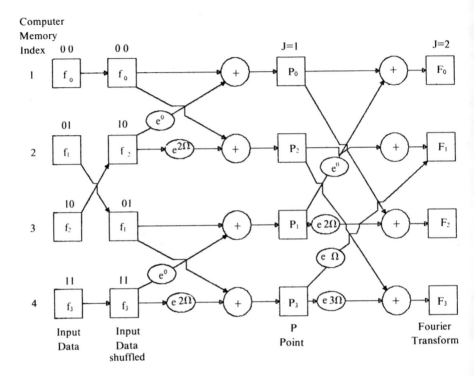

Figure A7.1 Flow diagram for a Cooley-Tukey fast Fourier transform with data shuffled in binary digit reverse order. The diagram is for N = 4 data points and the symbols in the boxes are those used in the mathematical derivation. The binary indexing and the computer symbols in the algorithm given in section 3.8 are shown outside the boxes. Multiplication by exponential factors is shown inside the ellipses. Note that $e^0 = 1$ and $e^\Omega = -1$. The boxes are memory locations. Computation is arranged with some temporary storage locations so that vectors f, P and F occupy the same permanent space in the computer memory.

The final exponential in 7.13 is unchanged. Substituting into the discrete Fourier transform of 7.12 reduces the equation as follows. The sums are for T_0, T_1 to T_{n-1} equal to 0 to 1.

$$F(W_{n-1}, W_{n-2}, \ldots W_0) = \sum \sum \cdots \sum f(T_{n-1}, T_{n-2}, \ldots T_0)$$
$$exp\,[\Omega 2^{n-1} T_{n-1}] \cdot exp\,[\Omega(2W_1 + W_0)2^{n-2}T_{n-2}] \cdots$$
$$\cdot exp\,[\Omega(2^{n-1}W_{n-1} + 2^{n-2}W_{n-2} + \ldots + W_0)T_0] \qquad 7.15$$

Each sum is computed separately as a P point Fourier transform with sums from 0 to 1.

Cooley-Tukey fast Fourier transform

$$P_1(W_0, T_{n-2}, \ldots T_0) = \sum_{T_{n-1}} f(T_{n-1}, T_{n-2}, \ldots T_0)\, exp\,[\Omega 2^{n-1} W_0 T_{n-1}]$$
$$W_0\ T_{n-2}, T_{n-3}, \ldots T_0 = 0, 1 \qquad 7.16$$

$$P_2(W_0, W_1, T_{n-3}, \ldots T_0) = \sum_{T_{n-2}} P_1(W_0, T_{n-2}, \ldots T_0) \cdot$$
$$exp\,[\,\Omega(2W_1 + W_0)2^{n-2}T_{n-2}]$$
$$W_0, W_1, T_{n-3}, \ldots T_0 = 0, 1 \qquad 7.17$$

$$P_n(W_0, W_1, \ldots W_{n-1}) = \sum P_{n-1}(W_0, W_1, \ldots T_0)\, exp\,[\Omega(2^{n-1}W_{n-1} +$$
$$2^{n-2}W_{n-2} + \ldots W_0)T_0] \qquad 7.18$$

Finally the binary digit output is reversed.

$$F(W_{n-1}, W_{n-2}, \ldots W_0) = P_n(W_0, W_1, \ldots W_{n-1}) \qquad 7.19$$

In the computer algorithm the vector **f** in 7.16 is replaced by P_1 using equation 7.16. Next P_1 is replaced by P_2 and so on until **F** is finally computed by equation 7.19. It is convenient to shuffle the input data into a binary digit reverse order as an initial step so that the final answer is in the correct decimal index order. The computational stages for N = 4 are shown in figure A7.1. Because W_j and T_j can take on values of 0 or 1, equations 7.16 to 7.19 must be computed N times. Therefore there are Nn complex additions and Nn complex multiplications although many are by factors of 1 or -1. The algorithm in section 3.8 may be modified easily to perform an inverse Fourier transform or to compute the Fourier transforms of two data sets simultaneously. The theory as well as the references are given in Chapter 3.

Author index

ABELS, F., 281, 297
ADAMS, J. A., 173
AGARWAL, R. G., 12, 18, 20, 56, 61
AHMED, I., 252, 279
AKAIKE, H., 142, 143, 155, 162, 173, 174
ALBERSHEIM, W. J., 97
ALPASLAN, T., 310, 318
ALSOP, L. E., 6, 18
ALTERMAN, Z., 6, 19
ANDERSON, C. W., III, 16, 20
ANDERSON, D. L., 9, 19
ANDERSON, G. L., 55, 61
ANDERSON, N., 154, 168, 173, 174, 175
ANDERSSEN, R. S., 409, 410
ANDREWS, C. A., 392, 398
ANDREWS, H. C., 398
ANSTEY, N. A., 91, 97, 383
ARCHAMBEAU, C. B., 328, 335, 341, 344, 353
ARTHUR, C. W., 353
ARYA, V. K., 233
ASBEL, I. JA., 7, 19
ASELTINE, J. A., 73

BACKUS, G. E., 6, 9, 10, 19, 399, 406, 411
BACKUS, M. M., 21, 291, 293, 294, 296, 297, 299, 301, 304, 318, 319, 324
BAILEY, J. S., 46, 610
BAILEY, R. C., 410, 411
BANKS, S. C., 235, 327
BARANOV, V., 55, 61, 281, 290, 297
BARBER, N. F., 2, 19, 20, 141
BARRODALE, I., 162, 174, 232, 233
BARRY, K. M., 383
BARTLETT, M. S., 97, 99, 109, 130, 144, 174, 459, 460, 462
BASHAM, P. W., 335, 343, 353
BATH, M., 186, 197
BAYER, R., 411
BAYLESS, J. W., 230, 233
BELL, L. W., 383
BENIOFF, H., 5, 19,
BERKHOUT, A. J., 233, 323, 326
BERNOULLI, D., 31, 61, 434
BERRYMAN, J. G., 163, 174, 411
BHANU, B., 360
BHATTACHARYYA, B. K., 55, 61
BIEHLER, S., 7, 20
BISHOP, T. N., 145, 163, 167, 175
BLACKMAN, R. B., 1, 19, 35, 61, 97, 99, 109, 112, 117, 118, 130, 456, 462
BLOCH, S., 318

BLOOMFIELD, P., 165, 174
BODE, H. W., 1, 19, 225, 233, 361, 371, 441
BOGERT, B. P., 355, 359, 360
BOIS, P., 392, 398
BOLONDI, G., 323, 326
BONILLA, M. G., 324, 326
BOORE, D. M., 2, 19
BORCHERDT, R. D., 291, 297
BORN, M., 328, 332, 353
BOSTICK, F. X., 136, 141, 233
BOUGUER, P., 54
BOWYER, D. E., 398
BOX, G. E. P., 139, 143, 174
BRACEWELL, R. N., 371, 378
BRADLEY, J. J., 383
BRAILE, L. W., 411
BRENNER, N. M., 53, 61
BRIGHAM, B. O., 34, 58, 61
BRIGHAM, E. O., 230, 233
BRILLOUIN, L., 111, 117, 147, 174
BROMWICH, T. J., 63, 66, 73
BROWN, R. D., 398
BRUNE, J. N., 9, 20, 335, 353
BRYAN, G. M., 359, 360
BRYAN, J. G., 10
BUCHEN, P. W., 2, 289, 291, 295, 297
BUHL, P., 359, 360
BURG, J. P., 1, 19, 118, 130, 142, 147, 150, 155, 174, 180, 181, 183, 296, 299, 301, 304, 318
BURKE, M. D., 62
BUSLENKO, P. P., 7, 19
BUTTERWORTH, S., 253, 279

CADZOW, J. A., 174
CAGNIARD, L. 17, 19
CANTWELL, T., 136, 141
CAPON, J., 130, 142, 174, 177, 180, 183
CAPUTO, N., 138, 141
CARERS, D. A., 383
CAREY, M. J., 280
CARPENTER, E. W., 136, 141
CARSON, J. R., 63, 73
CASSANO, E., 326
CHAI, C. K., 279
CHAN, A. K., 279
CHAN, D. S. K., 279
CHAN, M., 297
CHEN, W. H., 33, 61, 153, 164, 174
CHILDERS, D. G., 360
CHURCHILL, R. V., 266, 279
CIZEK, V., 368, 371

CLAERBOUT, J. F., 2, 19, 34, 58, 61, 186, 197, 233, 297, 323, 326, 368, 371
CLARKE, G. K. C., 153, 175
CLAYTON, R. W., 97, 145, 162, 176, 234, 359, 360
CLOWES, R. M., 12, 19, 309, 318, 412
COLE, A. J., 266, 280
CONAWAY, J. G., 234
CONSTANTINIDES, A. G., 279, 280
COOLEY, J. W., 2, 19, 43, 59, 61, 122, 126, 130
COOPER, H. F. J., 291, 297
COPSON, E. T., 429
CORINTHIOS, M. S., 46, 61
CRAMER, H., 449, 455
CRAWFORD, J. M., 97
CRIBB, J., 411
CROOK, A. W., 281, 297
CRUMP, N., 230, 234
CUER, M., 411, 412

D'ERCEVILLE, I., 281, 297
D'MELLO, M. R., 235
DANIELL, P. J., 99, 109, 122, 130
DANIELS, R. W., 278, 279, 280
DARLINGTON, S., 97
DAVIES, D., 323, 326
DAVIS, J. M., 398
DER, Z. A., 399, 411
DEREGOWSKI, S. M., 234
DIMRI, V. P., 235
DIRAC, P. A. M., 430, 433
DOBRIN, M. B., 117, 299, 318, 383
DOHERTY, S., 323, 326
DOHR, G. P., 326
DOORNBOS, D. J., 323, 326
DORMAN, L. M., 411
DOTY, W. E. N., 97
DOUZE, E. J., 320, 326
DRAPER, N. R., 96, 97
DUESTERHOEFT, JR., W. C., 233
DZIEWONSKI, A. M., 7, 9, 19, 20, 299, 318

EGELAND, A., 328, 354
EINSTEIN, A., 448, 455
ELEMAN, F., 328, 354
ELLIOTT, R. D., 328, 353
ELLIS, R. M., 335, 343, 353, 360
EMBREE, P., 297, 301, 304, 306, 318
ERICKSON, R. E., 162, 174, 232, 233
EULER, L., 31, 61
EVENDEN, B. S., 383
EWING, M., 18

FAIL, J. P., 299, 301, 318
FARNBACK, J. S., 368, 371
FEJER, L., 31, 61
FILSON, J. R., 323, 326
FISHER, R. A., 99, 109
FLINN, E. A., 328, 335, 340, 341, 342, 353
FORD, W. T., 186, 197, 234
FOSTER, M. R., 295, 296, 297, 299, 318
FOURGERE, P. F., 165, 174
FOURIER, J. W., 31, 61
FOURMANN, J. M., 234
FOUTCH, D. A., 174
FOWLER, R. A., 328, 333, 353
FRANKLIN, J. N., 9, 20, 21, 399, 409, 411
FRASIER, C. W., 299, 304, 309, 318
FRYER, G. J., 174, 234
FUCHIDA, F., 54, 62
FUTTERMAN, W. I., 291, 297

GABOR, D., 371
GALBRAITH, J. N., 234
GANLEY, D. C., 160, 274, 292, 297
GARDNER, G. H. F., 371
GAZDAG, J., 323, 326
GEL'FAND, I. M., 410, 411
GENTLEMAN, W. M., 43, 62, 131, 141
GERSCH, W., 164, 174
GIBSON, B., 326
GILBERT, J. F., 6, 7, 9, 19, 20, 399, 406, 411
GILES, B. F., 299, 318, 320, 327
GIMLIN, D. R., 376, 383
GLENN, W. E., 411
GODFORTH, T., 299, 318
GOLD, B., 62, 278, 279, 280, 368, 370, 371
GOLDEN, R. M., 1, 20, 241, 266, 279, 280
GOOD, I. J., 43, 45, 62, 384, 387, 398
GOODMAN, N. R., 136, 141
GOUGH, D. I., 16, 20
GOUPILLAUD, P. L., 281, 297
GRAU, G., 301, 318
GREEN, W. R., 411
GRIFFIN, J. N., 335, 341, 343, 353
GRIFFITHS, L. G., 234
GUILLEMIN, E. A., 266, 278, 280
GUPTA, S. C., 383
GURBUZ, B. M., 97
GUTOWSKI, P. R., 143, 145, 174, 323, 325, 326, 412

HAAR, A., 384, 398
HADAMARD, J., 384, 398
HALPENY, O. S., 58, 62
HAMMOND, J. W., 295, 297

Author index

HAMON, B. V., 141
HANNAN, E. J., 99, 109, 130, 141
HARMUTH, H. F., 385, 398
HARRISON, J. C., 6, 20
HASSELMANN, K., 138, 141
HATTON, L., 326
HAUBRICH, R. A., 138, 141
HAVSKOV, J., 326
HEALY, M. J. R., 355, 360
HEARNE, J. W., 186, 197, 234
HEAVISIDE, O., 63, 73, 430, 433
HEISER, R. C., 236
HELSTROM, C. W., 455
HEMMINGS, C. D., 310, 318
HERRIN, E., 299, 318
HILDEBRAND, F. B., 429
HILTERMAN, F. J., 2, 20
HIRANO, K., 252, 280
HOOD, P., 323, 325
HOPF, E., 236
HUBRAL, P., 144, 175, 323, 326
HUREWICZ, W., 1, 20, 23, 26, 30, 186, 187, 197, 462
HURLEY, P. M., 10
HUSEBYE, E. S., 323, 326

INGALLS, A. L., 299, 318
IOANNIDES, G. A., 353

JACKSON, D. D., 399, 407, 410, 411
JACOBS, J. A., 17, 21
JACOBSON, R. S., 411
JAROSCH, H., 6, 19
JENKINS, G. M., 127, 130, 139, 141, 143, 144, 175
JENSEN, O. G., 173, 176, 360
JOHNSON, I. M., 410, 411
JOHNSON, J. B., 452, 455
JOHNSON, S. J., 168, 173, 175
JONES, R. H., 99, 109, 118, 130, 175
JORDAN, C., 28, 30
JORDAN, T. H., 9, 19, 20, 411
JUDSON, D. R., 323, 326
JUPP, D. L. B., 412
JURCZYK, D., 97
JURKEVICS, A. J., 235
JURY, E. I., 186, 197, 266, 280

KAILATH, T., 175, 232, 234
KAISER, J. F., 1, 20, 240, 241, 266, 279, 280
KALABA, R., 410, 412
KALMAN, R. E., 230, 234

KANASEWICH, E. R., 12, 18, 19, 20, 61, 309, 310, 318, 323, 325, 326, 328, 336, 354
KANE, J., 398
KAY, I., 410, 412
KAY, S. M., 175
KEILIS-BOROK, V. I., 7, 19, 20
KELLER, G. R., 411
KELLY, E. J., 323, 326
KELVIN, LORD, 4, 20
KEMERAIT, R. C., 360
KENNETT, L. N., 398
KHINTCHINE, A., 90, 97, 99, 102, 109
KIM, W. H., 97
KLAUDER, J. R., 93, 97
KNEALE, C. W., 383
KOEHLER, F., 297, 317, 319, 320, 327, 371
KOLMOGOROV, A. N., 1, 20, 187, 197, 225, 234, 361, 371
KONIG, H., 34, 51, 62
KORMYLO, J., 234
KORTMAN, C. M., 392, 398
KOSBAHN, B., 235
KOTCHER, J. S., 371
KOTICK, B. J., 328, 353
KREIN, M. G., 234
KREY, T., 323, 326
KROMER, R. E., 162, 175
KRUTKO, P. D., 24, 30
KULHANEK, O., 186, 197, 234
KUNETZ, G., 234, 281, 290, 297
KUO, B. C., 82, 197

LACOSS, R. T., 118, 127, 130, 142, 168, 175, 177, 180, 183
LAMB, H., 4, 20
LAMBERT, D. G., 335, 353
LAME, M. G., 4, 20
LANCZOS, C., 403, 412
LANDAU, L. D., 328, 353
LANDISMAN, M., 299, 318, 399, 411
LAPLACE, P. S., 28, 30, 186, 197
LARNER, K. L., 296, 297, 323, 326, 412
LASTER, S. J., 320, 326
LEBESGUE, H., 31, 62
LEE, D. T. L., 175
LEE, M. R., 97
LEE, Y. W., 97, 225, 226, 234, 451, 455
LEJEUNE DIRICHLET, G., 31, 62
LEVINSON, H., 157, 175, 208, 234
LEVITAN, B. M., 410, 411
LEVSHIN, A. L., 7, 20

LEVY, S., 412
LEWIS, B. T. R., 335, 344, 353
LEWIS, P. A. W., 34, 59, 61, 122, 130
LIFSHITZ, E. M., 328, 353
LIN, J., 326
LINES, L. R., 97, 234
LINTZ, R. R., 327
LOCKETT, F. J., 291, 297
LOEWENTHAL, D., 139, 141, 410, 412
LONG, J. A., 299, 318
LOVE, A. E. H., 4, 20
LU, C. H., 383

MACDONALD, G., 138, 141
MADDEN, T., 117, 137, 138, 141
MAL'CEV, A. I., 347, 353
MARTIN, M. A., 240, 280
MASSE, R., 411
MAY, B. T., 319, 326
MAYNE, W. H., 12, 20, 319, 326
MCCLELLAN, J. H., 360
MCCLOUGHAN, C. H., 62
MCCOY, J. F., 235, 327
MCDONALD, J. A., 368, 371
MCDONOUGH, R. N., 175
MCPHERRON, R. L., 353
MEANS, J. D., 353
MEDER, H. G., 230, 235
MENDEL, J. M., 234, 297
MERCADO, E., 326
MEREU, R. F., 231, 234, 236
MEYER, R. P., 335, 344, 353
MICKLE, M. H., 412
MILLER, G. R., 2, 20, 141
MIMS, C. H., 335, 341, 354
MITRA, S. K., 252, 280
MONTALBETTI, J. F., 12, 20, 319, 323, 326, 328, 336, 354
MOORE, E. H., 403, 412
MORF, M., 173, 175
MORREY, C. B., JR., 97
MUIRHEAD, K. J., 310, 318
MUNK, W. H., 2, 20, 134, 138, 141

NAHI, N. E., 297
NAIDU, P. S., 55, 62
NEGI, J. G., 235
NEIDELL, N. S., 320, 326
NESS, N. F., 6, 20
NISHIMURA, S., 252, 280
NOLET, G., 400, 412

NOLL, A. M., 360
NORTHWOOD, E. J., 383
NYQUIST, H., 110, 117

OAKES, J. B., 266, 280
ODEGARD, M. E., 175, 234
OLDENBURG, D. W., 412
OLIVER, J., 335, 353
OLSON, J. V., 354
OOE, M., 235
OPPENHEIM, A. V., 235, 278, 280, 357, 360, 370, 371
ORANGE, A. S., 136, 141
ORMSBY, J. F. A., 240, 280
OSSANNA, J. F., 360
OTIS, R. M., 360
OTT, N., 230, 235

PALEY, R. E. A. C., 187, 370, 371, 384, 390, 398
PANAGAS, P., 392, 393, 398
PAPOULIS, A., 97, 117, 130, 371, 460, 462
PARKER, R. L., 9, 20, 399, 410, 412
PARZEN, E., 130, 142, 163, 175, 455, 460, 462
PAULSON, K. V., 328, 333, 354
PEACOCK, K. L., 148, 175, 235
PEASE, M. C., 45, 62, 347, 354
PEEPLES, W. J., 411
PEIRCE, B. O., 280
PEKERIS, C. L., 6, 19, 20
PENDREL, J. V., 183
PENROSE, R., 403, 412
PETERSON, A. M., 252, 279
PETERSON, R. A., 117, 383
PHILLIPS, R. J., 411
PHILLIPS, R. S., 97
PISARENKO, V. F., 143, 175
PONSONBY, J. E., 173, 175
PORATH, H., 16, 20
PRATT, W. K., 398
PRESS, F., 6, 7, 19, 20
PRINCE, A. C., 97
PRINCE, E. R., 299, 318, 320, 327
PROTTER, M. H., 97

RABINER, L. R., 62, 279, 280, 368, 371
RADAR, C. M., 278, 279, 280, 370, 371
RADEMACHER, H., 384, 398
RAMSDELL, L., 62
READ, R. R., 370, 371
REED, I. S., 384, 398
REINING, J. B., 327

Author index

REISS, E. L., 291, 297
REITZEL, J. S., 16, 20
RICE, R. B., 1, 9, 21, 30, 223, 224, 225, 235
RICE, S. O., 28, 97, 455
RIEMANN, B., 31, 62, 66, 73
RIETSCH, E., 97
RILEY, P. L. 236
RISTOW, D., 97, 235
ROBINSON, E. A., 1, 10, 21, 25, 29, 30, 45, 58, 62, 141, 143, 145, 174, 175, 186, 188, 197, 233, 235, 237, 280, 281, 295, 297, 298, 361, 371, 455
ROCCA, F., 326
ROSENFELD, A., 393, 398
ROSTOKER, G., 17, 21
RUNGE, C., 34, 51, 62
RYU, J., 411

SABATIER, P. C., 410, 412
SABITOVA, T. M., 7
SAFON, C., 412
SAMSON, J. C., 17, 21, 347, 351, 354
SANDE, G., 43, 62, 129, 131, 141
SATO, Y., 318
SATORIUS, E. H., 175
SATTLEGGER, J. W., 327
SAVAGE, J. C., 291, 297
SAVELLI, S., 326
SAVITT, C. H., 383
SAX, R. L., 335, 341, 354
SCHAFER, R. W., 278, 280, 358, 360, 370, 371
SCHNEIDER, W. A., 10, 21, 296, 297, 299, 318, 319, 320, 323, 327
SCHREIDER, JU. A., 7, 19
SCHULTZ, P. S., 323, 327, 328
SCHUSTER, A., 2, 21, 98, 109
SCHWARZ, G. R., 398
SENEFF, S., 299, 318
SENGBUSH, R. L., 295, 296, 297, 299, 318
SERIFF, A. J., 97
SHANKS, J. L., 241, 280, 304, 309, 318
SHANNON, C. E., 147, 175, 225, 233, 441
SHARPE, D. R., 164, 174
SHEINGOLD, D. H., 383
SHEN, WEN-WA, 327
SHERIFF, E. E., 371
SHERWOOD, J. W. C., 186, 197, 281, 290, 298, 323, 326, 327
SHIBATA, K., 398
SHIMSHONI, M., 136, 141, 335, 354
SHOR, G. G., 411

SIEWERT, W. P., 62
SILVA, W., 291, 298
SILVIA, M. T., 197, 235
SIMONS, R. S., 344, 354
SIMS, G. S., 235
SINTON, J. B., 235
SLICHTER, L. B., 6, 20
SMITH, H. W., 96, 97, 233
SMITH, J. W., 376, 383
SMITH, M. K., 29, 30
SMITH, M. L., 9, 21, 399, 409, 412
SMITH, R. B., 360
SMITH, S., 6, 19, 136, 141, 335, 354
SMOLKA, F. R., 234
SMOLUCHOWSKI, M. V., 448, 455
SMYLIE, D. E., 153, 175, 183, 410, 411
SNODGRASS, F. E., 2, 10, 141
SOLODOVNIKOV, V. V., 362, 370, 371
SOMERVILLE, P. G., 360
SORENSON, H. W., 230, 235
STEGAN, G. R., 153, 164, 174
STIGALL, P. D., 392, 393, 398
STOFFA, P. L., 358, 359, 360
STOKES, G. G., 354
STOLT, R., 323, 327
STONE, D. R., 383
STRALEY, D. K., 319, 326
STRATONOVICH, R. L., 230, 235
STUMPFF, K., 34, 62
SUTTON, G. H., 18, 175, 234
SWINGER, D. N., 175

TANER, M. T., 297, 319, 320, 326, 368, 371
TATTERSALL, G. D., 280
TAYLOR, G. I., 2, 21, 84, 97
TAYLOR, H. L., 235, 327
THOMAS, J. B., 147, 175, 451, 455
THOMSON, W. T., 290, 298
THORNTON, B. S., 410, 412
TITCHMARSH, E. C., 362, 371
TOOLEY, R. D., 237, 280
TOTH, F., 326
TREITEL, S., 2, 21, 139, 141, 143, 145, 148, 164, 174, 175, 176, 186, 197, 235, 236, 281, 297, 298, 304, 309, 318, 370, 371, 412
TREMBLY, L. D., 234
TRIBOLET, J. M., 358, 360
TROREY, A. W., 2, 21, 186, 197, 281, 290, 291
TRUXAL, J. G., 245, 280
TSUBOI, C., 54, 62

TUKEY, J. W., 1, 43, 61, 60, 97, 99, 109, 112, 117, 118, 130, 141, 189, 3554, 360, 456, 462
TURIN, G. L., 299, 318

ULRYCH, T. J., 145, 153, 163, 164, 165, 173, 175, 176, 183, 232, 234, 235, 359, 360
URSELL, F., 2, 19

VALUS, V. P., 7, 21
VAN DEN BOS, A., 176
VAN DER POL, B., 63, 73
VAN NOSTRAND, R., 186, 197
VASSEUR, G., 412
VIENNA, A., 175
VOGT, W. G., 412
VOITH, R. P., 412
VOZOFF, K., 412

WADSWORTH, G. P., 10, 21, 236
WALKER, C. J., 235
WALKER, G., 176
WALSH, J. L., 384, 398
WANG, R. J., 236
WARD, R. W., 235, 327
WARD, S. H., 411
WATKINS, J. S., 235
WATSON, R. J., 295, 296, 297, 299, 318
WATTS, D. G., 144, 175
WEIDELT, P., 410, 412
WEINBURG, L., 253, 258, 264, 280
WEISINGER, R. C., 383
WELCH, P. D., 34, 59, 61, 122, 130
WHITE, R. E., 236
WIENER, N., 1, 21, 99, 102, 109, 137, 141, 187, 195, 225, 236, 361, 370, 371, 448, 455
WIGGINS, R. A., 2, 9, 21, 232, 236, 299, 386, 359, 360, 399, 404, 407, 412
WISECUP, R. D., 412
WOLD, H., 453, 455
WOLF, E., 328, 332, 353, 354
WOOD, L. C., 236, 384, 392, 395, 398
WOODS, J. W., 327
WUENSCHEL, P. C., 281, 298

YANOVSKAYA, T. B., 7, 19, 20
YUEN, C., 398
YULE, G. U., 143, 176

ZAANEN, P. R., 233
ZADRO, M. B., 138, 141*
ZAGUSTIN, E., 410, 412
ZEIDLER, J. R., 175

Subject index

AD-analog-digital converter, 372-377
adjoint, 348
AIC, 162-163
aliasing, 110-117
all zero process, 139
amplitude spectrum, 33, 64
analog recording, 372
analytic function, 65
angular frequency, 32, 41, 64, 415-416
AR-autoregressive process, 142-145, 161-165
ARMA process, 143, 164
asymptotic variance, 122
attenuation, 291
autocorrelation, 2, 84, 90
autocovariance matrix, 90, 219
autocovariance, 84-90, 104, 131, 165-167, 184, 206, 331, 440, 453

band pass filter, 237-279
band stop filter, 265
Bartlett's spectral estimate, 127-129
Bartlett's window, 127, 459-461
BCD-binary coded decimal, 381-382
beam steer vector, 181
beamforming filters, 229
Bessel filter, 277
Bessel type window, 240
BIC, 163
bicoherence, 138
bilinear transform, 266-269
binary codes, 377, 380-381
binary coded decimal code, 381
bipolar codes, 380-381
bispectral analysis, 132, 137-139
bispectrum, 137
bit, 374
body waves, 334, 341
Boolean algebra, 385
box-car function, 89, 118, 301
Brownian noise, 453
bubble memory, 378
Butterworth filter, 252-277

Cal, 385, 397
cascade filter, 246-247
cassette systems, 378
CAT, 163
Cauchy-Hadamard theorem, 186
causal system, 368
CDP-common depth point, 12, 319
central limit theorem, 450

cepstrum, 355-360
Chebyshev filter, 266, 279
chirp signal, 91, 116
coefficient of correlation, 83, 208, 332
coherence, 132, 137, 332
coherence interval, 328
coherency, 132-137, 319, 324
coherency matrix, 332-333
communication matrix, 289
complementary code, 382
complex cepstrum, 357-360
complex roots, 173
compression, 384, 392-396
convergence factor, 64
convolution, 22-30, 77-80, 184-197, 439
Cooley-Tukey fast Fourier transform, 465-469
correlation coefficient, 83, 337
correlation function, 2, 83-88
correlation matrix, 90, 96, 128
cosine bell, 120-121
cospectrum, 132-136
covariance, 73, 83-88, 294, 336
covariance matrix, 96, 336-337
Covespa, 323
cross correlation, 83, 132, 320
cross covariance, 85, 131, 206-207, 440
cross periodogram, 134
cross-power spectrum, 102-104, 132, 140
cross-spectral analysis, 131-134, 351
cumulative spectral distribution, 101-102, 188

Daniell spectral estimate, 121-124
Daniell window, 122-123, 132, 134
data adaptive spectral analysis, 142
data compression, 385, 393-396
decimal number system, 377-380
deconvolution, 198-236, 439
delay line, 30
deterministic deconvolution, 228
deterministic process, 102, 449
DFT-discrete Fourier transform, 34-40
diagonal matrix, 403
differentiating filter, 279
digital recording, 373
dipole, 188-195
Dirac comb, 34-35, 110
Dirac delta function, 35, 75, 79, 200-201, 223, 430-434
directionality, 338
discrete linear inverse, 399-412
dissipation, 290-291
downward continuation, 54-57

Subject index

duodecimal number system, 377
dynamic range, 373-376

eccentricity, 345
echoes, 355-358
eigenvalue, 32, 337, 349, 404
eigenvector, 337, 349, 404
elliptic filter, 277-279
ellipticity, 331, 334, 342
energy normalized cross correlation, 320
ensemble, 451
entropy, 146-147
envelope, 367
ergodic theory, 451
error squared, 200
Euler-Lagrange equation, 226, 438
excess three code, 381-382
expectation value, 146-147, 452

fan filter, 299
FFT-fast Fourier transform, 1, 31, 43-62, 462-469
filtering, 22-30, 184-196, 237-279, 462-469
filtre en eventail, 299
final prediction error, 155, 162
folding frequency, 110
forward method, 400
Fourier series, 31-34, 413-423
Fourier transform, 184, 423
Fourier-Stieltjes transform, 102
FPE-final prediction error, 162-164
free oscillations, 4-7, 139
frequency resolution, 122
Fresnel integral, 93

gain, 132, 136, 141, 185, 191, 195, 361, 371
gamnitude, 357
Gaussian function, 450
Gaussian process, 451
Gaussian white noise, 453
general linear inverse method, 9, 399-412
generating function, 28
ghost reflection, 294-296
Gibbs' effect, 240
Good transform, 385, 388
gravity waves, 139
Gray code, 380, 390-391
Gray scale, 375
group delay, 361

Haar function, 384
Hadamard matrices, 384, 387-390

Hadamard transform, 389
Hamming window, 456-459
Hanning window, 132, 457-459
hedgehog technique, 9
Hermitian, 332, 348
Hermitian transpose, 347
heterodyning, 138
hexadecimal number system, 377-380
high-pass filter, 238, 258-260
Hilbert transform, 362-372
homomorphic deconvolution, 232, 355-360
homomorphic filtering, 358-360

identity matrix, 349, 389
impedance, 283
impulse response, 24, 75-76, 189, 237-272, 302-305, 361, 370
information, 146
information density matrix, 406
instantaneous frequency, 115, 369
instantaneous phase, 115, 368
inverse filtering, 1, 199-233
inverse Fourier transform, 40-41, 423
inverse problem, 9, 399-412
inverse shaping filter, 216-223

Johnson noise, 453

Kalman filter, 230
kernel, 118
Klauder wavelet, 92-94
Kolmogorov condition, 188
Kronecker delta function, 41

Lagrange multiplier, 153, 409
Lanczos inverse, 405
Laplace transform, 63-73
Laurent series, 186
least squares deconvolution, 200-225
Levinson algorithm, 214-219
linear inverse method, 399-412
linear operator, 24, 74, 188, 226, 242
linear system, 74-75
Love waves, 328, 334, 344, 346
low-pass filter, 237, 254-258

MA-moving average process, 139-140, 142
magnetic disc, 378
magnetic pole, 55
magnetotelluric method, 17, 136
MAICE, 163
Markov process, 144, 450

Subject index

maximum delay wavelet, 194-195, 198
maximum entropy spectral analysis, 142-173, 183
maximum likelihood method, 96, 118, 177-183
mean square error, 200, 226
MEM-maximum entropy method, 1, 142-183
Mereu filter, 231-233
MESA-maximum entropy spectral analysis, 1, 142-183
micropulsations, 328, 454
microseismic noise, 127, 139, 293, 334, 454
minimum delay, 1, 184, 192, 199, 361-363
minimum entropy deconvolution, 232
minimum phase, 361, 370
mixed delay, 195-196, 362-363
MLM-maximum likelihood method, 177-183
modulus, 188
moment, first, second, 452-453
Monte Carlo technique, 9
Moore-Penrose operator, 405
moving average (MA) process, 139
multiple reflections, 299
multiplexer, 375
multiplexing, 134, 374
multivariate regression, 94-96

non-linear processes, 309-311, 358-360
non-linear velocity filter, 309
normal function, 432, 450
normalized wavelet, 192, 193
notch filter, 247-252
Nth root stack, 310-311
Nyquist frequency, 111, 237, 373

octal number system, 377-380
offset binary, 379-381
ones complement, 379-381
optimum linear filter, 1, 148, 225, 411
overconstrained system, 403

P waves, 10, 286, 292, 328, 334, 340
Paley transform, 384, 391
Paley-Wiener criterion, 188, 370
parallel filter, 244-246
Parseval's theorem, 80, 361
partial energy, 196-197
Parzen window, 132, 461-462
pearls, 139
periodogram, 1, 98, 118
permutation matrix, 48
phase lag, 33, 64, 185, 195, 369
phase spectrum, 33, 64, 134, 192, 195

physically realizable system, 226, 368
pie slice, 299
polarization, 134-136, 328-353
polarization filters, 328-353
poles of order n, 67, 169-170, 253-258
power spectrum, 1, 85, 87, 96-136
prediction error filter, 151
prediction operator, 149-151, 225-228, 446-447
predictive deconvolution, 228, 291
prewhitening, 126-127
probability distribution, 448-449
purely polarized state, 351

Q 291
quadrature function, 363
quadrature spectrum, 134-136
quefrency, 355, 359

Rademacher transform, 385
radix, 378
random process, 102, 450
random variable, 448
Rayleigh waves, 328, 334, 344, 346
realizable filter, 24, 188
rectangular window, 118, 188-119
rectilinearity, 336, 338-341
recursion filters, 217, 219, 240-246, 304-309
Reed-Muller code, 385
regressor matrix, 217, 219
rejection filter, 247-252
REMODE filter, 335, 341-344
resolution, 122
resolution matrix, 406
reverberation, 291-294
reverse of a wavelet, 184

s plane representation, 64-70, 186-187
S waves, 10, 328, 334, 341
Sal, 385, 397
sample and hold, 374
saphe, 358
self information, 146
semblance, 320
sequency, 385
Shannon-Fanno-Huffman code, 393-394
shaping filter, 216-223
shot noise, 453
shuffle matrix, 48-49
signum function, 365
similarity transformation, 349
smoothing, 225

sparse matrices, 48
spectral decomposition, 405
spectral window, 2, 118-126, 457-462
spherical harmonics, 4-7
square wave, 71-73
squared norms, 200-201, 205, 218, 404
state-space model, 230
stationary random series, 352, 450
stationary time series, 448-454
statistically normalized cross correlation, 320
step function, 72
stochastic process, 103, 449
subroutine BNDPAS, 274
subroutine BURGAR, 160
subroutine FASTF, 59
subroutine FILTER, 275
subroutine LEVREC, 216
subroutine SPECTR, 161
subroutine WALSH, 395
surface wave discrimination filter, 334, 344-347
synchronous tape decks, 378

taper, 120
third moment, 137
Toeplitz matrix, 154, 158, 208
transfer function, 80-82, 188, 192, 237
truncated inverse filter, 199-200
twiddle factor, 44
two-dimensional Fourier transforms, 54-58
twos complement, 379-382

uncertainty principle, 120
underdetermined case, 408
unit circle, 185-186
unit pulse, 75-76, 198, 223
unit step function, 27, 416, 431
unitary inner product, 348
unitary matrix, 348-349
upward continuation, 54-57

variance, 83, 96, 122, 336, 453
velocity filters, 299-311
velocity spectra, 319-324
Vespa, 323
vespagram, 323
VIBROSEIS, 91-94, 376

Walsh transform, 384-398
white noise, 448-456
Wiener-Hopf optimum filter, 1, 148, 225-228, 439-447
Wiener-Khintchine theorem, 98
windows, 96-111, 118-124, 457-462
Wold decomposition, 454
word, 373

z plane, 185
z transform, 1, 29-30, 71, 186, 199, 208-210
zero of order n, 68, 190-193, 247-250, 270
zero phase shift filter, 237, 246-247